£148-00

14

Culinary Arts and Sciences
Global and National Perspectives

FIRST INTERNATIONAL CONFERENCE ON CULINARY ARTS AND SCIENCES
GLOBAL AND NATIONAL PERSPECTIVES
ICCAS 96

CONFERENCE CHAIRMAN

P.A. Jones
Bournemouth University, UK

CONFERENCE ORGANISER

L. Sucharov
Wessex Institute of Technology, UK

LOCAL ORGANISING COMMITTEE

J.S.A. Edwards
Bournemouth University, UK

B.J. Pierson
Bournemouth University, UK

INTERNATIONAL SCIENTIFIC ADVISORY COMMITTEE

E.W. Askew	T. Lang
A. Babuchowski	D.H. Lyon
E.Th. Cassee	S. Larson
M. Franti	H.J.H. MacFie
N. Johns	H.L. Meiselman
	P. Sembury

Organised by:
The Worshipful Company of Cooks, Centre for Culinary Research, UK
Wessex Institute of Technology, UK
Bournemouth University, UK

Culinary Arts and Sciences

Global and National Perspectives

EDITOR:

J.S.A. Edwards
Bournemouth University, UK

Computational Mechanics Publications
Southampton   Boston

J.S.A. Edwards
School of Service Industries
Bournemouth University
Poole BH12 5BB

Published by

Computational Mechanics Publications
Ashurst Lodge, Ashurst, Southampton SO40 7AA, UK
Tel: 44 (0)1703 293223; Fax: 44 (0)1703 292853
Email: cmp@cmp.co.uk
http://www.cmp.co.uk/

For USA, Canada and Mexico

Computational Mechanics Inc
25 Bridge Street, Billerica, MA 01821, USA
Tel: 508 667 5841; Fax: 508 667 7582
Email: cmina@ix.netcom.com

British Library Cataloguing-in-Publication Data

A Catalogue record for this book is available
from the British Library

ISBN: 1 85312 399 4 Computational Mechanics Publications, Southampton

Library of Congress Catalog Card Number 96-083659

*The texts of the various papers in this volume were set
individually by the authors or under their supervision*

No responsibility is assumed by the Publisher for any injury and/or any damage to persons or property as a matter of products liability, negligence or otherwise, or from any use or operation of any methods, products, instructions or ideas contained in the material herein.

© Computational Mechanics Publications 1996.

Printed and bound in Great Britain by Bookcraft Ltd, Bath.

All rights reserved. No part of this publication may be reproduced, stored in a retrieval system, or transmitted in any form or by any means, electronic, mechanical, photocopying, recording, or otherwise, without the prior written permission of the Publisher.

Preface

The Worshipful Company of Cooks are a London livery company who can trace their origins back to 1311. During their history, they have always supported the catering (foodservice) industry, most recently by endowing a Centre for Culinary Research at Bournemouth University, to undertake research into Public Sector Catering. In order to progress this work, one of the activities has been to organise and establish an international conference to be held in alternate years. The International Conference on Culinary Arts and Sciences is the first of those conferences.

Historically, the arts, science and craft of the food industries have been regarded as separate entities, and in many cases deliberately kept apart to the detriment of the consumer. This Conference aims to redress that imbalance by providing an holistic approach to the subject. It is an attempt to provide, not only an interface between various academic disciplines, but also to bring together experts in their fields, those employed or involved in food related and catering industries in a forum where they can share their knowledge, experiences and expertise.

This book contains the edited proceedings of the Conference and has been grouped under the overall themes Food Service, Food Science & Technology and Food Habits. Although *art and science* have been grouped in this way for ease of reference, the Conference enabled participants to explore subject areas that they perhaps would not normally have dealings with in their daily routine. In this respect the Conference was also fortunate in having papers and delegates from over 20 countries providing a truly international theme.

The editors are grateful for the assistance and advice provided by members of the International Scientific Advisory Committee, colleagues at the Bournemouth University and the Wessex Institute of Technology for their help, and to the authors and participants for their support of the Conference which has made this publication possible. They are also grateful to the sponsors, the Worshipful Company of Cooks Centre for Culinary Research.

Special mention should also be made of Lynn Morton, the Conference Secretariat, whose hard work and dedication made these proceedings possible.

JSAE
June 1996

CONTENTS

INTRODUCTION

The Worshipful Company of Cooks of London: a brief history   3
From work by Past Masters A.W. Goodinge, P.F. Herbage
Adapted by J.S.A. Edwards

SECTION 1: FOOD SERVICE

Section 1.1: Menu Planning

Gastronomy, the importance of combining tastes   15
J.M. Van Westering

Menu Planning: what to do before menu engineering?   25
J. Kivela

Incorporating customer choice in menu planning   49
J. Eastham

Workplace caterers' attitudes toward providing healthier menus   59
M.J. Corney, A. Eves, M. Kipps, C. Noble

Catering for vegetarians - a challenge or a chore?   69
A.E. Sharples

Take the plunge: get the breakfast right   81
I.S. Folgerø, S. Larsen

Children's choices: variability in children's menus in restaurants in Northern Norway   91
L. Lervik, S. Larsen, T. Gustavsen

Section 1.2: Equipment and Systems

Multi-media development and optimization of concepts for a fast food restaurant   101
H. Moskowitz, A. Gofman, P. Tungaturthy

Catering systems revisited   117
W.G. Reeve, J.S.A. Edwards

Catering systems in use in the British Army   129
S. Ridley, P. Jones

The evaluation of trolleys used in hospital meal delivery systems   139
A. West

Section 1.3: Education and Training

Greening the food and beverage curriculum   151
J. A. Wade

Gastronomy - towards vocational reality   161
J. M. Schafheitle, P.A. Jones

Eating with your eyes: a teaching programme for Food and the Media   173
P. Hann, A. Colquhoun

Posters

Barriers to skills transfer of hotel management students and graduates in Hong Kong   184
J. Kivela

Hotel management education in Hong Kong at crossroads   185
J. Kivela

Quality improvement in foodservice management   186
L.R. Ferrone, P. Jones

SECTION 2: FOOD SCIENCE AND TECHNOLOGY

Section 2.1: Microbiology

The development and implementation of a hazard analysis system in a welfare feeding organisation   191
D. Worsfold

Microbiological safety of recipe mayonnaises   199
H.I. Dodson, A.S. Edmondson, M.A. Sheard

Section 2.2: Technology

Application of fuzzy mathematics in *sous vide* cook-chill food process development   211
G. Xie, H. Keeling, M.A. Sheard

Ensuring food authenticity in the food and catering industry   221
P. Creed

Two-stage beef roasting 231
M.A. Sheard, G. Xie

Improvement of the quality of typical Hungarian wines by membrane 241
filtration
K. Manninger, E. Békássy-Molnár, Gy. Vatai, M. Kállay

Demographic change and food product development: the challenge of an 251
ageing UK society
A. Colquhoun, A. Reid, P. Lyon

Posters

The effects of natural antioxidants on the quality and storage stability of 262
ready-to-serve foods
M. Karpinska, J. Borowski, M. Danowska

Application of yellow mustard mucilage in foods 263
W. Cui, N.A.M. Eskin, H. Liu, R. Pereira

Sedimentation phenomenon in canola oil 264
H. Lui, R. Przybylski, N.A.M. Eskin

Your roasted chicken quality problem solution - might be brine solution 265
S. Zalewski

A study of the functional properties of soy proteins for emulsion type 266
foodstuffs
I. Alexieva

A study of model emulsions stabilised with pectins 267
I. Alexieva

The influence of a plasticizing system and physical properties of formed 268
masses on the quality of "convenience food" production
A. Póltorak, A. Neryng

Influence of some physical parameters of dough on the quality of machine- 269
formed convenience foods
A. Wierzbicka, A. Neryng, E. Biller

The eating quality of the meat of four different slaughter masses of 270
Bonsmara cattle
I.C. Kleynhaus

Cooking variables and sensory quality characteristics of South African 271
venison with reference to springbok (*Antidorcas marsupialis marsupialis*)
D.M.J. van Rensburg

Section 2.3: Information Technology

Towards a model of computer use in the hospitality industry 275
M. Peacock

An information system for determining an optimal nutritional intake 285
G. Karg, S. Kreutzmeier

Poster

Computers and dietary assessment 294
C.E.A. Seaman

Section 2.4: Nutrition and Dietary Intake

Nutritional changes in Eastern Germany since 1990 297
G. Ulbricht

Dietary intake in the Czech Republic 307
A. Aujezdská, D. Müllerová, L. Müller

The development and characteristics of Bulgarian national cuisine 315
I. Alexieva

The nutritional quality of vegetables processed by traditional cook-serve 325
and meals assembly techniques
A. Walker, A. West, J. Lawson

Food and nutrition at environmental extremes - altitude 335
J.S.A. Edwards

SECTION 3: FOOD HABITS

Section 3.1: Food Habits

The relationship between post-war cultural shift, consumer perspectives 351
and farming policy
S.C. Beer, M.H. Redman

All in the garden is not dim sum. Trends and potential for the 363
development of western style food & beverage operations in the Peoples
Republic Of China
J. Sutton

The influence of cultural traditions and social change on domestic meals in a New Zealand context 375
J. Mitchell

Food habits: concepts and practices of two different age groups 387
S.S.P. Rodrigues, M.D.V. Almeida

Application of Kosher dietary laws by the food industry 397
N.A.M. Eskin, C. Grysman, G. Brojges

Supermarket shopping and the older consumer 405
C. Leighton, C. Seaman, M. McGlade

Class, income and gender in cooking: results from an English survey 415
T. Lang, M. Caraher, P. Dixon, R. Carr-Hill

Community meals in the next millennium 427
J. Kenny, B. Pierson

A consumer led approach to the development of community meals 437
N. R. Hemmington

Catering systems as laboratories for studying the consumer 447
H.L. Meiselman

Theories of consumer behaviour - their relation to the patient and the catering service in hospitals in the United Kingdom 457
J. Ervin, W. Reeve, B. Pierson

Determinants of repeated lunch choices in a cafeteria situation 469
L. Lähteenmäki

The effects of exposure to a food odour on food choice, consumption and acceptability. 479
L. R. Blackwell, B. J. Pierson

Physiological aspects of beer drinkability 489
T. Fushiki, H. Kodama, T. Yonezawa, K. Morimoto

Section 3.2: Marketing and Media

Service and guilt - a Norwegian view on service 501
S. Larsen, I.S. Folgerø

The marketing of the wine experience: an innovative approach 509
J. Fattorini

Poster

Destination North Cape: are German tour bus operators' needs and 522
expectations understood by the North Norwegian hotel business?
T. Gustavsen, L. Lervik and S. Larsen.

Index of Authors 523

INTRODUCTION

The Worshipful Company of Cooks of London: a brief history

From work by Past Masters A.W. Goodinge, P.F. Herbage
Adapted by J.S.A. Edwards
Worshipful Company of Cooks Centre for Culinary Research, Bournemouth University, Poole, Dorset, BH12 5BB, UK

Cooks have flourished for many centuries, but the first recorded mention is in Fritz-Stephen's (Clerk to Thos. A. Beckett) "Description of London" of 1170. Here he describes in great detail the cookshops on the banks of the River Thames which he thought the acme of civilization; "...At any time of day or night, any number could be fed to suit all palets and all purses...". Later in the 13th Century, the banks of the River Thames were taken over by the wine vaults of the vintners and the cookshops moved to Eastcheap and Bread Street.

Few people today would be able to stomach Mediaeval cookery, especially the more elaborate dishes. Everything was chopped or ground, seasoned and coloured beyond recognition. This was due to a number of reasons: little livestock could be kept through the winter; transport was slow, hence meat was salted; over-kept flesh needed to be disguised and cooks, like most specialists, made an elaborate mystery of their art; people's teeth fell out, thanks mainly to their diet; but above all, forks were not used at the table until 1600, hence food had either to be eaten in lumps with the fingers, or scooped up with a spoon.

The main meat was pork, with venison, game, hares and rabbit much more common than today. Fish and shell fish were eaten on Fridays and during Lent. Salmon was plentiful as were trout, eels, and sturgeon. Vegetables were limited to those in season mainly, onion, leek and cabbage, usually eaten raw. Fruits were apples, pears, plums, cherries, sloes, quince and strawberries. Dark rye bread was far cheaper than white made from wheat: pastry provided a great deal of the bulk which later came from potatoes. Sweetness came from honey whilst curnin, orpiment and other exotic spices were used. Large and heavy meals were washed down copiously with mead, ale or wine which was expensive.

Although the origins are lost in the mists of antiquity, the tendency of workers and townsfolk to organise themselves into fraternities, mysteries and guilds was one of the great social phenomena of the Middle Ages. Some of these organisations were religious, for example Parish Guilds, whilst others were for those engaged in crafts, trades or occupations. In the City of London separate Guilds were formed for each trade but in smaller towns these tended to be federated guilds. The first recorded redemptions in the Guildhall Records of Cooks are between 1309 and 1313 when 36 are recorded. The first edict from the Mayor was in 1312 but it is generally accepted that Cooks became a fraternity in 1311.

In the 14th Century there were cooks, pastlers (cooks or makers of pasties) and pie bakers. A pie at that time was meat baked in a dish with a pastry cover, whereas a pasty or pastie was defined as venison or other meat seasoned and enclosed in a crust of pastry and baked without a dish. In these early days each trade had its own fraternity or guild although there were two fraternities of Cooks, one in Bread Street and the other in Eastcheap. The earliest record of Masters being sworn in is in 1393 when the Master of the Cooks of Estchepe (Eastcheap) and Master of the Cooks of Bredstret (Breadstreet) were sworn in on the same day. The earliest record of a Master of the Piebakers is 1377. Amalgamation of Companies was not unknown in the 15th Century and it appears that these three Crafts fraternities, who had so much in common amalgamated about 1420 and became "The Cooks of London".

On 12th July 1482, King Edward IV granted the Cooks of London (the Company) the first of its eight charters, and from this date became incorporated. The charter is written in Latin, but whether it has been lost in one of the fires or whether is still in existence is a mystery. A copy of the charter does however exist in the Public Record Office. In the preamble it states "...the freemen of the Mistery of Cooks have for a long time personally taken and borne and to this day do not cease to take and bear great and manifold pains and labour as well at our great feast of St. George and at others according to our command...". In the charter granted by King James I in 1616, many services undertaken for the King by the Freemen of the Mistery of Cooks are again cited. "...as well at the Royal Feaste of our Coronacon at the intertayninge of our deere brother the Kinge of Denmarke, the Marriadge of my well beloved daughter the Ladie Elizabeth, our annual Feastes of Sainte George, as at the intertayninge of Forraine Princes and upon other occasions...". Today the Company acts under the charter granted by King Charles 11 in 1664 and Bye Laws approved in 1686 by the Lord Chancellor George Lord Jeffereys and Two Lords Chief Justice.

In these early days, the Masters and Wardens were granted the powers of supervision over the trade in the City and suburbs. This included the power to search, examine and scrutinise the business premises of cooks, to punish by seizure and fine bad cooking or selling unseasonable meat, and for breaking other ordinances of the Company. The other purposes of the Guilds were to: control the system of apprenticeship; ensure and maintain standards of trade;

regulations for holidays, hours, and to some extent wages and prices; pageants; and charities for widows, orphans and poorer brethren. Members accepted their positions primarily because the Charter or Ordinances were for the common good of the trade. Their disputes were settled by the Master and Wardens, which was probably better than civil litigation, as lawyers and judges at this time were often corrupt.

In 1500 the Company purchased a plot of land in Aldersgate Street, and soon afterwards built its own Hall. Little idea of the style of architecture of the Hall can he discovered. All that can be found is one description as "mean and ordinary" and another "...The Hall is more to be admired for its convenience than its eligence of building...". The measurements of the building and rooms were recorded and from these it appears that the Hall was built some way back from Aldersgate Street. The space in front was occupied by two houses and to the rear there was a long gallery and garden. In 1674 the long gallery was found to be dangerous and it was decided to erect a new Hall where the gallery and garden stood, but at the same time, retain the old hall.

For two and a half centuries, the Hall was the centre of the Company's activities which were, technical, charitable, educational, social and judicial. In order for so many functions to be undertaken in one place, a servant, the Beadle, had become indispensable. He acted upon the bidding of the Master and Officers kept a record of the Livery whom he summoned to courts, feasts or to church, attended searches, visited the poor, distributed alms, kept a check upon apprentices, ordered provisions and guarded the Master and Wardens. The Mace, which was his symbol of office, was also his first weapon of defence. Another official to be employed was a Clerk, but smaller Companies did not normally appoint an officer until regular meetings of the court demanded a factual record of the Company's affairs. However from records found so far, the first Clerk of the Cooks was in 1567 when Richard Tomson held the appointment, whereas the first record of a Beadle is 1665, and the first record of the staff is 1671.

Companies always cared more for the stomachs of their Liverymen than their health and have eaten and drunk well to the greater glory of God - a legend from the ancient parish fraternities. Feasts in the middle ages were grand affairs and Cooks Hall must have become quite a notable place as it was mentioned in a speech at the entertainment given by Dudley, Earl of Leicester to Queen Elizabeth at Kenilworth Castle in 1575. Although there are no records of Feasts at Cooks Hall, in 1425 the provision for an Election Feast of the Brewers included such 'delicacies' as porpoise, oysters, pike, lampreys, swans, plovers, geese, partridges, woodcock, larks, various spiced and sweetened breads, honey, figs, almonds and dates. Twelve gallons of cream and eight gallons of milk were used to make pastry and the whole was washed down with Red Gascony wine and ale.

The Hall survived the Great Fire of 1666, and for a period afterwards over a dozen Companies who had lost their Halls used Cooks Hall until their own were rebuilt. On 24th November, 1764 the New or Great Hall were destroyed by fire but the Old Hall escaped and the damaged Hall rebuilt. Unfortunately, on 8th August 1771 the entire building was destroyed by fire including many old documents and all the Company's Pewter. Luckily, the Company's silver, including the oldest piece, Cocoanut Cup, hallmarked 1588, seven chalices dated 1656, and a number of cups dated 1660-1680, were not in the hall at the time and have survived. After this fire the Court decided not to rebuild, and since then the ground has been leased out while the Company still retains the Freehold. The City Corporation have fixed to 10/12, Aldersgate Street a plaque reading "Site of the Cooks Hall destroyed by fire 1771". Recently a Roman mosaic has been discovered in the basement of the present building.

In 1609 King James I directed the City of London to undertake the Plantation of Ulster and £40,000 was raised by the twelve great Companies and forty four of the minor Companies. The senior Great City Guild, the Mercers, had associated with them the Cooks, Broderers and Masons and 3210 acres of land near Kilrea were allocated to them. This property was administered on behalf of the Associated Companies for nearly 300 years but finally sold in 1906. The traditional friendship between the Companies is maintained today by each entertaining the Masters and Wardens of the other Companies annually. At no time in the long history of is association has there been any written word, but the association has been based upon the mutual friendship and respect for each other's interests.

The Company maintained its powers over the trade in the City of London until about 1850, when in common with many other Companies, these were lost. At the end of the last century there were no recognised craft training and qualifications so the Corporation of the City of London and many of the City Guilds jointly formed the City and Guilds of London Institute to undertake craft examinations. The Cooks Company are proud to be a founder member of this Institute, whose qualifications are still recognised world wide.

The Company has always tried with limited resources to encourage various schemes, to teach the art and science of cookery. These have included the active support of teaching at schools in the City of London, the Shaftsbury Homes, prizes at food and catering competitions, and for City & Guilds. In 1952 the Company purchased a silver salver to be presented to the winner of the Grand Pieces Competition at the Salon Culinaire de Londres at Hotelympia. In 1975, following a comprehensive review of its prize schemes, the Company decided that, as the training of apprentices had been one of its original functions, its main efforts should be to encourage young people in craft training. Prizes awarded would be Caithness Goblets and Bowls engraved with the Company's Arms. Two years later, the Company instituted an annual competition for the London Technical Colleges. Members of the Company subscribed to a Trophy which is known as the Silver Jubilee Trophy, as the competition was instituted during HM The Queen's Silver Jubilee year. To encourage apprentices in the Army Catering Corps, (now Royal Logistics Corps) one of the Company's Caithness

Glass Goblets is presented to the Apprentice RSM of each Passing Out Parade. Prizes are also presented to other branches of the Armed Services.

The Company administers many old trust funds for charity, the earliest recorded benefactor was Walter Mangeard 1421. Among the many administered today is one from John Shield 1616 and another from Edward Corbett (Master 1664). Two children are presented to Christ's Hospital (Blue Coat School) under the will of John Phillips 1674.

The Company ranks 35th in order of seniority of the City Guilds, has a Livery of 75 of which 26 are members of the Court of Assistants. It is the only Company to have Second Master, and in addition to the elected Officers there is a Clerk and Beadle. Company Livery Medals were instituted in 1771 and the Officers' present robes and regalia were presented by various members in the last quarter of the 19th century. The Officers of the Company, Master, Second Master, Warden and Renter Warden, are elected annually on 14th September and sworn into office in November. This is followed by a banquet for the Livery, Masters and Wardens of the Associated Companies in order to meet the New Master, and all officers to be crowned, a ceremony which dates back to the 16th century. Another banquet is held, in May to meet the Lord Mayor and Sheriffs of the City of London. At every dinner the toast of the Livery is drunk in the traditional manner -

"The Livery, root and branch, may it flourish and thrive for ever".

The Worshipful Company of Cooks Centre for Culinary Research

The Worshipful Company of Cooks Centre have continued to support the catering industry with the established for Culinary Research at Bournemouth University at the beginning of 1994. The decision to locate in Bournemouth was taken following open selection and applications from 26 other Universities in the United Kingdom.

The aims of the Centre are to stimulate research in the fields of culinary knowledge and understanding in relation to the cooking process, technological advances, gastronomy and diet with particular reference on public sector catering.

Oversight of the Centre is provided by a Management Committee who are responsible for determining the policies, monitoring and oversight of the activities. Advice to the Management Committee is provided by a Steering Advisory Group consisting of eminent industrialist and academics whose remit is to provide a forum for advice, guidance and assistance to further the research activities and to better inform the Centre of the strategic issues affecting public sector catering.

The empirical research comprises a programme funded partly by the Company, partly by other research and consultancy contracts, and partly by the University. The central themes and where researchers are actively engaged are as follows:

Theme: Public sector catering

A Comparative Analysis of Catering Systems with Particular Reference to Hospitals and the Public Sector.

The purpose of this study is to provide an up to date critical analysis of the principle catering systems, i.e. conventional cook-serve, cook-freeze, cook-chill and sous vide and make recommendations for further research.

The Effective Factors of Consumer Satisfaction for a Hospital patient and Their Application to the Catering Service.

The purpose of this study is to identify the components of satisfaction for the hospital consumer (patient) which, taking into account their needs, desires, expectations, beliefs and attitudes, influence their perception of meal acceptability.

Feeding the Elderly Population Living at Home in the Next Century

The purpose of this study is to review and evaluate the factors to be considered when feeding the elderly population living at home and to propose solutions on how this could be most effectively achieved in the future.

An Evaluation of Organisational Structures and the Development of a Model on Which the Future Strategy for Army Feeding Could be based.

The purpose of this study is to develop potential strategies for the future organisational structure of Army Food Services, reflecting developments which have occurred within commercial and public sectors.

Theme: Food, diet and feeding strategies for the optimisation of health productivity and performance

Food Intake, Absorption and Nutrition at High Altitude - Exercise Sheer Hypoxia.

The purpose of this research is, using a Joint Services Expedition to Bolivia, to quantify what changes occur in the nutritional intake, intestinal absorption and other dietary related issue when climbing to altitude.

The Stability of Food Habits and How They Affect the Performance of Students Entering and Leaving University.

The purpose of this study is to monitor the food habits, food intake and performance of students entering and leaving university and assess how they change; the primary influences for these changes and what effects they might have on performance.

Feeding Strategies and the Optimisation of Productivity and Performance.

The purpose of this study is to establish how the timing, composition and quantity of food consumed can be utilised to optimise productivity and performance in the workplace.

Theme: Food choice and meal acceptability

The Combined Role of the Human Senses in Determining Food Choice.
The purpose of this research is to determine the interrelationships and interdependence of the human senses in determining food choice.

A Comparative Study of the Appropriateness of Sensory Evaluation Techniques Used to Measure Food Acceptability.
The purpose of this study is to evaluate and compare the suitability and applicability of sensory tests currently used in the laboratory and eating environments to measure food acceptability.

A Comparative Study of the Effects of Sous Vide Cook-chill Systems on the Sensory Quality of Selected Foods.
The purpose of this study is to determine the effects of sous vide cooking over a range of processing conditions on the retention sensory acceptability of selected food items.

A Study of the Factors Influencing Consumer Choice and Expectations When Eating Out.
The purpose of this study is to identify and quantify the elements in consumer choice when eating out and to construct an analytical model showing the inter-relationships between these qualitative and quantitative elements. The concluding aim is to test and validate the model in order to predict consumer choice in selecting a restaurant.

The Role of Olfactory Cues in Guiding Food Choice, Selection and Acceptability.
The purpose of this study is to establish the effects of visual, auditory, gustatory and textual cues on food choice and acceptability and to investigate the roles of olfactory cues in guiding food selection.

Biennial International Conference
An International Conference is to be held every two years, the first being this conference.

"Satisfying the Consumer Through Innovation"

FIRST INTERNATIONAL CONFERENCE ON CULINARY ARTS AND SCIENCES
Global and National Perspectives

25th - 28th June 1996

SECTION 1: FOOD SERVICE

Section 1.1: Menu Planning

Gastronomy, the importance of combining tastes

J.M. Van Westering
*Department of Management Studies, University of Surrey,
Guildford, Surrey, England*

Abstract

The paper's central issue is that of combining tastes - complementing food and beverages. This is a crucial area of gastronomy which, it is argued, has been substantially ignored to date, mainly due to historical rhetoric and to lack of research incentives. This has reinforced the notion that the combining of foods is based solely on an 'experiential art form' known only to the discerning chef's palate. The view that such valuable instinctive experience can be enhanced by 'the appliance of science' seems unvoiced or eclipsed by the mystique shrouding gastronomy. Although restaurants and retail outlets, combine food and drink for their consumers, alerting them to flavour, to taste combinations and to gastronomy, the information regarding combinations is scant and inconsistent, leaving the world of gastronomy bewildering and remote: the prerogative of the restaurant industry.

1. Introduction

Restaurants, and other food and beverage retail outlets are well equipped to make gastronomy available to all, yet they too conspire to perpetuate the myth that prevails, and are slow to inform their public. Food and drink are still mainly treated as separate items; how they might combine is rarely considered. The abundance of exotic foods and global wines render gastronomy a nightmare, one impossible to the consumer. Restaurants and shops need to simplify their approach and with this in mind, the paper addresses advances in food science and technology which can pertinently direct the consumer to a gastronomic experience, either in the restaurant or at home, with minimal risk taking.

2. Gastronomy for all

The term gastronomy immediately states good food, good wine; these alone are insufficient - *'the sum of the parts does not create the whole'*. Achieving the right combinations, considering what goes with what, lifts the combined tastes to the heights making the meal memorable, making *'the sum of the parts greater than the whole'*. The premise of this paper is to address the need to understand how food and drink can be combined to make eating, either at home or in the restaurant, a gastronomic experience, one within the reach of all today. Convincing the public is a difficult task. The term 'Gastronomy' retains exclusive connotations: historic rhetoric dominates these. The medieval delicacies of swan and dormouse, 'curiosities' of the times, were aimed at the rich[1]. Similarly, two decades ago, Beluga Caviar and Bolinger, ostentatiously consumed by Society's *'creme de la creme'* or the sleek heroes of popular novels, were viewed as unavailable to the less wealthy. To dine at the Ritz, to eat the best, was beyond the imagination of the majority. This is to presuppose that the best in terms of food was the best quality, and that it had to be seen to be eaten in opulent settings. An aura seemed to mythologize gastronomy, *taking it back to the gods,*[2] rendering it inaccessible to all but the rich. The belief, that gastronomy lacks general access, persists today.

Gastronomy does not rely on the most expensive or rare foods, the most elaborate of meals, or special occasions. Bode [2] captures its true essence, 'it is represented in a simple piece of bread, an apple, a piece of cheese - as long as the bread is fresh and crusty, the apple, ripe and juicy, and the cheese, well-matured and full of flavour - to which a glass of good wine could add perfection'. Lewis Carroll's *'Walrus and Carpenter'*[3] seem to epitomise this ideal. Their meal of fresh oysters, devoured on the beach at midnight, made their eyes stream with delight. In declaring: *'Now, if you're ready, Oysters dear, We can begin to feed'*, the accompaniments were minimal: *'A loaf of bread,' the Walrus said, 'Is what we chiefly need: Pepper and vinegar besides are very good indeed. No* extraneous conversation was required - only, *'Cut us another slice':* presumably this was washed down with the local *'vintage'* - seawater. The message here is that gastronomy is accessible to all. It is about choosing products, combining flavours, about 'pairing wine and food'.[4] Making choices and combining them in the meal experience has become increasingly difficult. The sheer volume of new wines and different food items available challenges even the most knowledgeable about appropriate combinations. Irrespective, retailers, the food and catering industry, need to be proactive in removing the aura surrounding the knowledge of food and wine pairing, the gastronomic culture. They need to communicate a wider accessibility to the public.

3. Combining Food and Beverages.
3.1 The rationale.
To reiterate, the term gastronomy immediately states good food, good wine and pleasurable eating. Given these, consumer expectations of an outstanding meal experience should be high. All too often, excellent cuisine and quality beverages fail to live up to their potential, fail to live up to customer expectations. What promised a gastronomic delight, a truly pleasurable, unforgettable occasion, even an adventure, falls short of its potential. Whilst a good meal was presented, an outstanding one was not achieved. The food and wine consumed, all excellent in

their own right, but together, when combined at the table, failed to raise the meal experience to greater heights. The combinations of food and drink just did not work fully, did not do justice to all the preceding hard work.

A chef spends infinite time in designing menus, in choosing ingredients, in discussing preparation methods and food presentation; such skills are finely balanced by scientific, technological and artistic inputs. Meticulous attention is paid to selecting quality beverages, but these efforts are often thwarted. The significance of food and drink taste combinations, throughout the meal experience, have not been fully appreciated.

It is contended here that the public are becoming more knowledgeable about food and drink and with greater awareness, consumers have become more adventurous, and open to change. Customers are moving away from the basic menu, main-course dictate of 'red wine with meat, white with fish'.[5] The average restaurant meal has three courses, each needs to be complemented with an appropriate beverage. Tastes must be combined to ensure a more pleasurable meal experience.

3.2 Understanding flavour

Focus should be directed at understanding flavours and tastes, and their combinations. Whether working with traditional dishes or experimenting with new recipes, the aim is to combine flavours which set off the ingredients' good qualities, and which in turn produce, even transform, a flavour into something perhaps more savoury, more tangy - unexpected. The palatability and acceptability of food are related to food flavour (a composite of texture, touch, temperature, common chemical sensitivity and smell, as well as taste). Food flavour, combined with sight and sound, contributes to the perception that delights the gourmet's senses. The four taste qualities, salt, bitter, sweet and sour, are crucial in determining the acceptance of different food combinations. It is their functional independence which plays a special role in taste modification. Traditionally they have been used to formulate general guidelines for compatibility in ingredient combinations.

Since the field of combining tastes and taste compatibility is extensive, the application of sensory profiling affords scientific rigour and a quantitative methodology to the task. This is a method of describing the sensory characteristics of foods (appearance, flavour and texture) and of quantifying them. It demonstrates how products differ in their flavour, and all importantly by how much. Profiling, graphically illustrates a product's sensory characteristics. It is carried out by people taught to be ultra sensitive to these characteristics, and extensively trained in the scales applied; they act as an analytical tool. Products initially need to be described in terms of their flavour and their intensity. Samples representing the extremes are scored individually (in terms of the intensity of their attributes), averaged, and the mean scores recorded. Such profiling enables the taste of the product to be noted individually, and then after different preparations. This points to consumer preference and indicates how the product might be changed in order to increase its acceptability. The data gained is widely used in product development, including product matching. With the ever increasing variety of food and beverage products on the market, such profiling can clearly benefit the restaurateur and the food retail industry, the more knowledge of the flavour of products, the better. Lawless[6] reinforces the need for testing the flavours of new market products. Pointing to the sheer volume of such products he questions, 'were the over 200 oat bran products launched in 1989 all optimised for flavour and texture appeal?' He answers by citing Shapiro[7] whose calculations indicated an '80-90% failure rate of new product introductions annually'.

The information gained from sensory profiling combined with other research can also facilitate menu construction and food combinations. Potatoes, for example, could be profiles to analyse attributes such as cooking time, peelability, flour content, smoothness, sweetness and smell. Variations are likely to be quite substantial, reinforcing the notion that there is a correct potato for the job. Chips, mashed, baked and roast, one potato variety does not suffice and the wrong choice can make or mar a meal. This is well-known but attention needs to be drawn to variety, choice and alternative flavours. Arguably such a choice is never offered. The BBC's *'Food File'* programme[8] drew attention to the demise of the British vegetable and the fact that far more varieties and flavours are available. However, European Union (EU) legislation (on seed propagation and importation restrictions) alongside 'protectionist policies' preclude wide-scale access to such potential. The profiling method could be applied to any food product and would be particularly helpful to those pursuing flavour - taste combinations and compatibility of food and drink.

Although knowledge in this area is increasing, little is known about the compatibility of different food products. While products can be analysed for their chemical components, how these interact often lacks clarity. Food and wine experts frequently cite, "the acidity of the wine cuts though the greasiness of the meat beautifully". Research in this area is scant and redress to the anecdotal guidelines is common. This point is taken up by Bartoshuk[9] who suggests that a 'systematic study of taste modification' is warranted to 'ensure that the necessary techniques to increase food palatability will be available when needed'. Arguably the scientist and the chef have similar objectives, hence the verve and inventiveness of the industry should not be undermined in the wake of scant research. The restaurant and retail industry are beginning to recognise the roles which they can play in informing their consumers.

3.3 The Role of the Restaurant.

The menu warrants more than a cursory glance. Clients, weaned on Rod Hull's *'Emu'*[11] albeit a 'distant' relative may balk at the idea of *'House Smoked Ostrich'*. Those with romanticised notions of the Australian Outback and 'baby-roos' will clearly 'skip' the *'Medallions of Kangaroo'*. Others sympathetic to the *'Crocodile Dundee'*[12] version of Australasia might be persuaded to sample the offerings. Whilst some might suggest the 'adjectival exotic' apt for the culinary proposals, Antipodeans will, of course, be all too familiar with such menu mainstays. Those equipped with the volume *'Recites ou Curiosites Litteraires'*[1] will doubtlessly point out that 'there is hardly a dish which has not been attempted already' referencing the practices (16th Century) at the 'Tour d'Argent', (Paris), in its serving of 'dormouse pastry, mixed dishes of snakes, porpoise, roast swan and crane stuffed with plums'.

While the menu provides vital information about the food available and how it is to be served, it fails to suggest appropriate plate combinations, to indicate tastes, or to recommend wine to accompany and complement the food. *'Blackened Swordfish Steak served with clementine chutney and aoili'* offers a combination of opposites as an accompaniment. What tastes might be anticipated from such a dish? Here the attention changes from the unfamiliar food content (the swordfish) to the vegetable, fruit and sauce combinations. A sense of challenge tempered by cautious restraint might reverently be felt. Julius Maincave[13] gave to the world, *'fillets of mutton with crayfish sauce, beef cooked in kummel, bananas with Gruyere cheese, sardines with Camembert cheese and herring soup with raspberry jelly'*. The latter appear to be highly unpalatable. Of course experimentation is necessary, otherwise many excellent combinations would be lost. It would be foolish to condemn the unknown as a matter of culinary principle, or to reject out of hand unfamiliar flavours. The menu clearly avoids the accusation of being stereotyped! Risks have to be taken, experiments have to be pursued in the advancement of any activity. If this were not so, the range of gastronomic enjoyment today would be exceedingly limited!

What dish, what wine, even what bread should precede the *'Blackened Sword Fish Steak'*? What should accompany it to render the meal a challenge, an adventure well repaid? Given good restaurant staff, the choice might not be insurmountable. In some restaurants the chef is often easily accessible. Customers are more likely to choose *'the known'* foods and beverages if help is not readily available. As a result, many an excellent meal experience has been avoided and lost to the public. Clearly restaurants wish to avoid this, yet often the simple, and more cost effective ways of proffering information are ignored. It is easy to understand why kangaroo, ostrich and crocodile[12] might be passed over if constructive advice is not available. Restaurants wishing to be adventurous must accept more responsibility for their sales if they are to encourage their customers to take full advantage of the courses offered.

Menus should offer more guidance to customers, for example, by recommending starter, main and dessert course combinations. The idea of 'complete menus within the menu'[4] goes beyond the food and its accompaniments, and focuses on complementary tastes. The inclusion of a recommended wine to further complement each course offers a pertinent suggestion to restaurateurs. McWhirter and Metcalfe[5] contend 'choosing wine with food is nothing to worry about' adding reassuringly 'few combinations are bad, and many wines will usually marry well with a number of food types'. However, they recognise that whilst 'many wines taste good with all sorts of food ... some combinations of food and wine flavours and characteristics go further than that' and link up into a perfect marriage'. The premise here is a nonsense, choosing wine is something to worry about, good food and good wine deserve the perfect marriage. Ryan argues vigorously that 'wine-and-food menus are flavor matches made in heaven for today's chefs and restaurateurs'[14]. Finding the right wine to complement a dish is a somewhat daunting task. Good restaurant staff spend much time guiding customers' food choice. While the menu connoisseurs often prefer to select dishes a la carte, lesser-informed mortals choose from the table d'hôte menu. Similarly, the wine buff, selects in an a la carte style from the wine list. Diners less familiar with wine variations would benefit from a recommendation alongside the meal course offered. If this were common practice in the restaurant, it would be economical to serve single glasses of the recommended wine with the meal rather than expect the consumer to purchase a full or half bottle. The latter, although a common practice, reinforces the false assumption of a *'wonder wine'* that can be drunk with everything. In acknowledging that the single glass of wine is usually available in restaurants[4] it is not necessarily seen as the wine which genuinely complements the meal, and as such is part of the meal package. If the wine is 'the correct' wine, the one which exactly complements the dish and allows all the flavours (food and drink) to blend fully, then other recommendations would be unnecessary. In this way, consumers would come to recognise the efforts of the restaurant in its attempt to render the meal something special.

3.4 The role of the retailer

Food and drink retailers must, like the restaurateur, take on greater responsibility for informing their customers. Supermarket shelves are stocked with a range of quality goods of puzzling shapes, sizes and colours. It seems that something 'new' arrives daily. The consumer is assaulted by choice. Squashes (Gem, Butternut, Spaghetti, to name a few), exotic fruits (kumquats, physalis, grandilla), vegetables of amazing shapes and sizes plus an alphabetical list of names fill the racks. The choice is most challenging and acceptable. What is unacceptable is the erratic flow of information or even total lack of product detail. Food and beverage outlets have a need to sell more than just their produce. Today's generation are consumer-wise, are articulate about what they want, about what they intend to do with their purchases and what they might be persuaded to buy. The sales of food and drink magazines, the popularity of television coverage, the stardom status of food and drink presenters, and high programme ratings, all reinforce the point that the general public want to know more about the produce they can buy[4] and how to eat it, how to enjoy and how to appreciate what they buy.

Arguably supermarkets only sell the product, although some are doing more. Retailers generally are failing to provide customers with sufficient information. That which is provided is offered in an erratic and inconsistent manner. However, some wise competitors are well aware of the power in addressing the public's taste needs. Recent television advertising exemplifies this point. While Tesco sought to sell the quality of individual food items ('*the parts desperately Seeking the Chicken*'), [15] Sainsburys[16] invested in putting the parts together (moves towards the quality of the whole, but not more than the whole). Their menu combinations were screened at prime time to attract sales. Waitrose have used the media pull of Raymond Blanc to offer his *'Menu of the Month'*, auspiciously marketed in their *'New Products'* booklet[17]. The latter recommends wines to accompany each course, and amazingly even a specific bread, and a selected cheese. In offering the *'complete menu'* Raymond Blanc has an eye firmly directed towards *'the sum of the parts being greater than the whole'*, as exemplified in the *'January, 1996'*[17] menu, Figure 1.

Figure 1. Raymond Blanc's Menu of the Month

WINTER SALAD WITH WALNUTS AND ROQUEFORT CHEESE
Recommended wine to accompany starter:
Waitrose Solers Jerezana Dry Amontillado
LA POULE AU PO
Recommended wine to accompany the main course:
Cotes du Roussillon
GRAPEFRUIT & ORANGE TERRINE IN A MANGO COULI
Recommend wine to accompany the dessert:
Ungsteiner Herrenberg Scheurebe Spatlese 1992

More recent interest in consumer needs has focused on informing the consumer about wine, resulting in an upsurge in sales. Consumers want to know about the wines they are choosing, what they should accompany, and how they should be drunk. Wine labels are now used to indicate the wine's smell, taste and food accompaniment. Wine sellers, such as Oddbins,[18] offer quite copious detail about their wines, as well as hosting frequent tastings. It no longer suffices for wines to be identified by label, grape variety, or as being dry or sweet; more accurate information regarding flavour, taste, and food combinations is required, for example, an Italian sauvignon, *'Sauvignon Bidoli',* is described as *'being full of citrus and tropical fruits - lovely with starters or fish'*[18]. Arguably, with the plethora of wines on the market even this lacks in specificity. Such information, however, reduces the risk-taking element in purchasing wines. If supermarket and wine chains were prepared to be even more specific, for example, what might complement a smoked haddock and spinach dish, rather than just a fish dish, purchasers might buy the dearer priced wines particularly if the taste can match their meal proposals, and their home-based dinner party ideas. All round benefits could be accrued from just a little additional information and concern for the consumer. Retailers are making positive moves to inform their public but arguably, as in the case of the restaurateur, more proactive incursions are warranted.

3.5 Conveying the product message

The written word seems to dominate marketing approaches and consumers bombarded with time-consuming literature. Market research today is recorded in highly presentable, entertaining computer-graphic formats, digestible at a glance. Potato research offers such example. British shoppers would probably only be able to name three different varieties of potato readily available. They would probably have scant idea that cooking time, ease of peeling, flour content, smoothness, sweetness and smell should be considered before purchase. This detail is revealing since most supermarkets convey the message that vegetables come in one variety only! The potato research demonstrates suitability in terms of food preparation, and in the flavour that each can contribute to the individual dish and to the overall meal. Here market forces are being used to target the use of the potato, indicating the right product for a specified purpose.

There is a need to present more sophisticated information to consumers. It seems appropriate for food producers and retail outlets to inform customers about the smell, texture, tastes, preparation and serving methods of the new products on the market. It is insufficient just to place the product on the shelves, identified by name and country of origin. Arguably many such sales will be low risk purchases[19] for the consumer, but once bitten, without the advantage of suitable information, the consumer may not buy again: again, a product lost,[7] an experience lost. People want to branch out, and try new things, the industry should be tempting them more positively. The technology to inform is there, yet the mainstay of such dissemination is still the pamphlet, the menu, product tastings and the brief video screening. These are not the only ways to inform, neither are they necessarily the most suitable. It is possible to offer consumers product image, smell and taste, and the changes brought about by differing preparations, at minimal cost.

British United Provident Association's (BUPA) 1995 marketing campaign[20] made use of smell panels, basically scented printing inks. The message from the panels was so precise that such a method could be similarly used to inform supermarket customers and restaurant clientele. Smell panels could be sited over a wide range of produce. The marketing approach invited 'rub this panel and breathe in gently. At this moment, millions of sensors in your nose will send messages to your brain. Your brain can consult a built-in library of over 3,000 smells. This process is so precise that more refined noses can even distinguish between wines grown yards apart'. If the less refined nose becomes more sophisticated, the introduction of smell panels might aid in the marketing of new products and indeed new menu combinations. If these are feasible then taste panels must similarly be a possibility. The smell panels proved effective even after twelve months use, a good shelf-life. The message here is surely applicable to the food industry. Customers having to choose between six pizza sauces would benefit from a strategic siting of a smell panel on the pizza sauce label and the choice could be simplified. It is clear that something about the flavour, smell, taste and texture has to be communicated to the public. Denny Bros[21], printers of the inks for the smell panels suggest, 'In 'A World Devoted to the Senses' Fix-a-form with Fragrants brings your product to life. Smell, Sight, Sound, Taste, Touch - all are used to influence potential buyers and today's consumer has an ever-increasing choice. Fix-a-form is the labelling solution with utmost versatility'. They claim 'Fragrants are available in a whole range of ready prepared aromas or can be formulated to meet individual requests'. Clearly there must be some potential for such pads in the marketing of food and drink.

In advocating that alternative product marketing methods to up-date consumers should be considered, the awareness of marketing technique failures must be borne in mind. Given the range of products available, it is clear that in order to promote 'the unknown' more detail must be forthcoming. It is the public who will be consuming such products, whether at home or in a restaurant, and as such the industry should be helping them to appreciate what they are eating. Again, in conveying the message regarding such combinations, the market forces must be more proactive.

4. Conclusion

The main thrust of this paper has focused on combining flavours, tastes, and on complementing foods and beverages, with the aim of enhancing the key activity of human life - eating and drinking, in the form of the meal experience. Because so much time is spent in eating and drinking, the meal experience should be pleasurable and prove a rewarding adventure moving food and people from the realms of the known to the unknown. It has been argued here that in terms of combining food tastes 'the unknown' dominates. The unknown here being the behaviour of food ingredients under different processing conditions and in a variety of combinations with other ingredients. Clearly a vast undertaking, and indeed one seemingly too difficult to conceive. However, with the bewildering array of produce readily available today, it is similarly inconceivable to rely on the knowledge of individual chefs to advance the boundaries of gastronomy. Research in this field is sorely needed. Lawless[6] suggests that 'it remains to be seen whether sensory evaluation matures into a discipline with the quantitative precision of engineering and the cumulative knowledge of science in general, or whether it remains a collection of technical methodologies'. He adds that in the meantime, research lies with the sensory and evaluation scientists 'who must continue to explore links between scientific findings and practical product applications'. Here, greater prominence should be given to combining food tastes. Clearly such research would benefit those concerned with producing good food and beverage combinations, those working in the field of gastronomy.

In addition it has been argued that gastronomy should be made more readily accessible to the public. The food and beverage industry, as a whole, should direct more energy into informing their customers. To further this concern, the dissemination of information in the restaurant and in the supermarket should be more consistent and more 'palatable'. Modern technology should be used more widely and ingeniously to provide customers with information on produce, and how they might be successfully combined with others. To reiterate earlier concerns 'there is an art and a science to combining the products of gastronomy and enology and bringing them together' - *'The time has come'*.[4]

References & Notes
1. Poiret, P. *'107 Recettes ou Curiosites Litteraires'*. In Encyclopaedia Britannica, A New Survey of Universal Knowledge. Encyclopaedia Ltd. London. Volume 10. 1961.
2. Bode, W.K.H. *European gastronomy. The story of man's food and eating habits.* Hodder & Stoughton, 1994.
3. Carroll, L. *The Walrus and the Carpenter.*
4. Beckett, F. *The Drink. Caterer and Hotelkeeper.* August 18th 1995.
5. McWhirter, K. & Metcalfe, C. *The Quick and Easy Guide to Choosing Wine with Food',* Macdonald and Co. 1989.
6. Lawless, H.T. *Bridging the Gap Between Sensory Science and Product Evaluation.* New York State College of Agriculture and Life Sciences, Cornell University, Ithaca, New York. 1990.
7. Shapiro, E. *New products clog groceries.* New York Times, May 29, D1, D17. 1990.
8. BBC *'Food File' Programme.* 15th February, 1996.
9. Bartoshuk, L.M. *Modification of Taste Quality*, In Birch, Brennan and Parker, National College of Food Technology Conference. Applied Science Publications. AS. 1977.
10. Anon *Menu, The Naked Turtle Restaurant.*
11. Rod Hull and Emu are a well-known children's ventriloquist act.
12. A reference to the ethos of the blockbuster film *Crocodile Dundee*.
13. Maincave, J. In: Encyclopaedia Britannica A New Survey of Universal Knowledge. Encyclopaedia Britannica, Ltd. Volume 10. 1961.
14. Ryan, N.R. and Stephenson, S. *Menu Opportunities.* In: Restaurants & Institutions. March 27, 1991.
15. A reference to the 1990's Tesco television advertising campaign.
16. A reference to the 1990's Sainsbury's television advertising campaign.
17. Anon, *New Products*, Waitrose, January 1996.
18. Oddbins. Reference to a High Street chain of beverage retailers.
19. Mitchell, V-W, and Greatorex, M. *Risk Reducing Strategies Used in the Purchase of Wine in the UK.* In: International Journal of Wine Marketing, Volume 1, Number 2, 1989.
20. Anon, *BUPA Advertising Campaign*, Reference 6429, 1995.
21. Denny Bros. *A World devoted to the Senses.* Denny Bros Printing Ltd. 1996.

Menu Planning: what to do before menu engineering?

J. Kivela
Department of Hotel and Tourism Management,
Hong Kong Polytechnic University, Hung Hom, Kowloon, Hong Kong

Abstract

Traditional approaches to menu planning and menu-item selection often ignore the information which reflects the organisational and market needs. In addition, concentrating efforts on a menu engineering-type of analysis highlights menu performance after the menu has been planned and put to use.

This paper reports on the concept and application of Menu Planning Qualitative Variables (MPQV) as an aid for menu planners. A Control Group and Experimental Group of menu planners were chosen for the study to test the scope and usefulness MPQV. The respondents (Control group A), were required to develop a menu using a traditional menu planning approach, whilst the Experimental group (B) used the MPQV approach. There are important differences in the outcomes between the Control and Experimental groups in terms of menu planning approaches. The Experimental group thought that the menu planning information given to them was more applicable to their on-the-job needs.

1. Introduction

Executive chefs are usually responsible for menu planning and the development of menu items. They identify those menu items that are performing well and encourage their sales. They also remove dishes which are not performing well. Menu planning is often based on the gastronomic knowledge of the chef, and in most organisations, on the sales performance of menu items. Menu planning, therefore, often relies on chef's experience and intuition, which in this paper will be referred to as the "traditional" approach.

Growth of the Food and Beverage (F&B) business, diversification of cuisine's, development of new cooking styles, changing trends, global transportation of food, changing markets, changing eating habits, and more aggressive approaches to merchandising of food and beverages, call for a move away from menu planning methodology that is principally based on gastronomic know-how, intuition and often, limited information. It can be argued, therefore, that menu planning needs to consider not only gastronomic variables, but also marketing, financial and merchandising variables.

Research about dining out which reports on the customers' perspective by Lewis[1], Williamson[2], Ewan and Ewan[3], Finklestein[4], Auty[5], and Kivela[6], overwhelmingly suggests that the great majority of customers choose to eat-out for pleasure-seeking purposes, and that the food quality and food type may not always, as is often thought, be the most important choice variables, but rather, other seemingly less important attributes such as ambience, location, service quality, price mix, menu item mix, menu typology and art work, trend/style/fashion, may be more important factors determining restaurant choice. From the restaurateurs' perspective, traditional approaches to menu planning often fail to provide the answers to some important questions. For example,

- What should be the ideal menu item mix be if the operation's financial objectives are based on high-volume-low contribution margin?
- What specific profit maximising techniques have been considered?
- Is the menu mix based on price popularity or volume popularity or both?
- What percentage of menu mix will be devoted to red meats, white meats, seafood, and vegetarian?

It is suggested that the menu should be an accord between aesthetic-gastronomic and production desirables, communication, marketing, merchandising, and financial variables. These aspects must be in harmony and balance if desired business outcomes are to be achieved, Kivela[7].

During the initial menu planning stages, consideration should be given to a different approach to the traditional approach when considering menu-item selection. Variables which usually affect menu-item performance, once the menu is put into production, must be known, or at least anticipated, and used to identify the most appropriate menu mix **before** the menu is put into production. This must be done so as to minimise the potential negative impacts of these variables.

This paper reports on the outcomes of an experimental research project which tested the utility of using Menu Planning Qualitative Variables (MPQV) to determine whether or not these variables can help to minimise the negative impact of extraneous factors when constructing menus during the initial menu planning stages.

2. Menu Planning

Menus are often referred to as the blue-prints by which foodservice operations organise their food and beverage procurement, production and service. Increasingly, menus are viewed as the core of the food product which directly influences demand and cost of supply, and profit. The problem with the menu planning, is that it is often a hit-and-miss exercise, which results in an end product (the menu), which is not in tune with the dynamics of the foodservice business environment. The other problem with menu planning is that many menu planners often rely entirely on menu engineering-type techniques to analyse menu performance, or the outcomes of the menu planning. It must be noted that menu engineering tends to highlight the outcomes of good and bad menu planning, and then only **after** the menu is put in production, by which time it is often too late to do anything about the problems anyway. Kivela[7].

In the last fifteen years menu analysis or 'engineering' literature by Kasvana and Smith[8], Miller[9], Pavesic[10], and Hayes and Huffman[11], has featured prominently in analysing menu-item performance. These analysis techniques measure menu item performance relative to their cost and profit, but they do so **after** the menu has been planned, and put into production. Each model gives the restaurateur insights into menu-item performance and in general, menu engineering specifically gives attention to three important menu elements, (1) the Contribution Margin (CM), which is the analysis of gross profit for menu items relevant to menu item price, (2) Menu Mix (MM), an assessment of customer preferences in a menu item group related to menu item demand, and (3) Customer demand forecasts, anticipated and actual total number of customers served. There are however, some inherent strengths and weaknesses of these methods which should be pointed out.

First, all of menu engineering methods require large amounts of data to provide reliable and valid analysis. Second, effective menu engineering depends on little or no variation in the menu item range over the time period analysed, hence historical data for new menu compositions, which do not use the same menu items, are not all that helpful. Third, it is necessary throughout the analysis period to monitor and update commodity costs, yields, and dish costs. Analysis can become cumbersome if, for example, material cost change frequently. Fourth, a series of analyses are required to ensure that the manager is reacting to real trends rather than short term changes. On the plus side, menu engineering is very useful for measuring the performance of menu items, once the menu is in production.

In recent years, menu engineering models which analyse menu item performance have achieved reasonable reliability and acceptability as indicators of how successful menu planning has been. However, there is scant literature about how to take into consideration the many variables that need to be considered during the initial stages of menu planning. So far, literature about menu item performance by Kasvana and Smith[8], Miller[9], Pavesic[10], Hayes and Huffman[11], and Beran[12], have dealt mostly with financial outcomes, vis-à-vis menu-item sales performance analysis. On the other hand, menu planning literature is in general, by Fuller[13], Kotschevar[14], Harris[15], and Miller[9], deals mostly with the gastronomic principles of menu planning and item selection, cost attributes, and marketing mechanics. Review of the literature identifies such menu planning variables as, marketing, finance, production, gastronomy, customer needs, as well as, knowledge, skills and experience. With the exception of Kivela's work, the majority of literature cited does not actually highlight the methodology as to how to apply some of these variables to actual menu planning before the menu is considered for production. The most likely outcome of inexperience and incomplete knowledge about these variables, is a menu structure where some items have an unjustifiable dominance over others, inconsistent sales performance, gastronomic imbalance, and inappropriate cost/price/contribution rations. As noted earlier, some menu planners rely solely on the outcomes of menu engineering analysis as a 'quick-fix', as a solution to their sagging sales and less than satisfactory menu item performance. This is because menu engineering will only yield 'good news' if the menu has been developed and structured well in the first place. In other words, the objective is to plan a sound menu first, rather than focusing on measuring whether the menu is sound via menu engineering.

3. Traditional approaches to menu planning

In most foodservice operations the task of menu planning is given to the executive chef. The parameters for this exercise vary from place to place. In many operations the chef is simply given a directive to produce the new menu. In some operations the chef is aided by additional data by which to construct the menu such as the past sales history. In most cases, executive chefs do a reasonably good job based on prior knowledge and experience, guided by a repertoire of recipes, and past experiences in working out the most suitable menu composition. But, what if the chef is not very experienced? What if the chef does not have all the relevant information at hand? Is this the right approach to menu planning? Furthermore, will the new menu fit the purpose it was intended for? In today's highly competitive food and beverage business environment, the restaurateur should have the answers to these questions.

Training young executive chefs and food and beverage managers during the past twenty years, the author has often questioned traditional approaches to menu planning, which resulted in menus that did not fit the intended operational and marketing purpose. The analogy to menu planning is that of a building construction. A building begins with a design criteria, (design variables), and an architect pens the drawings. This is followed by commencing the foundations, structural reinforcement, and then followed by the envelope. Finally the cosmetics-plant, equipment, and furnishings are completed. It seems odd that most experienced menu planners attempt planning menus, 'the blue print' (design variables) in reverse. They start by developing the cosmetics, try to put on the envelope, build the reinforcement, and then try to sink-in the foundations (organisational and marketing objectives). Then they try and assess whether or not the 'building' meets the design criteria. As mentioned earlier, young chefs and management trainees over the years have found the traditional or 'reversed' process to menu planning rather difficult and ad-hoc, especially when the menu planner lacks experience.

The alternative approach to menu planning, although not perfect, selects the most appropriate Menu Planning Qualitative Variables (MPQV). The data is derived from existing organisational records and usually include, financial attributes, market analysis, production and service records, and past sales records. A careful selection of MPQVs are then used to construct the menu framework. At this stage the menu framework is devoid of actual (food and beverage) menu items. These are usually selected later. The MPQV framework helps the menu planner to select the actual food items at some later date. Hence the building construction analogy.

4. The MPQV approach to menu planning

The MPQV process to menu planning begins with identifying variables which are most likely to have an effect on menu item development and selection. These variables, must reflect the current and future operational and marketing objectives. Menu planning should begin with identifying customer needs. It could then be followed by addressing the overall menu needs in relation to specific operational and organisational goals and objectives.

Whilst there are a plethora of variables that affect menu planning, from the review of literature, a set of variable categories emerge. These are, gastronomic, financial, marketing and merchandising variables, and within each category, there are of course, specific attributes that influence menu planning and menu planning outcomes. Apart from the financial category attributes, most of the other categories' attributes are qualitative in nature, thus the reference to 'qualitative variables' (attributes). Exploring the MPQV concept further, Tables 1, 2 and 3 show a list of some of the more important MPQVs. Menu planners should use these like a series of prompts to construct a menu framework.

(In this study the Experimental Group applied MPQV methodology to every menu item and every course development **before** any attempt was made to select the actual (food) menu items. However, it must be noted that MPQVs highlighted in this study are not exhaustive, and specific organisational and marketing needs may require more exacting MPQVs).

Table 1. Examples of possible gastronomic MPQVs

Menu Item, hot-cold	Menu Item, soft-hard	Menu Item, sweet-sour	Menu item, light-dark	Menu item, mild-strong
Menu Item, wet-dry	Menu Item, suitable if pre-prepared	Menu item, suitable only to-order	Uniqueness of dish	Menu Item, meat - white-red
Menu Item, fish - oily-white	Menu item, fish sea-fresh water	Menu Item, suitable for bulk production	Menu item, suitable for limited production only	Cuisine Type/classic/ nouvelle/California/ contemporary(other)/ oriental/Ethnic (type)
Menu Item, for special event	Uniformity and balance of culinary concept	Plate service - other service (specify)	Are specialised skills needed to prepare the menu item? (specify)	Is specialised service needed to serve the menu item? (specify)
What is nutritional composition needed?	Have you considered food groups	Is specialist equipment needed to prepare the menu item?	What is the occasion?	What are the cooking processes/s needed to prepare the menu item?
Required quality of materials	Are the ingredients in season?	What is the anticipated popularity of the menu item?	What is the anticipated popularity of the menu item by food group?	Is the item a popular dish for the chefs to cook?
What is the popularity of menu item by cuisine style	What is the popularity of menu item by flavour	What is the popularity of menu item by menu item presentation/ service	What is the percentage of convenience (value-added) as opposed to all in house produced of the menu item?	How much of the menu will be prepared on-site (if a centralised operation?
What percentage of menu items will be devoted to originality and artistic creativity?	What is the level of standardisation of quality and production methods?	Is the menu item difficult or easy to prepare with existing staff skills?	Does the menu item compliment the ambience and theme?	What is the level of standardisation of food cost?

Culinary Arts and Sciences   31

Table 2. Examples of possible financial and marketing MPQVs

Is the menu item, High cost - low cost?	Is the menu item based on high cost - high volume?	Is the menu item based on low cost - high volume?	Is the menu item based on high volume/low contribution margin?	Is the menu item based on high volume/ high contribution margin?
Is the menu item based on low volume- high contribution margin?	Is the menu item based on low volume- low contribution margin?	Is the menu item based on high contribution- low food cost percentage?	Is the menu item based on high contribution margin-high food cost percentage?	Is the menu item based on low contribution margin-low food cost percentage?
Is the menu item based on low contribution margin-high food cost percentage?	Is the menu item based on high volume low food cost percentage?	Is the menu item based on high volume-high food cost percentage?	Is the menu item based on low volume-low food cost percentage?	Is the menu item based on low volume-high food cost percentage?
Is the menu item based on popularity by menu item price?	Is the menu item based on popularity by volume of sales?	Is the menu item based on popularity by selling price? (price sensitivity)	Did you do a yield test for these menu items?	Have you taken into account the labour cost to produce this menu item?
Have you taken into account the cost of utilities to produce this menu item?	Have you taken into account your competitor's selling price attributes?	Have you calculated the break-even point for this menu?	Did you calculate the AP- (as purchased cost) for this menu item?	Have you considered profit maximisation techniques for these menu items?
Did you calculate the AS- (as served cost) for this menu item?	Did you calculate the EP- (edible portion cost) for this menu item?	Have you considered future food trends in relation to your menu item suggestions?	Have you checked with marketing about the appropriateness of your menu item suggestion to the market level	Have you considered the occupancy demand (peaks and lows) in relation to your menu item suggestions?

32   Culinary Arts and Sciences

Table 3. Examples of possible menu mechanics MPQVs

Have you considered the menu cover size and shape?	What quality of stock will you use for your menu cover and it is appropriate?	What typography will you use for your menu cover and is it appropriate?	Will your menu cover be a bookend-fold-four-fold, why?	Will your menu cover be a window-fold-six-panel, why?
Will your menu cover be a poster-style, why?	Will your menu cover be a letter-fold-six-panel, why?	Is your menu cover legible, did you test it under real conditions?	What language are you using on your menu card?	Are your menu items described correctly and accurately? Did you test this?
How descriptive does menu have to be?	Does you menu inform the customer? How did you test this?	How and who decided on the type of art treatment?	What merchandising treatment will you employ on the menu card? What is the objective and Why?	Will you use the boxing-in technique? Why or why not??
Will you use the highlighting technique? Why or why not?	Will you use the bolding technique? Why or why not?	How will you position your menu items on the menu card? Where and why?	Will the menu be in the written format only? Why?	Will the menu be in the pictorial format only? Why?
Will the menu be in the pictorial and written form? Why?	Have you or/and will you discuss printing attributes with your printer?	How large is the printing run?	What is the colour configuration (scheme) for your menu? Why this particular colour scheme?	What is the cost of final menu card?

An example perhaps best illustrates the applicability of MPQVs to menu planning. Since this example illustrates the construction of a simple dinner menu, it is important to first highlight the gastronomic composition (menu rhythm) for dinner menus. Normally a dinner menu (3-4 course) would be composed in a rhythm such as:

- light - heavy - heavy - light
- cold - hot - hot - cold
- dry - wet - dry - wet or wet - dry - wet - dry
- soft - hard - hard - soft
- mellow - robust - robust - mellow
- savoury - sweet

Example:

Simple Menu Brief

MENU REQUIREMENT	ADDITIONAL INFORMATION
Meal period	dinner
Season	Spring
Courses	3
Menu type	set
No. covers	250
Service style	plate
Item attributes	low fat, seafood, red meat, fruit. Hot, hot, cold wet, dry, wet
Cost/revenue preferred profile	Low-med food cost/high CM
Target market	Pre-theatre diners

Table 4 shows a simple first draft of a set menu composed of MPQVs on the left side and possible menu item interpretations on the right side.

Table 4. Sample draft menu using MPQVs

MPQVs	POSSIBLE (DISH) INTERPRETATION	WHY?
1st course Light/seafood/soft/wet/mild/ easy to prepare/low unit cost/poached/hot	**1st course** Poached Seafood Quenelles on light dill sabayon	**1st course** Follows rhythm commencing with 'light'. Quenelles are low-cost items (per unit) of otherwise expensive commodity. Low fat, wet in character (50% sauce), mild in flavour. Easy to prepare and quick cooking time.
2nd course Heavier/low fat/red meat/dry /medium flavour/easy to prepare/medium cost/ grilled/hot	**2nd course** Grilled Gum-smoked lamb fillet with herb butter	**2nd course** Follows menu rhythm with 'heavy'. Lamb fillet is a medium-cost item (per unit) of otherwise expensive commodity. Low fat (grilled), dry in character (30% sauce), mild in flavour. Easy to prepare and quick cooking time (customers have short dining time)
3rd course Light/fruit-based/wet/mild in flavour/pre-prepared/low-cost/chilled	**3rd course** Strawberry and yogurt mousse with fruit coulis	**3rd course** Follows rhythm following with 'light'. Strawberry and yogurt mousse is a low-cost item. Low fat, moist in character, mild in flavour. Prepared in advance. Helps digestion. Quick assembly time.

5. Methodology

The objective of this experimental study was to test whether the MPQV approach to menu item selection, where menu items were selected for their suitability against pre-determined specifications, was superior to the traditional approach, selecting menu items and retaining them unless menu analysis suggests that items are suitable.

Prior to this study, a group of food and beverage executives from a category five hotel in Hong Kong approached the author to stage a two-day, free, seminar on menu planning. This forum represented a unique opportunity to test the MPQV concept. However, because a small sample (one hotel and one group of food and beverage executives from the same hotel), would have severely limited the generalisation of any findings, a wider sample of food and beverage executives and hotels was sought.

From the 1994/95 Hong Kong Hotels Association listing of 76 hotels, 37 hotels (48.6%) were randomly selected for participation. Food and beverage executives who made the original request, were also included in the study. A wider representation of hotels would have been desirable; however, it was felt that a large number of respondents arising from such a wide sample would have been too difficult to manage, given the space limitations of the venue and the intended control-experimental groups arrangements

A letter of invitation to participate in the menu planning forum was sent to Food and Beverage Managers, Assistant Food and Beverage Managers, Executive Chefs and Restaurant Managers, in each of the 37 hotels. An assumption was made that these F&B Executives were the most likely people to be involved in some way with menu planning, although it must be pointed out that Executive Chefs and F&B Managers most probably have a greater role in the planning and development of menus than Restaurant Managers and Assistant Food and Beverage Managers. The invitation letter made no reference to the intended study.

For the purpose of clarity, in the next sections of the paper, these respective managers will be referred to as the "F&B Executive group/s"

Of the 37 hotels (148 invitations to various food and beverage executives), 20 hotels responded favourably (80 invitations). A telephone reminder to 17 hotels did not yield any additional interest in participating in the forum. The attributes of the 20 participating hotels are presented in Table 5.

Table 5. Hotel Sample

CATEGORY	NUMBER OF GUEST ROOMS	NUMBER OF F&B OUTLETS
5	513	5
5	517	5
5	573	6
5	592	4
5	605	4
5	600	5
5	862	4
5	509	5
4	875	5
4	591	6
4	723	4
4	500	6
4	665	6
4	440	3
4	400	2
4	590	5
3	166	2
3	194	2
3	496	3
3	245	2

The total F&B Executive sample comprised:

- 20 F& B Managers
- 20 Assistant F&B Managers
- 20 Executive Chefs, and
- 20 Restaurant Managers

The demographic profiles of the F& B Executives are shown in Table 6

Table 6. Demographic profiles of the F&B Executive

F&B Managers

Age	%	Years of hotel Experience	%	Months/years in current position	%
<25	0	<3	5	< 12 month	0
25-34	30	3-5	25	1-3 years	60
35-44	55	6-9	40	4-6 years	25
45>	15	10>	30	7> years	15

Executive Chefs

Age	%	Years of hotel Experience	%	Months/years in current position	%
<25	0	<3	0	< 12 month	10
25-34	35	3-5	60	1-3 years	55
35-44	45	6-9	25	4-6 years	30
45>	5	10>	15	7> years	5

Restaurant Managers

Age	%	Years of hotel Experience	%	Months/years in current position	%
<25	10	<3	15	< 12 month	20
25-34	70	3-5	20	1-3 years	15
35-44	15	6-9	45	4-6 years	25
45>	5	10>	25	7> years	40

Assistant F&B Managers

Age	%	Years of hotel Experience	%	Months/years in current position	%
<25	10	<3	15	< 12 month	5
25-34	60	3-5	55	1-3 years	60
35-44	25	6-9	20	4-6 years	35
45>	5	10>	10	7> years	0

All F&B Executives

Age	%	Years of hotel Experience	%	Months/years in current position	%
<25	8.8	<3	8.8	< 12 month	8.8
25-34	48.6	3-5	40	1-3 years	47.5
35-44	35	6-9	32.5	4-6 years	28.8
45>	6.3	10>	20	7> years	15

The first day of the menu planning forum dealt with matters of menu planning, only the second day was designated for work-shop (menu-planning) activities. No reference was made to his study on the first day of the forum.

To commence the workshop it was first necessary to break-up F&B executive from "clustering" within their own hotel group and to get them to "mix" with other executives. It was decided to use a "lucky draw" method as an ice-breaker and an element of fun for the participants. For each F&B executive group, the respondent's name and title was written on a piece of paper and placed in four boxes - Food and Beverage Managers, Assistant Food and Beverage Managers, Executive Chefs, and Restaurant Managers

The first draw from each box was designated as Group A (Control Group), second draw Group B (Experimental Group), third draw Group A, and fourth draw Group B, and so on. Respondents were not made aware of the existence of a Control Group and Experimental Group. This process facilitated an equal number of Food and Beverage Managers, Assistant Food and Beverage Managers, Executive Chefs, and Restaurant Managers (n=40) within both groups, and a random (lucky) mix of F&B executive from various hotels.

The next step was to establish the participants (both groups) perceptions about their expertise (and lack of it) in menu planning. A modified representative-grid was used to establish this. This was an important measurement because it established the groups' and individuals' congruency (or lack of) about their own planning competencies. To achieve this the respondents were required to write down on forms provided four statements about their "Best Menu Planning Skills", LEFT SIDE, identified as **"S" statements**, and four of their "Poor Menu Planning Skills". RIGHT SIDE, identified as **"s" statements**. The participants were then asked to pass their response to the person next to them, so that each participant ended up with someone else's response, which they then had to score. The participants were asked to score their colleague's statements according to how best each statement fitted with their own response. They were asked to rate "YOUR BEST MENU PLANNING SKILLS" (Left Side, **"S" statements** according to;

- (3) **"Identical or same as my reply"**,
- (2) **"There is some difference but is close to my own reply"**,
- (1) **"There are some commonalties but not close to my own reply"**, **and**
- (0) **"Completely different to my own reply"**.

They were then asked to rate "YOUR POOR MENU PLANNING SKILLS" (Right Side, **"s" statements**) by using the same scale.

Before any computations were made a qualitative analysis of all of the statements had to be made. There were in all 320 "S" statements and 320 "s" statements.

(N = 80 participants x 4 "S" statements and 4 "s" statements). These were then qualitatively analysed for "best fit", i.e. statement which although not exactly the same, were close enough to represent the group's overall agreement about best (poorest) menu planning skills.

After this, F&B Executives were split-up according to their respective groups (Control group A and Experimental group B) and were assigned to different rooms. Both groups were then asked to form 5 planning teams, each comprised of 2 F&B Managers, 2 Assistant F&B Managers, 2 Executive Chefs, and 2 Restaurant Managers. This aspect was not controlled for either group, and was left up to the respondents to decide planning team membership. Once this was done, both groups were given a Menu Planning Brief which they had to complete during the workshop session. In addition to this, both groups were given commodity price lists, and menu samples from a range of F&B organisations in Hong Kong.

The Control Group (A) was allowed to proceed with their menu planning brief without further instructions. The Experimental Group (B) was given a Menu Planning Brief which was developed to include MPQV components in some detail. In addition, a detailed listing of MPQVs was also given to the Experimental Group to use as a guide (Tables 1-4) during the exercise.

Both groups were given a 2 hour time limit to complete the menu planning. Straight after completion of the menu planning exercise, both groups were brought together in one room, and a short questionnaire was given to both groups to elicit their opinion about the utility, suitability, advantages and applicability of the information given to them in order to complete the Menu Planning Brief. Selected representatives of both groups were asked to present and defend their menu planning outcomes.

6. Results and discussion

The number of usable "S" and "s" statements is as follows. In all 87 per cent of "S" (n= 278, N=320) and 86 per cent of "s" (n= 276, N=320) statements were usable for the analysis. The "S" and "s" statements which closely resembled the "best fit" of all the respondents are shown in Figure 1:

Figure 1. Most generally perceived "S" and "s" statements

	S-type STATEMENTS
1	The ability to determine future eating trends
2	The ability to involve a team approach to menu planning
3	The ability to have menus developed to meet budgeted costs
4	The ability to plan menus with good profit returns
	s-type STATEMENTS
1	Lack of cooking or culinary expertise
2	Not knowing food costs and budgets before menu planning
3	Lack of expertise in market analysis
4	Lack of knowledge about how to organise food production

The scoring method used was as follows:

FOR "S" STATEMENTS
Score for each F&B Executive Group for each "S" statement
= ST score .
= statement scores (3) x actual number of usable statements (n)

Note, the highest possible score for each "S" and "s" statement is 60, i.e. (N= 20 x 3), hence in order to obtain a percentage of agreement for each Executive Group for each "S" and "s" statement, it is necessary to divide n (usable statements) actual score by N. Taking the "S" statement S1 in Figure 2 as an example, ST = 42/60 or 70% agreement.

Score for all "S" statements for all Food and Beverage Executive Groups
= GS(1,2,3, and 4) score
= an aggregate of all ST1 scores divided by the number of F&B Executive Groups (4)

Taking the ST1 scores in Figure 2 as an example, **GS1** = 42 + 46 + 44 + 43/4 = 43.5 or 73% agreement towards ST1 (S2, S3, and S4) between all of the F&B Executive Groups.

Score for all "S" statements (1-4) for each Food and Beverage Executive Group = GST score
= an aggregate of all ST (1-4) scores for each F&B Executive Group divided by number of statements (4).

Taking the ST1-4 scores in Figure 3 as an example, **GST** = 42 + 33 + 63 + 38 /4 = 41.5 or 69% agreement towards all of the "S" statements of each F&B Executive Group.

The same scoring procedure was used for the "s" statements.

The individual and group results about the "S" Statements"- Attitudes Towards Menu Planning Skills are shown in Figure 2. There was a sufficient level of group congruence about these. However, some differences in competency opinions did exist. For example, only 45 per cent of restaurant managers concurred that they had the skills to involve a team-approach to menu planning "S2", compared with 82 per cent for Executive chefs and Assistant F&B Managers. On the other hand, F&B Managers and Executive chefs scored highest, 88 per cent and 77 per cent respectively, for the ability to plan menus to meet budgeted costs "S3".

The individual and group results about the "s" Statements"- Attitudes Towards Menu Planning Skills are shown in Figure 3. Although there was a sufficient level of group congruence about these, there are a number of very specific incongruencies which exist between F&B Executives. For example, only 8 per cent of the Executive chefs thought that they lacked cooking or culinary experience "s1" and lack of knowledge about food production "s4", whilst F&B Managers, Assistant F&B Managers, and Restaurant Managers all showed high scores for lacking cooking or/and culinary expertise.

Furthermore, 88 per cent of Executive chefs thought that they lacked market analysis expertise "s3", whilst 88 per cent of Restaurant Managers suggested that not knowing food costs and budgets made menu planning difficult for them "s2".

Figure 2. Scores about "Attitudes Towards Menu Planning Skills"

"S" STATEMENTS RESPONDENTS

STATEMENTS S	SCORE 0-3	F&B MANAGERS N=20			ASSIS. F&B MANAGERS N=20			EXEC. CHEFS N=20			RESTAURANT MANAGERS N=20	
S1	0	0	0		0	0		0	0		1	0
	1	3	3		6	6		3	3		5	5
	2	3	6		2	4		10	20		4	10
	3	11	33		12	36		7	21		9	27
ST1		n=17	42	70%	n=20	46	77%	n=20	44	73%	n=19	42
S2	0	0	0		1	0		0	0		1	0
	1	1	1		2	2		0	0		3	3
	2	7	14		4	8		11	22		0	0
	3	6	18		13	39		9	27		8	24
ST2		n=14	33	55%	n=20	49	82%	n=20	49	82%	n=12	27
S3	0	0	0		0	0		0	0		1	0
	1	2	2		1	1		0	0		3	3
	2	3	6		5	10		5	10		7	14
	3	15	45		11	33		12	36		5	15
ST3		n=20	53	88%	n=16	44	73%	n=17	46	77%	n=16	32
S4	0	0	0		0	0		1	0		1	0
	1	2	2		3	3		1	1		2	2
	2	3	6		5	10		2	4		10	20
	3	10	30		11	33		10	30		6	18
ST4		n=15	38	63%	n=19	46	77%	n=14	35	58%	n=19	40

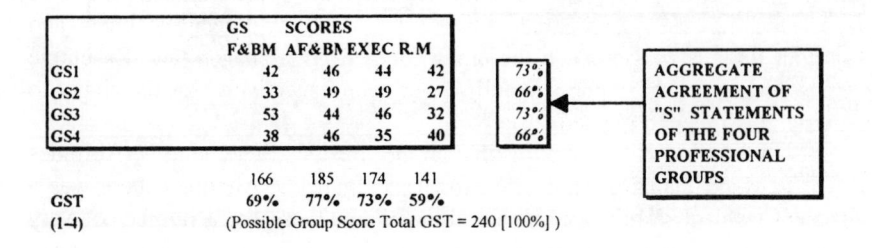

	GS SCORES					
	F&BM	AF&BM	EXEC	R.M		
GS1	42	46	44	42	73%	
GS2	33	49	49	27	66%	AGGREGATE AGREEMENT OF "S" STATEMENTS OF THE FOUR PROFESSIONAL GROUPS
GS3	53	44	46	32	73%	
GS4	38	46	35	40	66%	
	166	185	174	141		
GST (1-4)	69%	77%	73%	59%		

(Possible Group Score Total GST = 240 [100%])

USEABLE "S" STATEMENTS

ST1 n= 76 95%
ST2 n= 66 83%

42   Culinary Arts and Sciences

Figure 3. Scores about "Attitudes Towards Menu Planning Skills"

"s" STATEMENTS RESPONDENTS

STATEMENTS s	SCORE 0-3	F&B MANAGERS N=20		ASSIS. F&B MANAGERS N=20		EXEC. CHEFS N=20		REST. MANAGERS N=20	
s1	0	0	0	0	0	13	0	1	0
	1	1	1	3	3	3	3	1	1
	2	2	4	6	12	1	2	1	2
	3	11	33	9	27	0	0	16	48
st1		n=14	38 63%	n=18	42 70%	n=17	5 8%	n=19	51
s2	0	0	0	0	0	5	0	0	0
	1	10	10	1	1	10	10	0	0
	2	4	8	3	6	3	6	1	2
	3	2	6	8	24	2	6	17	51
st2		n=16	24 40%	n=12	31 52%	n=20	22 37%	n=18	53
s3	0	0	0	1	0	0	0	0	0
	1	1	1	3	3	0	0	0	0
	2	1	2	4	8	1	2	1	2
	3	13	39	11	33	17	51	13	39
st3		n=15	42 70%	n=19	44 73%	n=18	53 88%	n=14	41
s4	0	0	0	1	0	16	0	0	0
	1	0	0	1	1	1	1	0	0
	2	2	4	4	8	2	4	2	4
	3	18	54	13	39	1	3	15	45
st4		n=20	58 97%	n=19	48 80%	n=20	8 13%	n=17	49

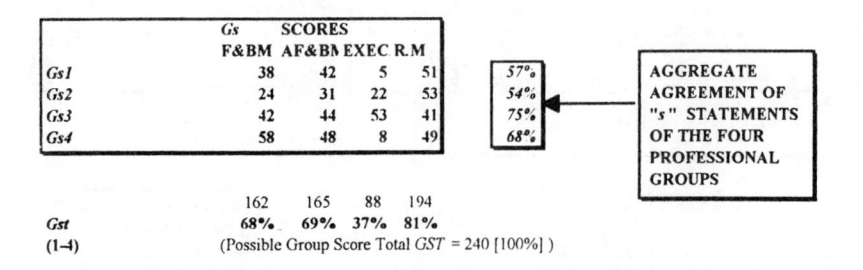

Gs	SCORES				
	F&BM	AF&BM	EXEC	R.M	
Gs1	38	42	5	51	57%
Gs2	24	31	22	53	54%
Gs3	42	44	53	41	75%
Gs4	58	48	8	49	68%

AGGREGATE AGREEMENT OF "s" STATEMENTS OF THE FOUR PROFESSIONAL GROUPS

	162	165	88	194
Gst (1-4)	68%	69%	37%	81%

(Possible Group Score Total GST = 240 [100%])

USEABLE "s" STATEMENTS

ST1	n=	68	85%
ST2	n=	66	83%
ST3	n=	66	83%

Culinary Arts and Sciences   43

Questionnaire responses

The feedback about menu planning information given to both groups, is shown in Figure 4, and reveals that some differences do exist between the Control and Experimental Groups about the scope and usefulness of menu planning information given to them for the menu planning exercise. As noted earlier, the Control Group only had limited information given to them to complete the Menu Planning Brief, whilst the Experimental Group had additional menu planning information which introduced the MPQV concept to menu planning.

Figure 4. Control group and experimental group feedback questionnaire results

QUESTION	SA		A		N		D		SD		n=	GROUPS
1	2	5%	7	18%	3	8%	24	60%	4	10%	40	CONTROL GROUP A
	7	18%	27	68%	1	3%	3	8%	2	5%	40	EXPERIMENTAL GROUP B
2	6	15%	3	8%	9	23%	14	35%	7	18%	39 #	CONTROL GROUP A
	11	28%	19	48%	2	5%	7	18%	1	3%	40	EXPERIMENTAL GROUP B
3	4	10%	8	20%	7	18%	20	50%	1	3%	40	CONTROL GROUP A
	9	23%	22	55%	0	0%	7	18%	2	5%	40	EXPERIMENTAL GROUP B
4	7	18%	13	33%	6	15%	12	30%	2	5%	40	CONTROL GROUP A
	15	38%	19	48%	3	8%	3	8%	0	0%	40	EXPERIMENTAL GROUP B
5	5	13%	11	28%	9	23%	14	35%	1	3%	40	CONTROL GROUP A
	9	23%	24	60%	0	0%	5	13%	1	3%	39 #	EXPERIMENTAL GROUP B

= 1 Missing case

Question 1.
Details in the Planning Brief were useful for me to be able to plan and develop an appropriate menu as required by the Brief.

There was considerable discord within the Control group, 60 per cent (D), 10 per cent (SD) about the range of menu planning information which was given to them for the menu planning brief. Overall, only 22.5 per cent of the Control group agreed that the range of information given was sufficiently useful. The Experimental Group on the other hand, were given the menu planning Brief information that incorporated MPQV attributes together with MPQV listings. The majority of the Experimental Group, 85 per cent (SA) and (A), thought that the range of menu information given to them was very useful for the menu planning exercise.

Question 2.
*The menu planning variables given in the Menu Planning Brief were **sufficiently appropriate** to the menu planning exercise*

Although the Control Group was reasonably divided about the suitability of menu planning variables that were given to them in the Menu Planning Brief 53 per cent of the group did register that they did not agree with the Brief information was suitable for the exercise. On the other hand, 23 per cent of the Control group thought that the Brief information was sufficiently appropriate for the menu planning exercise. The Experimental Group suggested that the variables given to them were sufficiently appropriate for the menu planning task, with a 75 per cent overall agreement.

Question 3
*The range of information given to me in the Menu Planning Brief gave me an **advantage** in the way I approached this menu planning exercise*

53 per cent of the Control Group disagreed that the menu planning information gave them advantage in the menu planning task, and commented that the Menu Planing Brief provided them with similar planning variables to what they usually experienced at work. However, 30 per cent did agree. The Experimental Group was more decisive and overall, 78 per cent of respondents felt that menu planning information provided, gave them an advantage for the menu planning task. Only 5 per cent thought that had not. The Experimental Group's non-menu planning "experts", i.e. F&B Managers and Restaurant Managers, also suggested that they were much more confident in attempting to plan menus using MPQVs, because the MPQVs provided them with necessary prompts by which to identify and address critical menu planning issues.

Question 4.
*The information given to me in the Menu Planning Brief is **readily applicable** to menu planning tasks where I work.*

The results between the Control Group and the Experimental Group show considerable response differences. Overall, 85 per cent of the Experimental Group thought that the menu planning information given to them in the workshops is readily applicable to their work, and thought that this approach to menu planning accounted for variables which they normally would have not thought about, but which were important. Only 7.5 per cent of the Experimental Group thought that these menu planning concepts will not be applicable to their work. On the other hand, 50 per cent of the Control Group thought that the information given to them was applicable to their jobs. Thirty per cent of the Control Group suggested that whilst the Menu Planning Brief information was useful, it did not necessarily foster better "applicability" to their jobs, because the information given to the was "nothing new" to what they already had at work.

Question 5.
*When possible, **I will apply** the menu planning approach which we used today to my menu planning tasks where I work.*

When asked whether they would actually use these menu planning methods on-the-job in their organisations, 37.5 per cent of the Control Group responded negatively whilst 22.5 per cent were not sure. Overall, 40 per cent of the Control group indicated that they would apply this menu planning methods on-the-job. However, they also argued that menu planning tasks are usually given to their executive chefs, and that they did not have the "expert" knowledge to be able to plan menus anyway. On the other hand, the Control group also responded positively, and argued that whilst it is their responsibility to plan menus from time to time where they work, they often do not have a more precise information about the variables which affect menu planning. The Experimental Group seemed more decisive and overall 82.5 per cent responded that they would like to use MPQV methods on-the-job. The Experimental group also suggested that given the exactness of menu planning variables (MPQVs), they felt they understood more about menu planning now than before, and were more confident to tackle the task of menu planing on-the-job. However, 15 per cent of the Experimental group responded unfavourably, and argued that even with the information given, they felt "under-qualified" to plan menus, and that this task should be given to the executive chefs only.

7. Conclusions

It can be concluded that the Experimental Group thought that they had a superior quality and scope of information about menu planning than the Control group. This was substantiated by the Experimental Group's unanimous accord about their choice of menu items during the presentation of menus. It can be further argued that using the MPQV process gave menu planners a set of more definitive and precise variables during the initial stages of menu planning. Interestingly enough, the Experimental Group presented the menu framework showing the MPQVs only which was followed by the actual menu (food) items and a justification why these menu items had been selected. The Control Group on the other hand, listed the menu items (food only) which they had selected without any framework. Some Group members however were not entirely in agreement with the Group's choice of menu items, and voiced their disagreement to the forum.

This study was an experimental undertaking, and needs to be followed-up by a more widely applied investigation. The study does however, provide insights into menu planning methodology that menu planners should consider during the initial menu planning stages. It is further suggested that MPQV planning methodology should be applied **before** menus are put in production (and before menu engineering). This study has also highlighted how the initial menu planning tasks can be significantly improved by providing the planners with more definitive menu planning criteria, that focus on operational and marketing variables, which reflect organisation's specific current and future F&B needs.

References

1 Lewis, R.C. Restaurant Advertising: Appeals and consumers' intentions, *Journal of Advertising Research,* 21.(5); pp 69-74: 1981.
2 Williamson, J. *Decoding Advertisements: Ideology and Meaning in Advertising.* Marion Boyas, London. 1978.
3 Ewan, S. and Evan, E. *Channels of Desire,* McGraw-Hill, New York. 1982.
4 Finklestein, J. *Dining Out: A Sociology of Modern Manners,* Polity Press Cambridge. 1989.
5 Auty, S. Consumer choice and segmentation in the restaurant industry. *The Services Industries Journal,* 12,(3); pp 324-339: 1992.
6 Kivela, J. Restaurant choice selection using determinant attribute analysis Unpublished research paper. Hong Kong: Department of Hotel and Tourism Management, Hong Kong Polytechnic University. 1994.
7 Kivela, J. *Menu Planning for the Hospitality Industry,* Hospitality Press Melbourne, 1994.
8 Kasvana, M.L., & Smith, D.I. *Menu Engineering: A Practical Guide to Menu Analysis,* Hospitality Publications, Lansing, Mich., 1982.

9   Miller, J.E. (1992). 3rd. (Edn.), *Menu Planning & Strategy*, Van Nostrand Reinhold, New York, 1992.
10 Pavesic, D.V. Prime Numbers: Finding your menu's strengths, *Cornell Hotel and Restaurant Administration Quarterly,* November, pp 71-77; 1985.
11 Hayes, D.K., & Huffman, L. Menu Analysis - A Better Way, *The Cornell Hotel and Restaurant Administration Quarterly,* February. Pp 64-70; 1985.
12 Beran, B. Menu Sales Mix Analysis Revisited: An Economic Approach, *Hospitality Research Journal,* 18(3), pp. 125-141, 1995.
13 Fuller, J. *The Menu, Food and Profit,* Stanley Thornes, London, 1991.
14 Kotschevar, L.H. 2nd. (Edn.), *Management by Menu,* Wiley, Lansing, Mich., 1987.
15 Harris, N. *Service Operations Management,* Cassell, London, 1989.

Incorporating customer choice in menu planning

J. Eastham

School of Leisure and Food Management, Sheffield Hallam University, UK

Abstract

This paper examines the value and possible impact of in-house sales transaction information on menu planning and explores classifications based on social context and methodologies to generate additional information on the consumer. This information can then be utilised to assist in menu planning.

Introduction

Information on sales transactions may now be generated much more rapidly as a consequence of the growth of computers, both in the form of Electronic Point of Sales (EPOS) and computerised accounting procedures. Information on items sold, turnover, ingredient costs, and gross profit, can provide managers in large organisations greater knowledge of day to day financial activity.

More sophisticated menu planning techniques established in the 1970's Lyons[1], and 1980's Kasavana[2] have provided the foodservice manager with a means of obtaining higher levels of information on the customer transaction in terms of sets of items chosen and their value.

By building on concepts of situational factors, Sinha[3], and other current trends of thought relating the context to choice, Dickson[4], Foxall[5], Thaler[6], and on the social status of food, Farb & Armelogos[7], both particularly significant in an environment in which contexts for an individual are potentially diverse and variable, menu planners could utilise information on the nature of the meal function and social context as a means of understanding consumption patterns. This would involve collecting information on social contexts of groups, individuals within the context and their respective choices, by those closest to the point of sale, i.e. the waiting staff.

In an environment in which there are increasing trends of mass customisation, the benefits of such a system are seen to in the development of more customer specific menus which should lead to higher profitability.

Aims

The aims of the research were to:

1. Explore a means of classification based on the social context of meals eaten in foodservice establishments.

2. Propose a methodological protocol for the generation of more detailed information on customer choice for the purpose of menu planning

Methods

Following a pilot study, research was undertaken to classify customers into market segments based on social context and function, and methods by which this information might be collected in two disparate food environments; a 3 AA star seaside hotel, and a popular food style Italian restaurant. Additional research into the feasibility of enriching the working practices of waiting staff took place in a carvery restaurant. Research sites have tended to be independent establishments primarily due to ease of access.

Data captured on the customer

To collect data the researcher took the role of a member of the waiting team. Although the waiting staff were aware of the research, and often participant, customers were not aware of being observed. Data collected consisted of customer profile data, data on social function and social context (figure 1), plus additional information such as time of arrival, relationship between individuals.

Once collected, information was linked to existing front and back of house systems which recorded information from guest checks/till receipts, recipe files, inventory files, invoices, and menu costings.

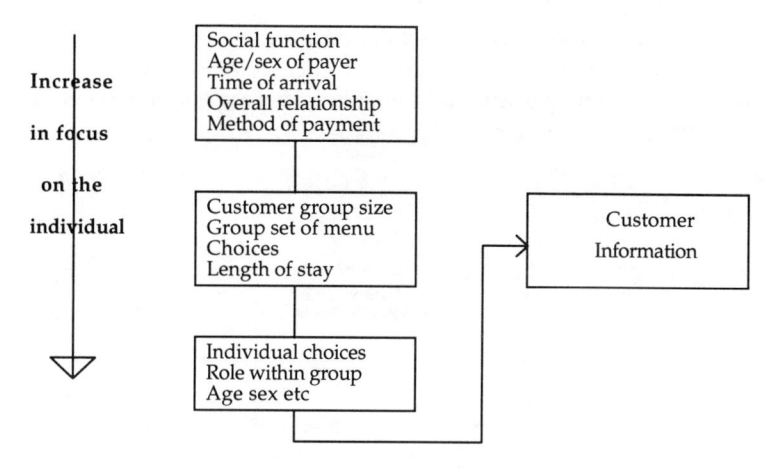

Figure 1 : Data to be captured on customers' characteristics

Data extraction

The nature of the information that could be extracted could be either on the basis of value and/or actual items chosen, and provide information on principle and subsidiary market segments as defined by social context (figure 2).

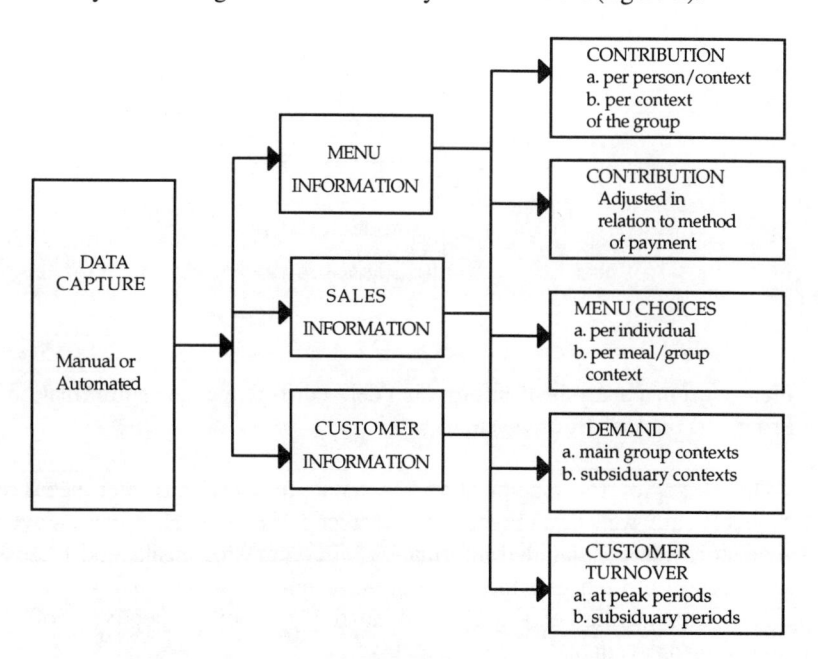

Figure 2: The nature of information which can be derived from customer information and Front and Back of House systems.

This information was then used to determine volume, revenue, contribution of each market segment, and their relative significance as well as their respective time of consumption, particularly significant if the planner should devise a system of differentiated pricing and menus designed to specific market segments (see figure 3).

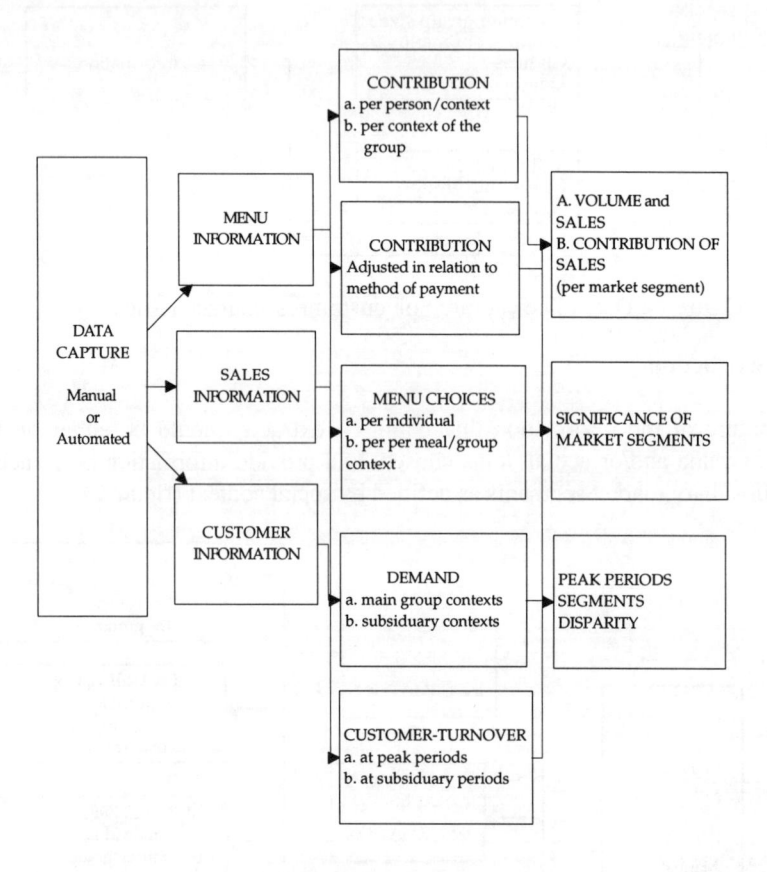

Figure 3: Final analysis of information derived from customer information, and Front and Back of House systems.

The strategies for manipulating the menu to meet customer needs more effectively, are well documented, e.g. Miller[8], Hayes & Huffman[9], Pavesic[10]. However, if there is detailed information on contexts of meals, and these were incorporated into the existing techniques, a more accurate analysis of current against potential market segments in relation to their respective profitability could be made available.

As seen in Table1 there were a number of different levels of extraction of data which could be used to determine the comparative value of the actual against the potential market.

Table 1. Possible ways of processing data (additional information e.g. volume of sales etc., would be recorded but is omitted here)

	Revenue	Contribution	Ratio	Significance of information
1. Per segment	Total Revenue	Total Contribution	Ratio based on total	Indicator of current profitability of segments
	Spend	Contribution		
2. Per market segment	1. Average Spend 2. Highest Spend	1. Average 2. Highest	1. Average 2. Highest	-benchmark
3. Individual customers within market segments	1. Average 2. Highest	1. Average 2. Highest	1. Average 2. Highest	- benchmark
4. Customer groups within market segments	1. Average 2. Highest	1. Average 2. Highest	1. Average 2. Highest	- indicators for potential marketing strategy
5. Individuals within market segment	1. Average 2. Highest	1. Average 2. Highest	1. Average 2. Highest	potential price and cost ranges

Information could then be used to set or adjust: *the menu content, pricing strategy* and *pricing structure*. However, the information may also enable the menu planner to devise *different pricing structures, strategies* or indeed *different menus* for different times of day or week in order to generate a more customised menu planning process and thus optimise profitability.

Results

Research site

A 3 AA star privately run hotel in an English seaside town offered 3 course table d'hote menus for lunch and evening meals (Table 2).

Table 2 The range of prices and contributions per meal type

	Range of contribution	Price range
Lunch	£ 3.15 - £ 4.88	£ 6.95
Sunday lunch	£ 5.87 - £ 6.85	£ 8.25
Dinner	£10.83 - £11.66	£13.50

Methods of analysis

Some 30 categories of customers were observed, ranging from; *holiday meals for a family; occasional meals with extended family* and *regular main meals for married couples.*

As indicated, there are a number of possible levels of extraction. Two are to be demonstrated; per segment and per customer group. These will enable us to make a comparison of profitability of actual /potential market.

Table 3 - Overall expenditure and contribution per segment

	Revenue £	Volume	% Volume	% of Total Revenue	Contribution	% of Total Contribution
Regular main meal (lunch) married couples)	244.94	20	15.4	8.6	165.7	8.0
Occasional meals Sunday lunch family	441.31	9	7.0	15.5	323.1	15.7
Regular meals for 1	144.44	25	19.2	5.1	94.51	4.6
Occasional meals married couple main meal	275.98	9	7.0	9.6	197.4	9.0

Within the research site 4 segments were considered most significant *regular main meals, occasional Sunday lunches for the family, regular main meals for 1* and *occasional meals for married couples).* These were selected on the basis of total revenue and contribution but as is evident from the volume and % columns, that already the menu planner would need to consider the relative value of market segments which generate a high volume of sales as against those with lower volume but higher contribution and revenue per group. There are a large number *of regular meals for 1,* those who tend to choose low cost, low price items, i.e. value and price, whilst family Sunday lunches are certainly much less value conscious.

Using this information the planner may decide to build on the largest market segments, and develop a marketing campaign to promote family Sunday lunches.

However, if information were extracted either on average / highest spends and contribution per group, and compared with the average and highest figures for all groups, a distinctly different picture of the respective importance of market segments may emerge. In Table 4, there is evidence that although *occasional Sunday lunch for family* is still recognisable as one of the most significant market segments both in terms of contribution and expenditure, it is less so than either *get together with friends* or *holiday dinner for family.*

In Table 4 categories of *regular main meals, regular meal for 1, occasional Sunday lunch for family* appear to generate significant total expenditures but per customer unit will generate considerably lower than that generated by other segments.

Table 4: Mean and highest spends and contribution per customer groups

	Contribution (mean) £	Revenue (mean) £	Contribution (Highest) £	Revenue (Highest) £
Special lunch for family	49.58	68.44	69.41	86.12
Holiday dinner for family	58.80	35.90	73.00	108.00
Occasional Sunday lunch for family	35.90	49.00	45.00	60.50
Dinner with friends	38.28	48.60	40.70	54.30

If managers were making decisions on the basis of information presented here, *holiday dinners for the family* and *special lunch for the family* may prove to generate the highest; highest and average spends and contribution per customer group. The spread of expenditure and contribution within these segments are also notably wider. This would indicate not only the potentiality of these market segments but also an appropriate pricing strategy.

Discussion

The purpose of this paper was to explore the possible value and impact of consumer specific information on the process of menu planning. The process of categorising profitability either by market segments, Jones & Hamilton[11], or by menu items choices[2,9,10], or differentiating menus and prices accordingly is certainly not original.

The originality of the research is the appreciation of the impact of the interrelationship between group and individual on choice. Restaurants can be seen as a micro-environment in which the culture of a society is reinforced, Finkelstein[12] Thus the same customers use the same restaurant for diverse occasions, Miller and Ginter[13]. Following existing research it is pertinent that: people encode value for money differently according to the social context, Fergueson[14], and specific types of meals and cooking methods are seen as symbolic to specific types of social event, Simoons[15]. Therefore in two types of social context, one would anticipate that there would be differences of; expenditure, value, and food, e.g. a business lunch and an evening with spouse, (the differences would be reinforced if the business person pays with a company credit card, on expenses).

Furthermore, in the case of individual customer choice, in context of the group and the function of the meal for the individual, it was evident that there were other important factors to consider. Role or desired role of an individual, Jerome, Kandel & Pelto[16], Devault[17], Conner[18], has been found to be a particularly significant influence; on food item choice, on value (cost-price) and on price of choice, in a number of research sites, whether in the context of business lunches, family meals,or romantic evenings out.

In additional dimensions to role, related to acceptable choices in relation to image, Foxall[5], and hidden benefits associated with the meal for the individual, Fergueson[14], i.e. choice patterns for married or committed women changed according to the nature of the occasion, less care was apparently taken to maximise ingredient cost and minimise price of items on Sunday lunch-time with the family than other identifiable occasions.

Further considerations

Further work is needed in the design of the system of collection and processing. Current employment practices in foodservice restrict initiatives in staff empowerment, as staff are largely part-time, poorly paid and of low status, Eastham[19]. Pilot studies have highlighted the fact that young staff with no prior experience of service adapt well to the enrichment of their task.

Further developments of the system are to incorporate a clustering technique rather than the apriori method proposed. The benefits will be in the additional option of yielding common choice patterns across segments rather than the existing narrow perspective.

Additional issues to consider relate to the economic feasibility of such a system. Will the financial benefits gained warrant the level of investment required? Although I won't presume to enter fully into the debate, the work is in early stages. The increasing demonstrated need of other manufacturing and retail industries, white goods and food alike, for market information could be difficult to ignore. Other industries have found other solutions: loyalty cards, retail cards, but these I believe to be inappropriate in the light of the individual versus group dimension of contextual choice in food service environments.

Conclusion and Recommendations

Issues considered above are felt to be of particular significance to the planning of menus in the light of current trends of increasing competition and movements of customisation. The knowledge of how customers choose as well as customer spend, food choice, and ingredient cost of items selected can facilitate the process of pricing, and promotion and product development. If the predominant set of social activities and groups at any single or combination of periods were also known, planners would have the opportunity of devising more specific menus for identified periods and market segments.

The method discussed is primarily a re-engineering of current systems employed in restaurants. Indeed, much of the information identified as required by the method is currently known and used by the menu planner. Yet in managed houses and franchised establishments although there is financial information available, there is not necessarily a means of obtaining qualitative information on the customer. The collection of such data could be relatively easy to incorporate into existing systems.

As previously indicated the system is being developed further. A statistical process which can be linked to a computer system needs to be established before the method could be confidently offered commercially.

However, even at this stage the value of the system is already evident. The results demonstrate more clearly than existing methods, the actual value of specific individuals, groups and segments to the foodservice operation, and in addition provide comprehensive information of individual choices in relation to the value of their sales.

REFERENCES

1. Lyons, H.M. The financial control of catering, Chartered Institute of Public Finance and Accounting Jnl Vol 1 No 10, 1974.
2. Kasavana, M.L. Computer systems for food service operations. Van Hostrand Reinhold N Y., 1984
3. Sinha, I. Conceptual model of the role of situational type on consumer choice behaviour and consideration sets, Advances in Consumer research 21, 1994, pp 477-482.
4. Dickson, P.R. Person-situation segmentation, the missing link, Journal of Marketing 46, Fall 1982, pp 56-64.
5. Foxall, G.R. The consumer choice as an evolutionary process: and operand interpretation of adopter behaviour. Advances in Consumer research 21, 1994, pp 312- 317.
6. Thaler, R. Mental accounting and consumer choice, Marketing Science 1985, Vol 4 no 3 summer pp 199-214.
7. Farb & Armelogos, Consuming Passions, Houghton, Mifflin company, Boston 1980.
8. Miller, S.G., Fine tuning your menu with frequency distribution, Cornell H. R. A. Quarterly, Nov 1988. pp 86-92.
9. Hayes, D.K. & Huffman, L., Menu Analysis a better way, Cornell H R A Quarterly, Feb. 1985, pp 64-70.
10. Pavesic, D. V., Cost/margin analysis: a third approach to menu pricing and design, International Journal of Hospitality management Vol 2 no 3 pp 127-134.
11. Jones, P. & Hamilton, D. Yield management: putting people in the big picture, The Cornell H.R. A. Quarterly, Feb 1992, pp 89-96.
12. Finkelstein, J., Dining out: A sociology of Modern Manners, Polity Press, Basil Blackwell Ltd, 1989.
13. Miller, K.C. & Ginter, J.L. An Investigation of situational variances in Product Choice Behaviour and Attitude, Journal of Marketing Research, 16 Feb 198 , pp 111-123

14. Ferguesen, D. H. Hidden Agendas in Consumer purchasing decisions. The Cornell Quarterly. May 1987. Vol 51 pp 1-10.
15. Simoons, F.J. Eat not this Flesh: Food avoidance in the old world, University of Wisconsin press, 1967.
16. Jerome, N.W. Kandel R.F. & Pelto G.H.(Eds), Nutritional Anthropology, (Ch1) Regrave publishing company, 1980.
17. Devault, M. Feeding the Family, University of Chicago press, Chicago, 1991. pp 227-243.
18. Connor, M.T. Understanding determinants of food choice-contribution from attitude Research, British Food Journal, 95, 9, 1995, pp 27-31
19. Eastham, J.F. Anthropocentric systems for the collection of market information,(Ed. P. Goodall) (page numbers currently unavailable) proceedings ALPS conference, Sunderland, September 1995.

Workplace caterers' attitudes toward providing healthier menus

M.J. Corney,[1] A. Eves,[1] M. Kipps,[1] C. Noble,[2]
[1] Department of Management Studies, University of Surrey, Guildford, Surrey, GU2 5XH, UK
[2] Applied Technology and Computing Department, Roehampton Institute, Roehampton, SW15 5PH, UK

Abstract

Analysis of a questionnaire-based study of workplace caterers' attitudes toward providing healthier menus is reported, along with differences related to attendance of an in-house nutrition training course and to the sex of the caterer. Correlation between the caterers' attitudes and their knowledge of government recommendations relating to consumption of items from the main food groups is also discussed. The data analyses reported should prove useful for planning strategies aimed at improving menu healthiness, where the caterers' active co-operation is required.

Introduction

The purpose of the study was to gain a theoretically-based understanding of caterers' views to inform government decision making. A primary objective was to identify the key beliefs that caterers hold regarding healthier menu provision. Very little, if any, psychological modelling of caterers' attitudes is to be found in the academic literature (e.g. Glanz, Hewitt & Rudd[1], Glanz & Mullis[2] & Poulter & Torrance[3]). Consumer studies tend to examine beliefs and nutrition knowledge of customers, along with the effects of the provision of nutrition information and other intervention strategies (e.g. National Consumer Council[4]).

The question of what attitudes are and how they relate to behaviour requires elaboration here. The layperson's definition of attitudes would probably be that they are beliefs about events or objects that lead to a more or less positive predisposition toward them. Psychological theory, as represented in a widely accepted model of attitude structure, generally supports common intuition, but stresses the importance of relating attitudes to specific behaviours, rather than generalities.

The most commonly used psychological attitude theory, certainly in health behaviour studies, is Ajzen & Fishbein's[5] Reasoned Action Model. The advantage that the model has over earlier theories is that attitudes are related to specific behaviours. The accuracy of the representation of an attitude in the model is measured by how well it predicts the behaviour to which the attitude is related, therefore the attitude measures must be behaviour specific. The shortfall of earlier theories was a lack of correspondence between attitudes and the behaviours they might be expected to predict. This was due to the relative generality of the attitude measures in comparison to the behaviours.

Providing vegetables means customers can make healthy choices *(probabilistic belief)*

Providing vegetables is *(attitude)*

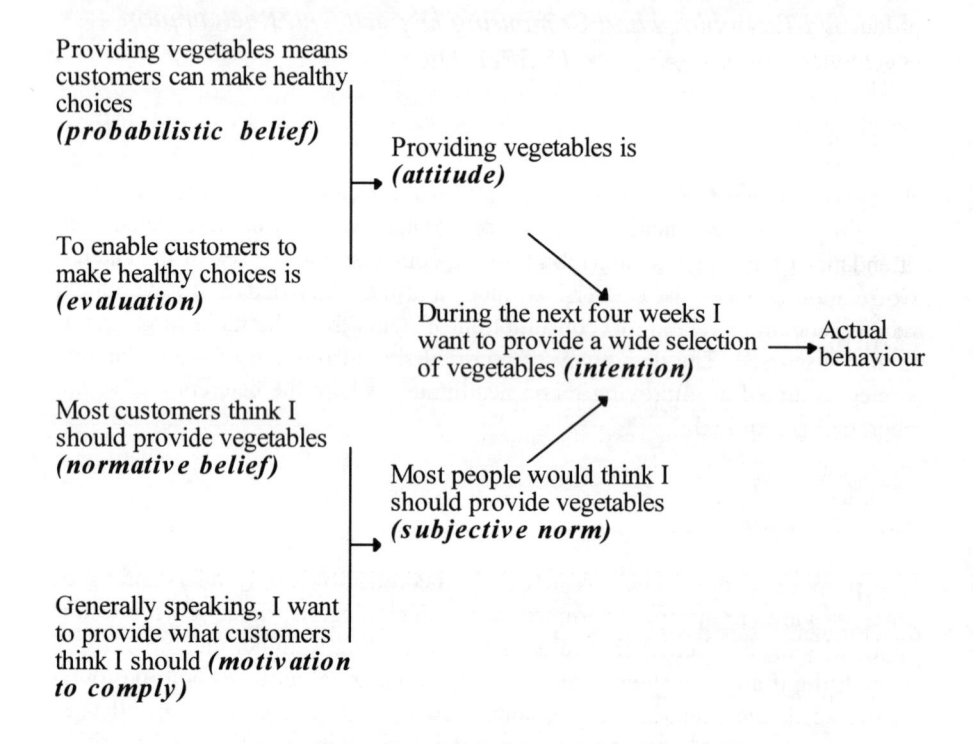

To enable customers to make healthy choices is *(evaluation)*

During the next four weeks I want to provide a wide selection of vegetables *(intention)* → Actual behaviour

Most customers think I should provide vegetables *(normative belief)*

Most people would think I should provide vegetables *(subjective norm)*

Generally speaking, I want to provide what customers think I should *(motivation to comply)*

Figure 1: The Reasoned Action Model.

In the model, attitude is seen to be made up of beliefs about performing a behaviour. In the study reported here the attitude in question relates to the provision of healthier menus. The behaviour relates to providing certain foods. Figure 1, shows that probabilistic beliefs are connected to attitude and attitude to intention and thence behaviour. The hierarchy of linkages indicates that the attitude formed leads to an intention and intention is the most direct predictor of behaviour.

The intention also receives input from subjective norm. Subjective norm is the perception of what other people think about the behaviour. Like attitude, subjective norm consists of specific beliefs. The normative beliefs reported here consisted of what caterers thought customers generally, slimmers, vegetarians and others would think about them providing more of the foods associated with healthier menus.

Both probabilistic and normative beliefs are multiplicatively weighted by additional measures. Multiplying each probabilistic belief by perceived outcome means the scores used to predict attitude reflect not only what is believed to be true, but also how good or bad the outcome is perceived to be. Similarly, multiplying each normative belief by a measure of motivation to comply produces a result that more fully reflects the significance of each belief to the behaviour.

Healthier menu provision was defined by creating an index of behaviours based on current government guidelines. Guidelines recommend the increased consumption of starchy carbohydrates, fibre-rich foods, fruit and vegetables, and the consumption of less fat.

Method

Participants

A Surrey based catering management company provided access to its staff. From 200 staff, 169 returned the questionnaires that they had been mailed, a return rate of 85%. Of these, 94 were female and 75 male; ages ranged from 19 to 58 years, with a mean age of 34 years. The company provides an in-house nutritional balance training course, 97 participants had attended this, 69 had not and three did not provide information on attendance. Participants completed the questionnaires in work time and were entered into a prize draw. Questionnaire development was assisted by preliminary discussions with two groups of caterers, comprising 16 caterers in total who were attending a training course at the company's head office.

Procedure and materials

The pilot discussions were used to ascertain generally held beliefs about providing the items on the healthier menus index. The pilot session groups completed a preliminary questionnaire asking about the perceived advantages and disadvantages of providing the foods, and who would approve and disapprove. Beliefs occurring most frequently were incorporated into the main questionnaire. Probabilistic beliefs were as follows:

'Providing starchy carbohydrates (e.g. potatoes, bread, rice and pasta) means customers can...' '...make healthy choices', '...choose filling foods', '...means variety on menus', '...can be a problem because they dry out'.

'Providing vegetables (e.g. peas, leafy green vegetables or salad vegetables) means customers can...' '...make healthy choices', '...means variety on menus', '...means waste because customers don't choose them', '...means extra expense'.

'Providing fruit and fruit juice means customers can...' '...make healthy choices', '...means variety on menus', '...means extra expense', '...means waste because customers don't choose them'.

'Providing oven baked rather than fried foods (e.g. chips) means customers can make...' '...healthy choices', '...means less fatty food for customers', '...means extra cooking time is needed'.

'Providing fibre rich foods means customers can make...' '...healthy choices', '...means waste because customers don't choose them'.

The normative beliefs were of the form 'Most ...think I should provide...'. For starchy carbohydrates 'most' were 'athletic types' and 'customers'. For vegetables they were 'vegetarians' and 'customers'. For fruit and fruit juice they were 'slimmers' and 'customers'. For oven-baked rather than fried foods they were 'health conscious customers' and 'manual workers'. For fibre-rich foods just 'customers'.

Responses were made by participants on fully labelled seven point Likert scales. Scale labels ranged from 'very unlikely' to 'very likely', with 'neither' as a mid-point, except for the evaluation and attitude measures which ranged from 'very bad' to 'very good'. 'Quite' and 'slightly' were used as intermediate values on all scales.

Knowledge of current nutritional guidelines was obtained using a multiple response checklist asking whether 'current guidelines on healthy eating advise eating more, less, or the same of the following nutrients?'. The nutrient list used consisted of fat, starchy carbohydrates, fibre, saturated fat, sugar and fruit and vegetables. Participants could check either 'more', 'less', 'same' or 'don't know' boxes for each nutrient.

The weighted scores were used in data analyses, so where probabilistic or normative beliefs are referred to they have been multiplied by the corresponding evaluation or motivation to comply score. Comparison of mean scores between groups, by sex of participants and attendance of in-house nutritional training, was made using the non-parametric Kruskall-Wallis ANOVA procedure, as the data were negatively skewed. For the same reason, non-parametric Spearman rank testing was used for correlation analyses that tested links between attitudes and in-house nutritional training. Principal components analysis was used to reduce the number of variables for the central attitude study. The components produced were used in multiple regression analysis. They were used to predict each caterers' overall intention on the index of healthier menu provision. The sum of the five intention scores was used as a measure of their overall intention.

The principal components analysis reduced the number of variables by examining the interrelations between variables. Variables that are highly correlated may be explained by an underlying component. The procedure grouped highly correlated variables so that high correlation within a group was matched by low correlation between groups. The difference between groups were maximised by rotating the axes that account for maximum variation within the data of each group of variables, until the best contrast between groups was obtained. Every variable is given a loading on each of the derived components. Loadings below 0.5 were discounted as this makes the task of interpreting components much easier. The components were then subjectively interpreted and labelled.

Loadings for the variables on each component were saved and used in the subsequent multiple regression analysis. The multiple regression analysis revealed the relative importance of each of the derived components to the caterers' intentions to provide healthier menus. A measure of the usefulness of the prediction obtained is indicated by the multiple correlation value, R. The value of R varies between zero and one, with one indicating perfect prediction of the dependent variable. The relative importance of the components is shown by the value of the beta weights.

Results

Analysis of male and female caterers' beliefs showed significant differences on three beliefs relating to baking rather than frying and providing fibre-rich foods (see table 1). The female caterers were consistently more likely to believe that their customers think that these items should be provided than their male counterparts.

Table 1: Significant differences between scores from male and female caterers.

	Male caterers' mean scores	Female caterers' mean scores	p
Most manual workers think I should provide baked rather than fried food	-5.15	0.47	***
Most customers think I should provide baked rather than fried food	1.08	7.64	***
Most customers think I should provide fibre rich foods	3.08	7.01	**

*** $p<0.001$ ** $p<0.01$

The in-house nutritional training clearly produced desirable results. Those who had attended training were consistently more positive in their views toward providing the foods and they were more likely to perceive that their customers want them. Their overall intention to provide healthier menus was significantly higher than those who had not attended the training.

Table 2: Significant differences between scores from those who had attended in-house nutritional training and those who had not.

	Training not attended	Training attended	p
Providing starchy carbohydrates can be a problem because they dry out	-2.15	-0.23	**
Providing fruit and fruit juice means extra expense	0.01	1.02	*
Providing fibre rich foods means customers can make healthy choices	5.93	7.48	**
Customers think I should provide starchy carbohydrates	2.12	5.13	*
Customers think I should provide vegetables	12.87	16.26	***
Customers think I should provide fruit and fruit juice	13.44	16.67	***
Customers think I should provide fibre-rich foods	2.63	6.82	**
Total score for probabilistic items on providing fibre rich foods	6.88	9.23	**
Total score for normative items on providing fruit and fruit juice	27.09	30.89	**
Total subjective norm score	7.14	9.14	***
Total intention score	8.65	10.80	***

*** $p<0.001$ ** $p<0.01$ * $p<0.05$

Table 3: Principal components analysis of the probabilistic and normative scores with loadings shown for each variable (explaining 67% of data variation).

Health		Costs		Variety		Baking	
Fruit & Juice provides healthy choices	0.88	Providing vegetables means waste	0.72	Providing vegetables means variety	0.82	Customers want baked food	0.88
Baking provides healthy choices	0.85	Providing fruit & juice means waste	0.71	Providing carbohydrates means variety	0.74	Manual workers want baked food	0.81
Vegetables provide healthy choices	0.83	Providing fibre rich food means waste	0.70	Providing fruit & juice means variety	0.69	Customers want fibre rich food	0.61
Fibre rich food provides healthy choices	0.77	Providing vegetables means expense	0.62	Carbohydrates are filling	0.58		
Carbohydrates provide healthy choices	0.74	Providing fruit & juice means expense	0.57				
Baking means less fatty food	0.62						

Table 3: Continued.

Fruit juice & vegetables		Carbohydrates		Not significant		Cooking time	
Customers want fruit & juice	0.80	Customers want carbohydrates	0.72	Carbohydrates tend to dry out	-.60	Baking means extra cooking time is needed	0.75
Customers want vegetables	0.65	Athletic types want carbohydrates	0.62	Vegetables provide variety	0.53		
Slimmers want fruit and juice	0.60						

Principal components analysis reduced the 27 questionnaire items to a set of eight underlying components. The amount of variation explained by each component is reflected in the ordering, as shown in table 3. The component labelled 'health' explained the most variation (23%) and the component labelled 'cooking time' the least (4%).

The amount of variation explained by a component does not necessarily indicate its predictive capacity however. Although the 'fruit juice and vegetables' component, consisting of items relating to customers' and slimmers' perceived needs of these, explained much less data variation than the 'health' component (5.4%), it was the most predictive of intention (see figure 2). The predictive capacity of the other components was generally matched by their importance in the principal components analysis.

The results indicate that the caterers' perception of their customers desire for fruit, fruit juice and vegetables, taking into account the caterers' willingness to provide them, is a key factor in determining their intention to provide healthier menus. Somewhat less important, but still significant, are the 'health' and 'costs' components. The positive beta weight given to the costs factor indicates that the caterers did not perceive costs to be an obstacle to their intention to provide healthier menus.

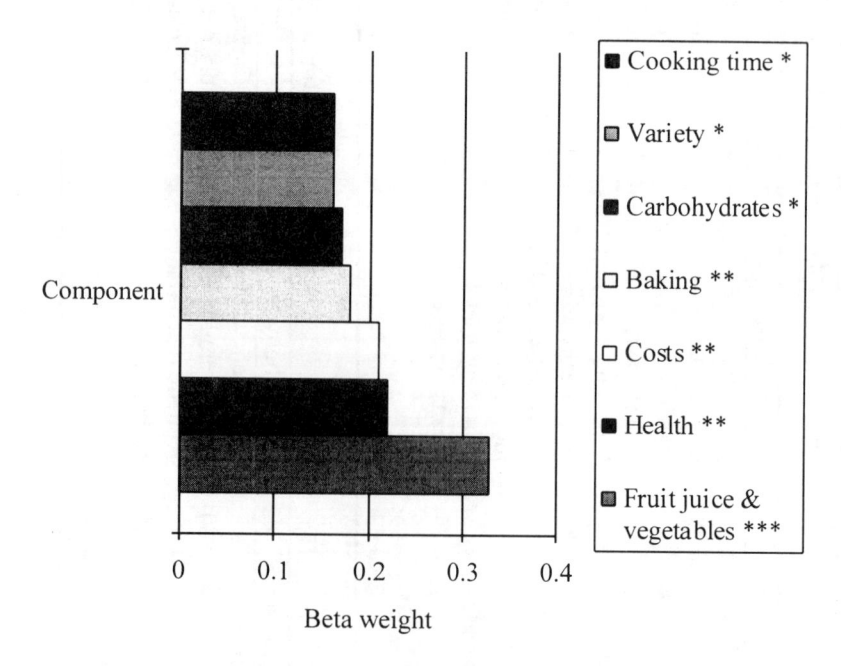

Figure 2: Relative importance of the components making up caterers' intentions, showing coefficients (beta weights) from stepwise multiple regression analysis (R.56). *** $p<0.001$ ** $p<0.01$ * $p<0.05$

There was a significant ($p<0.001$) positive correlation (0.33), between the caterers' attitudes and their score on the nutritional knowledge test. Separating those who had in-house training from those who had not, revealed the link was twice as strong with those who had not had training (0.22, $p<0.05$ versus 0.46, $p<0.001$). This result, taken together with the similarity between both groups' attitude and knowledge scores would suggest that the differences listed in table 2, showing significantly more positive beliefs, views and intentions, are due to factors other than attitude or knowledge.

There are three reasons that might explain the more positive intentions from those who had received training: First, the significantly more positive intentions of attendees' would have been influenced by the commitment shown by their employer through their provision of training. This would help to reinforce the belief that providing healthier menus is important and worthwhile. Second, attitudes were already generally positive, therefore the training was more helpful in increasing beliefs that their customers would want healthier menus. The analysis of attitudes showed that perceptions of customers' wants, and motivation to comply with them, are foremost in the intention to provide healthier menus. Third, the training provided culinary expertise, raising attendee's' skill levels and perhaps their perception that they could successfully prepare healthier menus. This self-efficacy factor has been found important in numerous health-related studies (e.g. Contento & Murphy[6] & Schwarzer & Fuchs[7]).

Conclusions

Perception of the customers' desire for fruit, fruit juice, and vegetables and motivation to provide them, was the factor which best predicted intention to provide healthier menus. Intervention strategies should therefore concentrate on convincing caterers that their customers want healthier menus. Caterers' motivation should be assisted by nutritional training that emphasises the importance of healthy eating and provides the skills necessary for caterers to be able to produce healthier menus. A recent survey has shown that many caterers enter the profession ill-equipped to provide more healthy menus, owing to limited (or no) nutritional training in catering related courses (Eves, A., Corney, M.J., Kipps, M. & Noble, C.[8]). Caterers' training should increase their knowledge of current dietary guidelines, whilst showing them that their adoption will not lead to increased costs or waste. Training provided by the employer is particularly valuable as it provides directly applicable skills and increases the perceived importance of their use.

Acknowledgements

This work was funded by the Ministry of Agriculture, Fisheries and Food. The authors are also grateful to the catering management company for enabling us access to their staff and to the caterers for their voluntary participation.

References

1. Glanz, K., Hewitt, A. & Rudd, J. Consumer behaviour and nutrition education: An integrative review, *Journal of Nutrition Education*, 1992, **24**, 267-277.

2. Glanz, K. & Mullis, M. Environmental interventions to promote healthy eating: A review of models programs and evidence, *Health Education Quarterly*, 1988, **15**, 395-415.

3. Poulter, J. & Torrance, I. Food and health at work - a review. The costs and benefits of a policy approach, *Journal of Human Nutrition and Dietetics*, 1993, **6**, 89-100.

4. National Consumer Council. *Consumer Information Needs Concerning Food Sold Through Catering Outlets*, HMSO, London, 1993.

5. Ajzen, I. & Fishbein, M. *Understanding Attitudes and Predicting Social Behaviour*, Prentice-Hall, Englewood Cliffs, 1980.

6. Contento, I. & Murphy, B. Psycho-social factors differentiating people who reported making desirable changes in their diets from those who did not, *Journal of Nutrition Education*, 1990, **22**, 6-13.

7. Schwarzer, R. & Fuchs, R. Self-efficacy and health behaviours, Chapter 6, *Predicting Health Behaviour: Research and Practice with Social Cognition Models*, eds Conner, M. & Norman, P., pp 163-196, Open University Press, Buckingham, 1996.

8. Eves, A., Corney, M.J., Kipps, M. & Noble, C. Nutrition education of caterers in England and Wales, *International Journal of Hospitality Management*, 1996, in press.

Catering for vegetarians - a challenge or a chore?

A.E. Sharples
School of Leisure and Food Management, Sheffield Hallam University, Sheffield, UK

Abstract

This paper presents an overview of the current catering provision in a number of selected eating-out venues in the Sheffield area, in order to evaluate their level of commitment to catering for vegetarians and to assess the expertise held within the organisation regarding the production of vegetarian food.

The paper concludes that the industry is in a transitional stage, and whilst some establishments have reshaped their catering polices to cater for vegetarians more responsibly and innovatively, others have been more hesitant in their approach. The results of the study revealed that economic and cost implications are key factors in establishing catering policy and any attempt to improve the standards of vegetarian catering must bear this in mind.

The results of the investigation also indicated that some caterers may need to improve their knowledge about vegetarian requirements if they are going to take full advantage of this increasing market opportunity.

1 Introduction

The relative success or failure of a catering business often hinges on the ability of the management team to respond effectively and responsibly to changing market needs. In recent years, the caterer has had to cope with a rapidly changing environment, tackling the increasing popularity of fast food, the development of 'grazing' habits in preference to set meal times, and a growing concern about the healthiness of the foods we eat.

The increase in the number of people adopting a vegetarian diet or significantly reducing their meat consumption is well documented (Realeat Survey[1]) and presents the industry with yet another important challenge but one which is not always straightforward. This is due to the fact that there are several 'confusing' issues which can have a major impact on the way that the

caterer anticipates and copes with this new market need.
- There is a wide spectrum of people who describe themselves as 'vegetarian', ranging from strict vegans to demi-vegetarians. This typology is discussed in various studies (e.g. Beardsworth and Keil[2]). Research has also indicated that it is not unheard of for some vegetarians to have occasional lapses in to meat consumption in certain situations, for example when vegetarian choices are not readily available (Beardsworth and Keil[2,] and Wright and Howcroft[3]). This situation could create difficulties for the caterer in attempting to make accurate predictions about the demand for vegetarian food.
- People become vegetarian for widely differing reasons (Beardsworth and Keil[2] and Leatherhead Food Research Association[4]) and this can have a major impact on food selection patterns and the acceptance or otherwise of certain commodities. For example, if a person becomes a vegetarian for health reasons they may be happy to eat battery produced eggs, but a person driven by moral grounds may not. This creates a considerable task for the caterer in attempting to provide a menu and recipes to suit everyone.
- The whole issue of vegetarianism is a highly emotive subject and it is likely that both caterer and customer will have their own strongly held viewpoint about where they stand in the 'pro-meat' or 'anti-meat' debate. These personal opinions are bound to have some effect on the commitment of the meal provider and the tolerance of the consumer.

For these reasons the caterer could be confused about the scale of their responsibility and in an attempt to be 'all things to all people' plays the safe card in providing the standard 'vegetarian choice' on the menu. Although this gesture may have been done with good intent, some vegetarians may view this as the caterer paying little more than lip service to this market need. Caterers are often torn between concentrating on the main bulk of the customer base who want an omnivorous diet with the risk of alienating the vegetarian customers completely, or investing some time and effort into attracting more non-meat eating clients, in a hope that they will also bring their families and friends. The question could be asked - 'how much energy should be directed towards 5 - 10 % of the customer base'?

The data collected during this study will help in establishing the current state of play within the industry, and act as a catalyst for implementing good practice for vegetarian catering.

2 Methodology

The data collection technique used in this study consisted of eighteen in-depth interviews held with catering managers within the Sheffield area. Sheffield was chosen because despite ranking as one of the five largest cities in the country, it has only one specialist vegetarian restaurant, suggesting that many vegetarians in the area rely on main-stream catering operations. In an attempt to draw some comparisons between different sectors of the industry, interviews were held with managers of six hotel restaurants, six independently-owned restaurants and six pub restaurants. The half hour interviews were semi-structured, being based around a series of questions, but kept deliberately flexible to encourage free discussion. The interviews focused on the current menu provision within the outlet, the food production techniques used, the awareness of the staff, and the perceived advantages/disadvantages of providing vegetarian options. Due to the intricate and complex nature of the information that was collected, it was inappropriate to carry out purely statistical analysis, but themes were focused on using content analysis.

3 Background

There is now a wide body of knowledge emerging that indicates a consistent upward trend in the number of people describing themselves as vegetarian within the UK. The series of surveys carried out over the last twelve years by Gallop Poll Limited (Realeat[1]), provide the most consistent source of data on the growing vegetarian market. Their studies are supported by a host of other reports, for example Jones Rhodes Associates([5]) which fundamentally echo the Realeat findings. Principal conclusions of the Realeat Study include:
- that as many as 4.5% of the UK population may now be vegetarian, which is more than double the figure of ten years ago;
- there has been a significant upward trend in the number of people who have consciously reduced their red meat consumption. The study suggests that as many as 26% of the population may now eat red meat only 'occasionally';
- that the female population lead the way in picking up the vegetarian banner and it is estimated that there are currently twice as many vegetarian women as vegetarian men. Growth is particularly noticeable amongst the 16 to 24 year age group where it is estimated that as many as 12.4% of females are completely vegetarian;
- that vegetarianism appears to be most popular in the south east of the country which has a 5.2% vegetarian population, higher than the UK average;
- that growth of the vegetarian market has been recorded amongst all social grades, but the higher socio-economic groups lead the way with 6.2% of 'A/B's' classifying themselves as vegetarian.

An examination of meat consumption figures within the UK reveal a lack of agreement between the 'expert' bodies but it is safe to conclude that the amount of red meat being consumed in the home has declined in recent years and the consumption of poultry has increased (National Food Survey[6] and MAFF Statistics[7]). The Meat and Livestock Commission (MLC) however, are quick to point out that catering sales of red meat are 'on-the-up' (Dixon[8]), suggesting that people still like to have their steak when eating out.

Another indicator of the growing popularity of vegetarian food is the rapid growth of the vegetarian ready meals market, which in 1994 recorded an annual turnover of £100m in the frozen meat-free product sector (Ross Young Report[9]). The brands which have been most successful are the Linda

McCartney label (25% total sales) and the Birds Eye Vegetarian Cuisine label (22% total sales).

The reason for people adopting a vegetarian diet are personal and complex and influenced by an individual's experiences and beliefs. In 1993 Beardsworth and Keil([2]) proposed that the moral/spiritual path to vegetarianism was still the most significant, but the most recent Realeat([1]) survey of 1995 cites 'health reasons' as the primary motivation for eating less meat.

The violent clashes at Shoreham in 1994 over the live export of veal calves brought home the message of animal welfare in a very emotional and public way, and undoubtedly caused many people to re-examine their own feelings about modern farming methods. The recent scares over BSE have added more fuel to the fire and created a tense situation for both the government and the MLC. The crisis is far from resolved (Dealler and Kent[10]) and whilst the experts continue to discuss and debate the issues it is safe to assume that some of the population have decided to reduce the beef consumption until they have more concrete information. The debate over safe meat production is likely to continue until a more comfortable balance is reached between producer and consumer, and in an effort to promote meat as a safe and healthy commodity, the meat industry has implemented several quality assurance schemes (Dixon[11]) and advertising campaigns (The Grocer[12]).

The claim made by some that a vegetarian diet is more healthy and could prolong life is another case that is not clear cut. A study carried out into the nutrient intakes of vegetarian and non-vegetarian women, (Janelle and Barr[13]), concluded that current nutritional recommendations could be attained by both sectors as long as they were health conscious. Another report, published in the British Medical Journal in 1995 (Thorogood, et al[14]) concluded that vegetarians could suffer 30% less heart disease, 40% less cancer and 20% less premature mortality than meat eaters. The report stressed, however, that diet is one factor affecting health alongside other factors such as smoking and alcohol intake. These studies were supported by the British Nutrition Foundation in a Briefing Paper([15]) which recognised the possibility of achieving good and bad vegetarian and non vegetarian diets. It is also vital to remember that those who decide to adopt a vegetarian, and in particular a vegan diet could also be prone to a range of health risks, eg. mineral, vitamin deficiencies, if they do not pay sufficient attention to their nutrient intake.

When the Cranks restaurant opened in London in 1962 the name of the restaurant reflected how people viewed vegetarianism at that time. Now, of course, *"the lonely furrow that Cranks then ploughed has become the broad highway for a great many"* (Swann[16]). The current vegetarian market is a very different one from that of 25 years ago, with vegetarians no longer perceived as oddballs of society but as mainstream consumers who want to dine out with family and friends and be able to choose from a variety of different eating experiences. As caterers, maybe it is up to us to be asking 'how' rather than 'why'?

One positive move in this direction was the recent formation of Caterveg, an organisation whose brief was to educate and inform caterers about responsible vegetarian catering. Several schemes have been implemented by this organisation within the industry. These include an approval system for restaurants who have attained certain standards of vegetarian catering and an accreditation scheme for college courses which include vegetarian catering in their cirriculum.

4. Findings

The results of the study will be discussed under four main headings, namely; current menu provision, food production methods, staff awareness and the perceived advantages/disadvantages of providing vegetarian foods.

4.1 Current Menu Provision

The menus at all of the restaurants were scrutinised in order to identify the proportion of the menu that was devoted to vegetarian dishes and the types of dishes that were being provided.

Out of the six pubs that were visited, four of these used a static menu supplied by head office and two outlets used a blackboard system that was changed daily. All of the pubs operated under a brand. Five of the hotels used in the study operated under a brand and all of these used a static menu of some type. The other hotel was owned by a smaller company but was of a comparable size and rating to the other hotels. This hotel used a table d'hote menu. Two of the restaurants used a blackboard system, and the remaining four, used printed menus that were changed periodically. Discussion with the managers who used menus that changed on a daily basis confirmed that the menus on the day of the visit were representative of their normal provision.

4.1.1 Normal Menu Item Versus Special Provision
Several key points were noteworthy:
- sixteen of the eighteen restaurants visited included vegetarian options as a standard choice every day and only two of the restaurants stated that they would prefer advance notice if vegetarians planned to dine there. It was interesting to note that these two outlets were both from the 'independently-owned' sector where the proprietor/manager undoubtedly had more control over the menu composition than in a branded operation;
- provision for vegan consumers was considerably more limited, with only six of the restaurants expressing confidence about catering for this market sector without prior warning. The majority of the outlets felt that they had the ability to cater for vegans but would need notice in order to obtain suitable commodities. This was particularly important in outlets that served primarily 'bought-in' menu items, many of which contained dairy products;

- twelve out of the sixteen outlets which provided vegetarian choices on a daily basis had made the decision to include these dishes as part of their normal menu, but four of the operations preferred to offer a separate vegetarian menu. One restaurateur stressed the importance of ensuring that vegetarians are made to feel part of the normal customer base rather than customers requiring a specialist menu, but another manager suggested that a separate vegetarian menu could possibly offer more choice;
- Half of the operations visited thought that it was a good idea to highlight the menu with symbols, e.g. (V), so that vegetarian customers would feel confident about making the right selection, but the remainder thought that this was unnecessary and four of the managers said that this action could result in isolating vegetarian customers. It was also identified that meat-eating customers may feel less inclined to choose a vegetarian option if a V symbol was present, with the thought that 'it must just be for vegetarians'.

4.1.2 Proportion of Vegetarian Dishes on the Menus

4.1.2.1 Pub Sector All of the outlets had at least two vegetarian starters on their menus and several had as many as four. This obviously varied depending upon the size of the menu, but between 25 to 33% of dishes on the starter sections were suitable for vegetarians.

All of the pubs had at least two vegetarian main courses on their menus and one outlet had a choice of twelve! The data showed that between 10 to 50% of dishes on the main course sections were vegetarian. Several of the pub managers revealed that because they were branded operations, 'head office' had stipulated that a certain proportion of their menus must be devoted to vegetarian food. The manager of the pub which included twelve dishes on the menu had introduced a specialist menu because she felt that her 'softly branded' menu provided by head office was too limited for her clientele.

4.1.2.2 Hotel Sector All of the hotels had at least two vegetarian starters on their menus with some having four choices. The proportion of the starter sections that was devoted to vegetarian dishes ranged between 25 and 60%.

The main course selection, in the hotels however, appeared to be less extensive than in the pubs with three of the six hotels only offering one vegetarian main course. The proportion of the main course sections devoted to vegetarian dishes ranged between 7 and 25%. This was suprising as it is anticipated that some customers dining in a good quality hotel would expect a better choice than in a pub. It is possible that many hotel restaurants are targeted towards businessmen as their main client-base who may prefer a traditional meat-based meal rather than a vegetarian option.

4.1.2.3 Independent Restaurant Sector This sector was far less consistent in its approach. All of the restaurants visited had a least one vegetarian starter on their menus, with one outlet having a choice of four. The proportion of the starter sections that was devoted to vegetarian foods ranged between 17 and 50%. The main course selection however, was far more limited, with the managers of two of the outlets stating that they preferred advanced notice to cater adequately. Both of these managers remarked that they sometimes put vegetarian dishes on their menus but that they could not guarantee its inclusion. Three of the other restaurants only offered one vegetarian main course on their menus, but it must be remembered that in many of these smaller restaurants this single choice is from a total of six dishes. The implication of only offering a one selection is that a party of vegetarians dining together would be forced to select the same dish, which can prevent sharing and positive discussion about the meal. The remaining restaurant, which was targeted towards a younger and less affluent clientele offered a range of at least eight vegetarian main courses each day. The proportion of the main-course menu which was dedicated to vegetarian food ranged in this sector between zero and 40%.

4.1.3 Type of Dishes Provided

4.1.3.1 Pub Sector The most popular starters that featured on the menus were deep fried or garlic mushrooms which were offered by all of the six outlets on the day of the interview. Another starter which featured widely was melon which was present on four of the menus. Other starters that were available in more than one pub were soup and deep fried potato skins.

The main courses offered by the pub sector were a far cry from the quiche or vegetable lasagne that was the main stay of many vegetarian menus a few years ago. Pasta dishes featured strongly with all of the pubs offering a baked pasta dish of some description. Other popular choices which were available in more than one of the outlets were rissottos or paellas, mushroom strogonoff, vegetable crumbles/bakes and vegetable pies/pastry dishes. None of the main courses available contained any meat substitutes, e.g. soya, Quorn or tofu.

4.1.3.2 Hotel Sector This sector presented a less consistent service to the vegetarian consumer, with several outlets displaying a more innovative use of ingredients whilst others offered menus which were more predictable.

Melon still featured strongly being available on four of the six menus in this sector, but the dish had been upgraded from the pub specification to contain exotic fruits, sorbets and liqueurs. Deep-fried and stuffed mushrooms were again popular with three of the six outlets including them on their menus. There was also evidence of more exotic dishes available such as vegetable and nut terrines, deep-fried cheeses and stuffed peppers, illustrating an attempt to provide a more up-market eating experience in this more expensive sector.

Several of the hotels also demonstrated a degree of skill and interest in the main course sections of their menus. One hotel included filo parcels, an aubergine bake and a Quorn and mixed bean casserole, whilst another offered a fricassee of wild mushrooms, a broccoli and ricotta strudel and a spicy couscous dish. Other hotels however, were far more conservative in their menu provision, serving 'safe' dishes such as pasta with a tomato and basil sauce.

4.1.3.3 Independent Restaurant Sector This sector presented a more limited choice to vegetarian customers, although some of the dishes offered in two of the restaurants were innovative and demonstrated a high level of technical skill.

Melon, and other fruit based starters, were again popular in this sector and to reflect the higher price range in several of the restaurants these were presented with coulis or liqueurs. Home-made soups were also evident on several of the menus, and goat's cheese served with salad or brioche was available on two of the menus. The main courses were generally less imaginative than some of those served in the hotel sector but two of the restaurants offered pastry tartlets filled with seasonal vegetables and accompanied by interesting sauces. Another outlet offered a pastry dish that was based around sun-dried tomatoes. In several of the outlets the vegetarian main course was significantly cheaper than the meat or fish choices signifying a lack of parity between the quality of the food offered to vegetarians and non-vegetarians. The outlet which had a less expensive menu offered pasta dishes, salads, spicy casseroles, vegetable bakes, pizzas and quiches.

4.2 Food Production Systems

Thirteen out of the eighteen outlets visited stated that they prepared most of their vegetarian dishes on site. The remaining five used 'bought-in' frozen or chilled dishes. In all of the outlets the food provision was consistent for both vegetarian and non-vegetarian items, i.e. if the outlet produced home-made food for meat-eaters then it also produced home-made food for vegetarians. The outlets that used 'bought-in' meals were primarily those which operated under a brand.

All of the managers when questioned about the use of commodities in vegetarian dishes were aware of the importance of using vegetarian stocks, fats and oils in their dishes. All of the outlets which used 'bought-in' products expressed confidence in their suppliers and stated that they could produce detailed dish specifications if requested by the customers.

Out of the eighteen managers, sixteen believed that their customers were not concerned about the use of vegetarian/non vegetarian cheeses in the preparation of dishes and several managers expressed the opinion that their clientele would not be willing to pay a higher price for the use of vegetarian cheeses. Only two of the managers stated that they used 'free-range' eggs in there dishes with 'quality' being cited by both as the primary reason. Only one restaurant used organic produce on a regular basis, but this was driven by a desire to provide a top quality product with a high flavour profile.

Only four of the restaurant managers expressed the view that the nutritional content of their food was important to their customers. The majority of the interviewees thought that most people when eating out felt the need to have a 'treat' and the managers did not believe that this attitude was any different between meat eaters and non meat eaters.

4.3 Staff Awareness

The interviews revealed that two thirds of the managers were aware of the increase in vegetarianism over the last few years, and were also aware of the demand pattern for vegetarian food in their own outlet. Three of the outlets estimated that as many as 10 to 15% of their customers chose vegetarian options. Six restaurants estimated their demand to be between 5 and 10%, eight restaurants quoted 5% and one restaurant stated that demand was as low as 2%. Outlets in the pub sector showed consistently higher demand than in the other two sectors. This may be due to the fact that the pubs appeared to have been reasonably proactive in their approach to vegetarian catering. Several of their managers expressed the view that a good vegetarian menu was capable of attracting business by providing interesting choices for both non-vegetarians as well as vegetarians. The restaurant where demand ranked only 2% had 'set its stall' by providing a menu centred around meat and fish and the manager had no intention of changing this policy, although he was happy to cater for vegetarians by specific request.

Eight of the managers said that they had witnessed a vast increase in the demand for vegetarian food in the last two years, five managers reported a slow but steady increase and four managers had not discerned any particular trend. One manager reported a slight reduction over the last year but recognised a general upward trend in demand. One manager also reported an increase in the demand for vegan meals. When asked about the reason for the increase, thirteen managers cited health reasons, four managers thought that BSE had been an inportant factor and one manager thought that moral reasons were still the most relevant. Other factors which were mentioned were the cost of meat and the desire to follow a trend.

Ten of the interviewees felt that both they and their staff were sufficiently knowledgeable to cater for vegetarians adequately, but the remaining eight managers felt that chefs should receive more input in their training to be able to cater for vegetarians more successfully. One manager thought that some chefs did not like to prepare vegetarian food and that they needed encouragement to use their skills in preparing more imaginative food for this sector.

4.4 Perceived Advantages/Disadvantages of Providing Vegetarian Food

4.4.1 Perceived Difficulties
Ten of the managers were reasonably happy about providing vegetarian food and could not identify any major problems. The remaining eight however, made reference to several negative issues that they had recognised. As can be seen in Figure 1, the most important factor is that this type of food can sometimes be more labour intensive to produce, particularly in small quantities. Financial risk was also cited, particularly amongst the small operations where the managers

were reluctant to prepare special dishes for their menus which could then be wasted. Two managers also commented that they did not want to purchase special ingredients which would 'sit on the shelf'. Two managers stated that

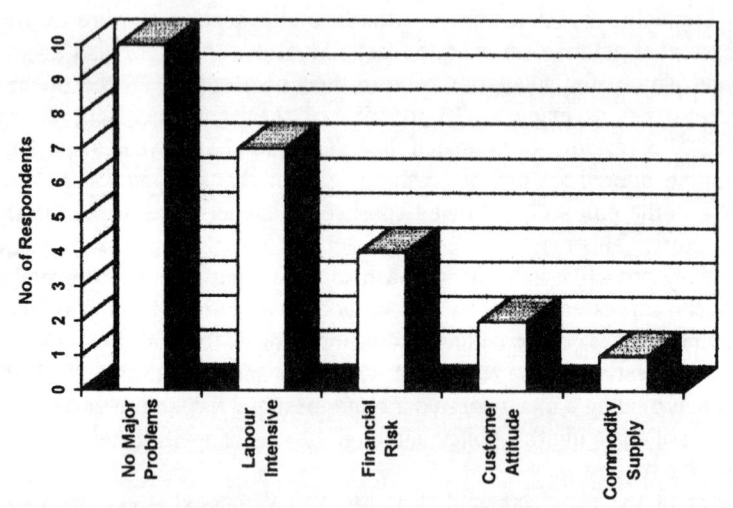

Fig 1: Difficulties in Catering For Vegetarians

some vegetarian customers could be over demanding and another stated that obtaining certain ingredients was sometime difficult.

4.4.2 Perceived Benefits

Eight of the interviewees could not see any major advantages in providing vegetarian food, but the remaining ten reported some positive aspects that they

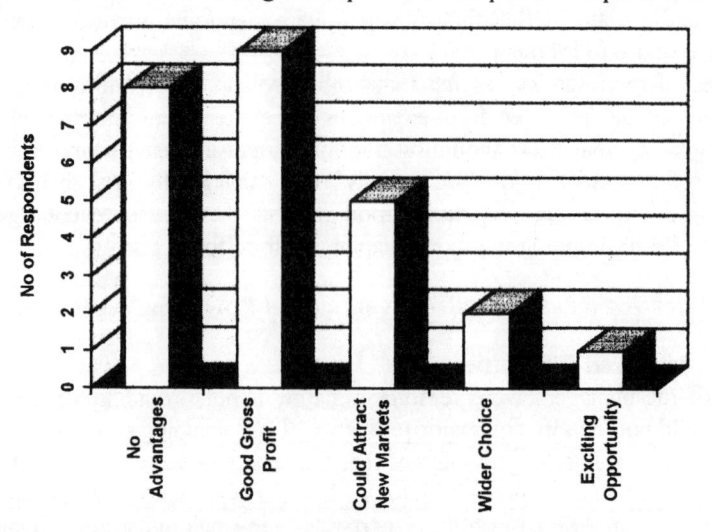

Fig 2: Advantages of Catering For Vegetarians

had considered. Fig. 2 illustrates that nine of the managers recognised the potential of being able to secure a good gross profit percentage from vegetarian food if it had been carefully produced and marketed. This opportunity is not always available to those operations who choose to use expensive 'bought-in' products. Five of the managers commented that a good vegetarian menu was capable of attracting a new market sector which could boost their business. Several managers had identified that vegetarian customers often dictate the choice of venue when dining out in a group of meat and non-meat eaters. Several managers also recognised that vegetarians are not the only people who like to eat vegetarian food. A vegetarian menu widens the menu and allows all of the customers the opportunity to break away from the 'meat and two veg' meal if they wish. One manager commented that the preparation of vegetarian food could be interesting and challenging and gave them the opportunity of working with different commodities.

5 Conclusions

The results of the study indicate that many outlets within the catering industry appear to have made significant steps in re-positioning their menus to cater for vegetarians more successfully. From the research carried out, it is clear that a lactovegetarian wanting to dine out within the Sheffield area should be able to choose from a number of different venues without having to give advance notice, and to feel reassured that the food has been prepared responsibly. The study indicates however, that there is disparity between the meal provision a vegetarian would expect to experience in a pub, a hotel or an independently-owned restaurant.

The pub sector appears to have taken an extremely positive approach in the way they have responded to this new market need. The vegetarian meals on offer are often simple but in keeping with other dishes on the menu and generally providing good value for money. A branded pub has the advantage of operating with the support of head office and a national network of outlets from which to draw experience. Pub restaurants also rely on operating an informal meal setting with a high through-put of customers so that the financial risk element of experimenting with new dishes on the menu is reduced.

Some of the hotel restaurants appear to have risen well to the challenge of vegetarian catering and provide interesting menus that are in keeping with their overall image. Other hotels however, include vegetarian dishes on their menus which are less inspired than those provided for meat-eaters.

The independently-owned restaurants are often run by chefs or enthusiasts who have the freedom to develop their own style of menu focused on particular food types or service style. Consequently, vegetarian provision is dependent on the owner's interest and commitment to this style of food.

The industry has now recognised that the need for vegetarian food is not a passing phase and many feel that this market is likely to develop further. If caterers are to take full advantage of this developing market it is essential that they start to view vegetarian catering as an exciting opportunity, rather than a chore or a duty they feel obliged to perform. This drive should not be seen as an effort to displace meat from its place on the menu, but to provide vegetarians and non-vegetarians with the opportunity of trying different and interesting dishes. The success of securing this market niche will be dependent on many factors, but undoubtedly those caterers who endeavour to enhance their skills and knowledge regarding vegetarian catering will surely reap the benefits.

References

1. The Realeat Survey Office, *The Realeat Survey 1984-1995: Changing attitudes to meat consumption in London,* Realeat Foods Ltd. 1995.
2. Beardsworth, A.D. and Keil, E.T. Vegetarianism, Veganism and Meat Avoidance: Recent trends and findings, *British Food Journal Vol. 93 No. 4 1991,* pp 19-24, MCB University Press, 1991.
3. Wright, G. and Howcroft, N *Vegetarianism : An Issue for the Nineties,* Hatan Publishing Ltd., 1992.
4. Leatherhead Food Research Association *Vegetarianism and the Consumer : Analysis of Consumer Attitudes and Behaviour,* Focus Report, 1994.
5. Jones Rhodes Associates *The National Health Survey,* Jones Rhodes Associates, Nottingham, 1988/1994.
6. Ministry of Agriculture, Fisheries and Food *National Food Survey,* 1995.
7. Government Statistical Service, MAFF *Quarterly Supplies and Total for Domestic usage of Meat in the UK.,* 1995.
8. Dixon, S. Meat and Four Million Veggies, *Hospitality Magazine,* August 1994, pp 30-31 H.C.I.M.A.
9. Ross Young *Market Report,* March 1995.
10. Dealler, S.F., and Kent, J.T. BSE: An update on the Statistical Evidence, *British Food Journal,* Vol. 97, No. 8, pp 3-18, MCB University Press, 1995.
11. Dixon, S. Quality Cuts, *Caterer & Hotel Keeper,* 23 February 1995 pp 62-65.
12. Marketing News MLC picks Doctor for Sales Growth Strategy, *The Grocer,* 23 September 1995, pp50.
13. Janelle, K.C. and Barr S.I *Nutrient Intakes and Eating Behaviour Scores of Vegetarian and Non-vegetarian Women,* Journal American Dietetic Association, February 1995, Vol. 95. pp 180-186.
14. Thorogood, M. (et al). *Risk of Death from Cancer and Ischaemic Heart Disease in Meat and Non-meat Eaters,* British Medical Journal, 1994, 308, pp 1667-71.
15. British Nutrition Foundation *Vegetarianism Briefing Paper,* Feb. 1995.
16. Swann, D. *Crank's Puddings and Desserts,* Guiness Superlatives, 1987.

Take the plunge: get the breakfast right

I.S. Folgerø,[1] S. Larsen[2]
[1]*School of Business Administration, Cultural and Social Studies, Stavanger College, N-4004 Stavanger, Norway*
[2]*Department of Tourism, Finnmark College, N-9500 Alta, Norway*

Abstract

Whereas the hotel guests may take lunch and dinner almost anywhere, the majority breakfast in their hotel. The guests' (lack of) satisfaction may have a deep and lasting influence on her/his impression of the hotel in general, and thus be a deciding factor the next time the guest is to choose a hotel or hotel chain. Still, this meal has received surprisingly little attention, and is often treated as a Cinderella in hotel and restaurant planning and management.

This article focuses on the guests' expectations and reactions regarding different food items and other breakfast stimuli. Through a qualitative study of hotel guests (observation and interviews) in an international chain hotel, the article identifies and describes the minimum requirements for customer satisfaction, and reveals features that may lift the breakfast experience above the customers' expectations: The exotic, the naughty, and that which is too complicated or time-consuming to prepare at home. This is done within a cost/benefit framework. Suggestions for further research and for practitioners are given.

Introduction

The hotel breakfast is a unique opportunity to impress the guest and increase sales. If the guest checks out of the hotel with the warm afterglow of a satisfactory meal, s/he is more likely to make a mental note of returning to the hotel. If the guest is staying another night, the likelihood of her/him patronizing the hotel restaurant in the evening is much higher if the breakfast was satisfactory than if the guest found this meal wanting. So rather than regarding the breakfast as a costly but necessary evil, the wise hotelier should grab this marketing opportunity and make the most of it.

Making the most of it does not necessarily imply spending a fortune on a vast and impressive menu or buffet spread. As always, one should aim at the highest possible customer satisfaction at the lowest possible cost. Learning how to optimize the breakfast buffet was the goal of this study. We wanted to know what the customers' basic needs and expectations are, and what distinguishes a memorable breakfast from an ordinary one.

The study

We performed this qualitative study in a four-star, international chain hotel centrally located in a Norwegian town. The hotel caters mainly to business travellers. Breakfast is included in the room price. The meal is served between 6 and 10 a.m., in the form of a large self service buffet with a vast selection of hot and cold food, plus a la carte upon request.

We interviewed a total of 14 breakfast guests. They were all male, aged between 28 and 55, clocking up between 8 and 60 hotel nights per year. Seven were Scandinavians, five British, and two from the US. Demographically speaking they represent a reasonably fair cross-sample of the hotel's guests. All the interviews were performed within a week in late August 1995.

The selection method was as follows: We would sit at a table in the restaurant, enjoying a cup of coffee and pretending to be ordinary guests, while observing the guest(s) at the table next to ours. Their selection of food was noted down, and as they seemed to be near the end of their meal, they were approached for an interview. Only two prospective respondents refused to participate, both claiming lack of time as the reason. Two others (nos 4 and 11) agreed to be interviewed later in the day, for the same reason. The respondents seemed happy to participate in the study.

We showed the respondents our notes of their meal, and asked them the following (open) questions:
> Is this what you normally have for breakfast at home?
> If not, what do you normally eat for breakfast?
> Would you consider this your typical hotel breakfast?
> Which, if any, items are to you absolute musts on the breakfast buffet?
> Which items were you most pleased to find on the buffet?
> Do you feel the teensiest bit guilty about anything you've eaten for breakfast today?

Results

The analysis of the data showed that two closely related concepts seemed to be underlying the breakfast guests' selection of food: Risk aversion and sensation seeking. Further, we found four criteria to be of importance in the selection process: Familiarity, exoticism, complex or time consuming preparation, and sinfulness.

People prefer an optimal stimulation level; not too much and not too little stimulation (Argyle[1]; Deci[2]; Horton[3]; Steenkamp and Baumgartner[4]). The optimum level is highly individual (Deci[2]; Steenkamp and Baumgartner[4]). Deviations from the optimum stimulation level lead to attempts to rectify the situation by reducing or increasing stimulation (Ellis[5]; Steenkamp and Baumgartner[4]), although *small* variations around the optimum stimulus level are perceived as pleasurable (Horton[3]).

Risk aversion

The wish to avoid or limit the number of unreasonable and/or unnecessary risks is of importance in most aspects of daily life. The restaurant guest is at risk in many ways: There is the possibility of food poisoning, of injury caused by own or other people's carelessness with hot, sharp or heavy items, of wasting one's resources (mainly time and money), and, not least, of embarrassment. The restaurant is a scene, where the customers' behaviour is shaped by the "set", the other "actors", and the directions given directly and indirectly via signs, cues and collective expectations (see Shelton[6]). To be shown to be incompetent, a beginner or a klutz, constitutes a loss of face that most people would rather be without.

To a large extent, our respondents seemed to play fairly safe at breakfast. They were afraid of ruining the day:
> "The last thing I want is an upset stomach. I'm here for a very important meeting, and I want to be able to concentrate, not having to run to the loo half the time or worry about embarrassing rumbles or even wind..."

They wanted predictability:
> "I want to be hungry again about one o'clock. The lunch is sometime after midday, and I don't want to be desperate for food at ten. So I eat what I know will keep me going for a few hours."

They wanted simple solutions:
> "In the morning I want what I'm used to, I don't want to have to think too much. Strange breads and flavoured teas and cold fish - you can keep that for me. I want my toast, I want something I can just eat, and get on with my day."

And they wanted to seem confident and well travelled:
> "Well, to be honest, I was dying to taste the pancake-like thingies people make over there, what do you call them? Waffles? Don't look like waffles to me... Such a gorgeous smell. But I have no idea of how to cook them or how to eat them, and I don't want to play the fool in front of the others. I mean: Even little children seem to know how to do it, and here I am, the managing director, useless..."

Familiarity

Thus they all wanted to find "the usual" on the breakfast table. Bread (or toast), cereals, butter, jam, milk, juice, tea and coffee constituted the absolute necessities for this particular sample of hotel guests. If any one of these items were missing, they would all regard it as reason to complain:
> "If there was no coffee? No coffee for breakfast? I'd be out of here this very morning!"

> "No bread?? That's impossible. No, I wouldn't be happy with just eggs and bacon and potatoes. You can't have breakfast without bread. [...] So I didn't actually eat it. Well, that's up to me, isn't it? It's got to be there, that's all there is to it."

But finding "the usual" on the buffet was the *minimum* requirement. All the respondents agreed that whereas this selection on its own might be considered sufficient in a cheap little back-street lodging house, they expected quite a lot more from a four-star hotel: Something interesting, something fun, something "extra".

Sensation seeking

A hotel meal may be an excellent opportunity to get that little extra spice into one's life. Nutrition is, of course, central to eating; but so is symbolism and hedonism. Consumers want to experiment, to seek variety, to fulfil fantasies, to play and have fun (Holbrook & Hirschman[7]).

Health, fitness and slimness are continually in media's focus, and creep frequently into everyday conversations. The emphasis is on being "thin, moral, & admirable" (Counihan[8]). Feelings of guilt and shame may also play an important part in the restaurant experience (Larsen & Folgerø[9]). We know we ought to be good; we know what is (currently regarded as) healthy, what the norms and expectations are; what foods are socially acceptable. These are, however, not necessarily the foods we want to eat.

The recent trend towards healthy eating may indeed be a passing fashion. The strong interest in fat-substitutes in the US tells us that people don't want to avoid eating fatty foods, they want to avoid gaining weight. Real food may be making a come-back. In Norway, "light" food is losing ground quickly; sales of low-fat cheeses are down dramatically, and sales of light soft drinks are also falling (Stavanger Aftenblad 09. January 1996, p.16). And the restaurant may *not* be the right place for self-denial and restraint, even for the most health-conscious.

All our respondents professed to enjoy that little bit of excitement in the morning. Something out of the ordinary; the little extra that marked the occasion, created what five of them termed "a real hotel breakfast". Actually *eating* the less usual items wasn't always necessary:

"Three out of four days I only eat my cereal, some fruit juice and my coffee. That's what I have at home. But the *last* morning in a hotel I always have a full breakfast, with eggs and bacon and yoghurt and a croissant and all the works. Yes, I look forward to it every time..."

"It's the *possibilities*, isn't it? To know that I can have sausages and cherries and even hot chocolate if I want to. To tell the truth, I don't much like any of that, so I never actually eat it, but I like the feeling... the choice, the luxury... knowing that I *could*, if I wanted to."

When the respondents talked about the food items they enjoyed in hotels but didn't normally eat at home, their descriptions fell in three broad categories: Exoticism, complex or time consuming preparation, and sinfulness.

Exotic

Most respondents liked to try something slightly exotic. This can be a lot of things; for some, a new yoghurt flavour is exciting enough. Other factors that increase the exotic value of a dish can be its "inappropriateness", as when "full-flavoured, ethnic style food" (Conan[10]), generally considered lunch-time fare, is served for breakfast; or when the chef makes the best out of the season's local produce; or even an unusual presentation, for example a well-arranged plate of fruit crudities with yoghurt dip.

When abroad, our respondents wanted something local, something typical:
"All of these hotels are the same, you know. And we never get to visit the locals at home. So we could be just anywhere, couldn't we? No way to tell. I like that little bit of local flavour on the breakfast table. Papaya in Kenya, fish in Norway. Makes me feel I'm abroad, like."

A hotel is a good place for trying something new. The cost is low; you can have as little or as much as you like; there is enough ordinary food to fill you up; and you don't have to finish what you don't like, the waitress won't tell you off for leaving food on your plate. It is also a good learning ground, where one can develop knowledge and a satisfying competence regarding food one has heard about, but never tried at home:
"Last time I went to the States I had pancakes, it was brilliant. Well, to tell the truth, I didn't think all that much of them, far to filling and sweet for my taste. But I'd seen it so many times in films and on TV, I was just dying to try it out, and then the waitress asked... She had to teach me how to do it, where to pour the syrup and all that, it was great fun."
"I like trying new things. I like to know about food, know what different dishes taste like and how you eat them and so on. And international table manners, like how to eat a Scandinavian open sandwich. It is really useful, sooner or later you'll need to know."

Complex or time consuming preparation

Some well-loved items are too complex or time consuming to prepare at home. Very few people can afford the time to buy fresh bread rolls or prepare a cooked meal in the morning. These items are mainly of the home-cooked, old fashioned kind; foods one may dream about in a nostalgic way, but hardly ever get to eat. Crisp strips of golden bacon; sensuously soft, scented cinnamon rolls... The National Restaurant Association (US) reports that breakfast orders for freshly baked goods grew an average of 2% yearly in the years 1989 to 1993. For busy executives, "home-cooked" food may be the ultimate luxury, and in a hotel one may enjoy these rare treats:

"I loved visiting my grandmother. She would always serve me fried potatoes and bacon for breakfast, and sometimes hot bread rolls with honey. My wife just won't do it, you know..."

"What I really like is the pretty display. Like a Sunday brunch... At home, we just prepare the food by the counter, then carry the plates over to the table. Don't have the time to lay the table properly."

"Hashbrowns! That's what I want. Takes forever, I believe. Never tried making it myself. But my mother would, once in a while."

"A fresh fruit salad, I think. But I mean *really* fresh, not yesterday's. *That's* real luxury."

Sinful

Lastly, the aspect of sin. Every respondent felt a little bit guilty about something he had eaten for breakfast that morning. The reasons for this were many: General health awareness, doctor's advice, calories, cholesterol levels, family rules, even expense (the hotel's, not their own!). The most common guilt-inducing objects were sugary goods (Danish pastries; honey; jam; real sugar in the tea), salt, eggs, fat (butter, bacon, sausages), and expensive items like smoked salmon, exotic fruits and half-eaten yoghurt cartons:

"Well, you can't lose weight while staying in a hotel. It's impossible."

"I know I shouldn't, my heart is a bit dodgy. But I enjoy it so. And *one* egg can't kill me."

"Yes, I guess I was a bit greedy. I should have finished that salmon, I know. But we pay enough for the room anyway, and I only had cereals yesterday..."

But mostly, the feeling of wickedness seemed to be important and enjoyable to the respondents; an ingrained part of the overall hotel experience. When they talked about it, we were reminded of big boys on adventure, gleefully away from mummy's watching eye:

"I know I shouldn't. But it's only once in a while, you know, and no-one will ever know about it."

"Goodness, no, I wouldn't feed this to my children. But really, I'm old enough to decide for myself, aren't I? And I'm not going to tell them, anyway."

"Yes - a croissant with butter *and* honey! Not bad, not bad... My wife would kill me, you know. Not to mention the doctor."

Healthy food seemed to counteract the bad effects of the no-no's in the guests' minds:

"I had a Danish pastry, yes, but I also had a lot of fresh grapefruit and some melon. So it wasn't all that bad."

Conclusion

The individual optimum stimulation level is related to curiosity-motivated behaviour, variety seeking, and risk taking (Steenkamp and Baumgartner[4]). Most guests may want their breakfast to be based on simple, healthy, everyday food. But they also want the little extra; the small helping of something unusual, old-fashioned or sinful, to brighten up the experience.

Modern western dietary habits - and ideals - may prompt hotel guests to state in surveys that what they want is healthy, low calorie food for breakfast. Our observations support this, but also found that the guests enjoy their little sins when given the opportunity. And even the rushed, health-conscious, habitual cereal eater may hanker for the occasional plateful of eggs and bacon. The customers' satisfaction with the overall hotel experience may be bolstered by a well planned breakfast buffet, and slow moving items may be far more important to customer satisfaction than many chefs realise.

This does not, however, mean that the hotel must supply every food item imaginable. A well planned selection can provide excitement and enjoyment for the customer at a moderate cost for the hotel. The magic words to remember may be "something old, something new, something borrowed and something blue": If the hotel provides its target group(s) with a breakfast comprised of something old fashioned, something exotic (i.e. local cuisine), something borrowed from their everyday breakfasting, and something deliciously sinful, the relationship should be secured.

Further research

Ours was an explorative, small-scale study. We believe the concepts of risk avoidance and sensation seeking in the hotel/restaurant setting are well worth further study. Of particular interest are the relationships between perceived risk; perceived cost; and memorability of a restaurant experience. National and social differences in breakfast eating habits, likes and dislikes, also merit further study.

Advice to practitioners

It may be tempting to concentrate on offering only the most popular breakfast items. Modern stock control software makes it easy to pinpoint the slow movers; a limited selection is easier to plan, prepare, and serve; fewer items ensures higher turnover of goods, which again may improve freshness and overall quality; and savings on raw materials may be used to improve the service level. Such a strategy may, however, be a serious mistake. The rare helping of bacon and fried potatoes, or the tiny portion of a high calory "sin", may be what lifts the breakfast from the ordinary to an experience worth remembering.

Every hotel should aim to find its own perfect breakfast mix. The selection of food should reflect the market segment's nationalities, social backgrounds, and expectations. Most hotels would benefit from even quite rudimentary studies of their guests' breakfast habits. Simple observations and/or interviews of breakfast guests may be all that is needed to improve the selection.

Many large international hotel chains are in favour of central menu planning and identical breakfast buffets, arguing that the traveller will enjoy the safe, no-hassle "seen-it-all-before" standardized international hotel breakfast. The idea behind this presumably being that the traveller can't be bothered to "learn" a new breakfast buffet every other month; or that a well-known menu will provide a welcome stress-free start to the day for the busy, jet-lagged, newspaper-reading executive with his mind on the clock. We believe this is wrong.

We believe the breakfast is far more important to the traveller than generally assumed. The sporadic guest wants something special, a "real hotel breakfast"; the frequent traveller gets bored if he/she finds the same old selection on offer in Vienna, London, Hong Kong and New York. Yes, keep the basics. But add some local flavour to your breakfast table.

References

1. Argyle, M. *The Social Psychology of Everyday Life*, Routledge, London, 1992.

2. Deci, E.L. *Intrinsic Motivation*, Plenum Press, New York, 1975.

3. Horton, R.L. *Buyer Behaviour: A Decision-Making Approach*, Charles E. Merrill Publishing Company, Ohio, 1984.

4. Steenkamp, J-B.E.M. and Baumgartner, H. The Role of Optimum Stimulation Level in Exploratory Consumer Behaviour, *Journal of Consumer Research*, 1992, 19 (Dec), 434-448.

5. Ellis, M.J. *Why People Play*, Prentice-Hall, Englewood Cliffs, N.J., 1973.

6. Shelton, A. A Theater for Eating, Looking, and Thinking: The Restaurant as Symbolic Space, *Sociological Spectrum*, 1990, 10, 4 (Oct-Dec), 507-526.

7. Holbrook, M.B. and Hirschman, E.C. The Experiential Aspects of Consumption: Consumer Fantasies, Feelings, and Fun, *Journal of Consumer Research*, 1982, 9 (Sept), 132-140.

8. Counihan, C.M. Food Rules in the United States: Individualism, Control, and Hierarchy, *Anthropological Quarterly*, 1992, 65, 2 (April), 55-66.

9. Larsen, S. and Folgerø, I.S. Service and Guilt - a Norwegian View on Service, *Proceedings of ICCAS 96: Culinary Arts and Sciences, Global and National Perspectives*, Bournemouth, England, 1996.

10. Conan, K. Bold and brazen breakfasts, *Restaurant Business*, 1995, 10/94 (July), 66-78.

Children's choices: variability in children's menus in restaurants in Northern Norway

L. Lervik, S. Larsen, T. Gustavsen
Department of Tourism and Hotel Administration, Finnmark College, N-9500 Alta, Norway

Abstract

Due to a change in behavioural patterns, an increasing number of families seem to be in demand of hotel- and restaurant services. Nevertheless, the business itself does not seem to have recognised the particular demands of the infant or child customer.

The present study addressed the variability of dishes offered to children in foodservice operations in the largest municipality in Finnmark, Northern Norway. All menus from all the foodservice operations in the municipality of Alta were examined.

As expected, results indicate that compared to standard adult menus, children's choices are fairly limited. The variation compared to the standard menus both within and between different foodservice operations is small.

Suggestions are made as to how one could improve the menu-quality for families with children. It is argued that a change in attitudes would be necessary in order to achieve such an increase in service quality for children and their families.

1. Background

Children, along with families with children and the single female traveller seem to be the forgotten groups in the hotel- and restaurant industry. Indeed, the phlegmatic hotel-saying that "two things do not belong in a hotel, dead flowers and living children" indicates an extremely negative attitude towards (families with) children in the hotel- and restaurant industry. This impression seems to be confirmed when surveying current literature, since most text-books covering hospitality management rarely ever mention the effective handling of children and families with children. The potential for developing high quality products aiming at these groups seems to be blatantly absent in the literature. In addition, literature search in the standard data-bases brought 0 hits for the combination children/restaurants or children/menus.

The aim of the present project was to investigate the variability in children's menus in a municipality in Northern Norway. For a long time it has been our suspicion that cafeterias and restaurants alike do not provide adequate menus for children, neither in the amount of variation offered nor in the nutritional content of current menus. It was expected that relative to the meal and beverage menus offered to adults, the variation in terms of number of meals and drinks offered children would be much less. In addition it was expected that children's courses would be less healthy and less expensive than that offered adults.

2. Methods and materials

All 17 of the foodservice operations in the municipality of Alta were included in the study. Of the 17 foodservice operations, 7 were recognised as restaurants, while the remaining 10 were cafeterias. The categorisation was based on the following definitions: "a restaurant is a foodservice operation, which has a kitchen that prepares it's own food and is open for orders most of the operating hours. The restaurant further has a menu and a varied and satisfactory dinner service" (Oslo municipality)[3], and a cafeteria "is a foodservice operation where we serve ourselves different cold courses and cakes from a buffet or food display, while the hot food is served, on plates, by the cafeterias' employees" (Nestande).[2]

Alta is a small township, with some 16.000 inhabitants, situated far to the north of the arctic circle. Tourism represents a cornerstone in the local economy, in addition to Alta being a growth centre in the county containing the northernmost college in the world.

When conducting the study, all available menus available to the guests without obstacles (i.e. without extra efforts on the part of the customer) were registered. Menus were understood consistently with Mooney's[1] definition, which suggests that a menu can, among other thing, mean "the piece of literature or display used to communicate the product range to the customer." Restaurant or cafeteria employees or staff were not contacted or interviewed. This implies more objective data, but may naturally cover that some of the foodservice operations offer more to children than what is evident from the mere observation of the menus. Consequently one might argue that the study has concentrated on immediate availability of children's menus.

The number of appetisers, main courses and desserts were recorded as well as number of (i.e. variation of) drinks. The number of drinks for children was calculated as "total number of drinks offered minus number of drinks containing alcohol." In addition, courses listed as small dishes on the menus were counted. If a special children's menu was offered, or if it was explicitly mentioned on the main menu that child-sized portions were available, this was considered a special offering for children and registered as such. If not, all courses were considered to be for adults.

Prices were also included in the registration form, and mean prices for adults' and children' menus (appetisers, main courses and desserts) were calculated. An analysis of whether the main course was served with or without vegetables was also done. This information was used to construct a "health-index"; percentage of main courses containing vegetables.

Differences were calculated using t-tests.

3. Results

Table 1 shows the number of dishes offered children and adults in cafeterias and restaurants respectively.

	Average number of children's courses	Average number of adults' courses
Cafeterias	9.50 **	39.00
Restaurants	3.43 **	30.71

** = $p<.01$

Table 1 Number of Dishes offered Children and Adults in Cafeterias and Restaurants.

As can be seen from Table 1, children are offered a significantly less varied menu in both cafeterias and restaurants. Naturally this holds true for appetisers, main courses and desserts alike. Children's choices concerning courses, do not vary from cafeterias to restaurants. The variation of drinks is much higher in restaurants than in cafeterias, an observation very well predictable from Norwegian alcohol laws. Nonetheless, the relative variation (percentage of drinks suitable for children) in drinks offered children is higher in cafeterias (60 %) than in restaurants (14 %).

Table 2 shows mean prices for adults' and children's menus.

	Average prices for children's menus	Average prices for adult's menus
Cafeterias	18.29***	54.39
Restaurants	22.01**	100.52

*** = $p<.00$, ** = $p<.01$

Table 2 Average Prices (NOK) for Children and Adult's Menus

As can be seen from Table 2, it is approximately twice as expensive to eat in a restaurant as compared to a cafeteria for adults, but there are no price differences between restaurants and cafeterias concerning children's menus.

The present research effort also studied the relative healthiness of the courses offered children and adults respectively. Preliminary results of this analysis are presented in Table 3.

	Percentage of healthy children's courses	Percentage of healthy
Cafeterias	20.00*	64.24
Restaurants	37.00***	62.77

*** = $p<.001$, ** = $p<.01$, * = $p<.05$

Table 3 Relative Healthiness of Children's and Adults' Main Courses

For adults, the main courses are approximately equally healthy in restaurants and cafeterias, whereas for children the main courses offered are about twice as healthy in restaurants as compared to cafeterias. On one hand, the variation in main courses is lower in restaurants. On the other hand, the main courses actually offered seem by this simple measure to be higher concerning healthiness in restaurants than in cafeterias.

4. Discussion

The results indicate that children's menus are less varied and less healthy than menus offered adults in cafeterias and restaurants studied in this report. In addition, the results indicate that children are offered more healthy, yet less diversed menus in restaurants than in cafeterias for more or less the same price. However, compared to adults' menus, children's menus are far less nutritionally balanced because relatively fewer courses are served with vegetables. As expected, cafeterias charge less for an average meal for adults than restaurants do, but somewhat surprisingly, no price differences between cafeterias and restaurants concerning children's menus were found.

The results indicate that children's menus generally are less varied and less healthy than menus offered adults. A major source of error in the present study is linked to the methodological approach of the study. Only written or otherwise displayed menus, were studied, implying that some of the foodservice operations probably would offer more to children than what is evident from the mere observation of the menus. Nevertheless, to the extent that the customer has to make a special request concerning children's menus, an obstacle is introduced.

The results further indicate that children are offered more healthy, yet less diversed menus in restaurants than in cafeterias for more or less the same price. As expected, cafeterias charge less for an average meal for adults than restaurants do, but it was surprising that no price differences between cafeterias and restaurants concerning children's menus were found. Nontheless, compared to adults' menus, children's menus are far less nutritionally balanced in as much as it was observed that relatively fewer courses are served with vegetables for children than for adults. Some might argue that the child customer demand unhealthy food (french fries and chicken or sausages) when they accompany their parents in such settings. On the other hand, one might as easily turn this argument to it's opposite, and state that children demand the products that are offered. In other words, our hypothesis is, that given the opportunity, children would appreciate more diversed and interesting courses. The negative attitudes towards children reflected in the limited and unhealthy choices offered to them when they are customers in restaurants and cafeterias function, in our view, as an obstacle to children's opportunities to develop a more refined taste.

It was surprising that children's menus on average cost the same, irrespective of the kind of foodservice operation studied. A natural prediction would be that restaurants would charge more and cafeterias less, a finding made for adults' choices. It may be that restaurants have approached the "problem of the child customer" more efficiently, in as much as their prices are not an obstacle for families when they bring their children to dine. It is also noteworthy that the relative price difference for families when dining out, depends on the adults and not the children, because children's menus cost the same in cafeterias and in restaurants. Children's menus are also more healthy in restaurants than in cafeterias. This implies that restaurants are preferable to cafeterias when having the children's interest best at heart.

All in all, our study has demonstrated that children's choices are more limited and less healthy than adults' menus in cafeterias and restaurants alike. These results probably indicate negative attitudes towards children among restaurant and cafeteria staff and management. Children are probably considered a less than welcome group in many of these settings. No doubt, to establish that such negative attitudes prevail, would require new studies. To the extent however, that negative attitudes towards the child customer are found in restaurant settings, one should aim at developing programmes in order to change them. Families with children represent an important target group in innovative marketing designs for the future. Not least, concerning capacity management; as families with small children usually have dinner earlier than other customer segments, e.g. business people, foodservice operations suffering from excessive capacity should look into the needs and potential of the family segment.

References

1. Jones, P. and Merricks, P. Management of Foodservice Operations, Chapter 3, *Planning and designing the menu*, ed S. Mooney, England. Cassell, 1994.

2. Nestande, K. *Hoteller og restauranter: en introduksjon*, Universitetsforlaget, Oslo, 1995.

3. Restaurant- og bevillingskontoret. *Skjenke-permen*, Oslo Kommune, Oslo, 1995.

Section 1.2: Equipment and Systems

Multi-media development and optimization of concepts for a fast food restaurant

H. Moskowitz, A.Gofman, P. Tungaturthy

Moskowitz Jacobs Inc., White Plains, New York, 10604, USA

Abstract

The past decade has seen three key advances in concept development of key interest to restaurant development, which are beginning to be used more widely:
a) <u>Stimulus Presentation Tools</u> - ability to show concepts in multi-media format, so that these concepts develop a reality beyond simply word/picture description. For instance, video clips can now be incorporated into the concept work.
b) <u>Experimental Design</u> - Recognition `that the creative act of developing concepts need not be an isolated instance, but can arise through diligent work, statistical layouts of study design, and the combination of hitherto conventional elements into new concepts, to create new ideas
c) <u>Rapid Consumer Feedback And Real-Time Modification Of Concepts</u> - Ability to speed up the study so that hitherto undreamed of speed of data acquisition, modeling, optimization, and re-evaluation can be achieved by straightforward procedures.

This presentation shows the approach, from beginning to end, illustrating the methods by means of a case history of a restaurant called "O'Steers". The presentation shows the types of stimuli, analyses, modeling, and concludes with a demonstration of the approach for designing the restaurant.

Introduction

What do consumers want in a restaurant? How can the restaurateur determine what the restaurant concept should be without spending undue amounts of money doing so? This paper focuses on new consumer-research techniques which allow the restaurateur to understand consumer needs and wants in an actionable fashion, and which provide a database from which to create new concepts.

Concept Testing - Traditional Techniques

In traditional concept testing the marketers develop ideas about restaurants. These ideas are fully formed, dealing with the entire restaurant or with specific features (as dictated by the particular problem). We call these ideas "gestalts". The elements work together, the concept is integrated, and the concept stands by itself. There may be a number of such gestalt concepts, depending upon who is creating them. These concepts may be in relatively rough form or may be polished to resemble advertisements. The bottom line, however, is that these are complete ides. The researcher exposes consumers to each of these gestalt concepts and obtains ratings of interest, frequency, and other attributes. (These other attributes are called "diagnostics". They show reasons for concept acceptance / rejection, or measurement of the degree to which the concept communicates convenience, price/value, etc.). At the end of the day the research tallies up the ratings assigned to the concepts, and selects the winning concept. The diagnostics show the reasons why concepts win or lose. Table 1 shows an example of concept test results. The data comes from a screening of four different restaurant concepts, each rated by consumers for interest, frequency, uniqueness, attractiveness of menu, and attractiveness of decor.

Concept	Overall Interest	Frequency (Per Year)	Unique	Interest In Menu	Interest In Decor
Scale Anchor 1	Hate	Actual	Ordinary	Hate	Hate
Scale Anchor 9	Love		Unique	Love	Love
A - Fish concept	6.5	8.0	4.5	6.7	5.3
B - Mexican concept (modern)	6.3	5.5	6.7	7.6	6.8
C - Mexican concept (traditional)	5.7	6.5	5.6	6.4	5.4
D - Home cooking concept	5.3	9.3	5.4	6.3	5.1

Table 1 - Example of results from concept evaluation, for four concepts

Although conventional concept testing has been accepted by many researchers and finds wide use in consumer package goods and in food service, many modern thinkers have begun to recognize that the traditional methods leave a lot to be desired. For one, the researcher only obtains a general measure of acceptance but does not know exactly what drives the acceptance or lack thereof.. There are certainly many ways to cut the data (e.g., by frequent Vs infrequent patrons, males Vs females) but these analytic techniques deal with the respondents, not with the stimuli. It is the stimuli, the ideas that must be optimized. The respondents can be optimized in terms of who will be the target for the advertising, Second, the data does not reveal to the researcher exactly what to do in order to improve acceptance and generate more consumer interest. Is it the name of the restaurant? The menu? The decor? The pricing? The service? It is all well and good when the concept scores well. If the concept scores poorly, however, then the researcher must be in a position to recommend alternatives.

Experimental Design In Concept Development

Statisticians and researchers have, for many years, preached that when the researcher systematically varies the physical characteristics of a product by an experimental design, then the research can determine what specific characteristics drive consumer acceptance versus diminish acceptance. The characteristics must be present in different combinations, so that the consumer evaluates multiple stimuli[1,2]

Following this logic market researchers have recognized that they can apply the same approach to concepts[3]. Concepts comprise elements which can be varied in order to create different combinations. If the elements are systematically varied and presented to consumers, then some combinations will be liked and some will be disliked. Since the elements or components of the concepts appear in a systematically varied fashion against multiple background, the researcher can trace the response to the concepts to the presence / absence of the elements, much as the statistician in the hard sciences can trace the response to a stimulus back to its components.

Concept researchers use experimentally designed concepts when they employ the method of conjoint measurement. Conjoint measurement combines concept elements into well defined combinations, with the property that the elements are statistically independent of each other. The combination are insured against being self-contradictory because part of the design consists of a matrix of concept elements, and another matrix of pair-wise combinations of elements that are restricted. It makes no sense, for instance, to combine elements dealing with "fast food" and elements dealing with "waiter service". Once the elements are chosen, the researcher combines these elements according to an experimental design. The design yields specific combinations. The consumer evaluates each concept monadically (one at a time) on a variety of response attributes. The ratings are then processed statistically by means of regression analysis to reveal the part-worth contribution of each element to the consumer ratings. The design, analysis and interpretation become extremely easy once the researcher has experience even with one study where the components are varied by experimental design. [Most of the difficulty encountered in designed experiments comes from the reluctance of the researcher to change research paradigms, and even consider a designed experiment instead of the evaluation of gestalt concepts].

Conjoint measurement has enjoyed a twenty year history of application in market research, beginning with Paul Green and his associates at the University of Pennsylvania, and continuing both at the university level and at the practitioner level[4,5]. This paper will present an adaptation of conjoint measurement designed for the complexities and speed encountered in real world problems.

The IdeaMap System For Conjoint Measurement

The pace of business has increased, and so has the need to understand consumer wants. In its typical mode conjoint measurement has been a long and drawn out affair, taking months. Moskowitz[6] has suggested that the pace be speeded up to weeks, and the scope of the elements expand from a maximum of 30-40 different elements (or things that can be said in concepts) to 300+. The revised approach to Conjoint Measurement is known as IdeaMap[7]. IdeaMap, now in it's fifth year, comprises the amalgamation of basic theory of concept measurement and statistics with real-time data acquisition and analysis to suit the needs of business.

This paper will present the IdeaMap System as used to design a new restaurant concept, O'Steers. The data is taken from an actual case history. The specific elements in the study are disguised, however, to retain client confidentiality.

IdeaMap comprises a sequence of nine choreographed steps, ranging from ideation to concept creation through to consumer evaluation, analysis and optimization. The steps follow, illustrated by the O'Steers database.

IdeaMap Step 1 - Ideation To Create Elements
In contrast to the traditional methods which require the development group to create complete, gestalt concepts, the IdeaMap system requires that the development group create elements. These elements are snippets of ideas, dealing with name decor, menu, etc. There may be dozens or even hundreds of these elements. The elements are created in focus groups and ideation sessions, as well as created by looking at the entire competitive frame of restaurants.

Culinary Arts and Sciences 105

Typically ideation sessions generate hundreds of these elements, which are then polished into short, easy to understand phrases. The ideation session may even enable consumers to create or select graphics elements, as well as text phrases. At the end of the session the participants will have generated a large number of potential elements that can be used in the new restaurant concept. Thus far the elements are simply raw material that has been polished. At this very early stage of development many different elements may deal with the same restaurant feature, but express that feature in different ways. The researcher should not attempt to prune these elements, other than to eliminate clearly redundant ones.

IdeaMap Step 2 - Classification Of Elements Into Categories
For research purposes it is important to classify the elements that are developed from the ideation session. The method for doing so is straightforward -- the research group looks at the different elements obtained in ideation, identifies common themes (categories), and then sorts the elements into the appropriate categories. Table 2 presents some of the concept elements created from ideation, classified into categories. In total there were 136 elements tested in the study. With 136 elements the researcher can assess many basic ideas, as well as nuances of these ideas. As we shall see, the 136 elements will suffice to create new concepts as well.

Name	Harvest & Beef
	O'Steers
Position	Ready here, or delivered to your home
	Relax and enjoy--you know you're eating well
	Wholesome and hearty meals, fresh from the grill
Foods	Our specialty is a variety of delicious roasts
	Featuring hearty main course specialties like great grill beef, meat
	Foods prepared using the freshest ingredients
Sides	Freshly baked breads to complete your meal
	We make our own ice cream

Table 2 - Examples of categories (left) and elements (right)

IdeaMap Step 3 - Create Pairwise Restrictions
When creating concepts the researcher must insure that illogical combinations do not appear in concepts. In order to prevent these illogical combinations the researcher uses a small group of consumers to identify those pairs of concept elements from different categories which do not logically go together. Sometimes researchers also identify combinations of elements which make sense logically from the consumer's viewpoint, but do not make sense from a business or technical viewpoint. In the end, however, the result is a file of pairs of elements which will never appear together in concepts. These are called restrictions. When the concepts are created by experimental design later on, the

restriction file will be consulted to insure that the concepts never comprise incompatible elements in the same concept.

IdeaMap Step 4 - Dimensionalize The Elements
Step 4 consist of locating the concept elements on a set of non-evaluative semantic scales. The dimensionalization will be used for three distinct analyses:

a) Estimate utilities (contributions) for each element untested, by an interpolation method. Moskowitz[7] described the procedure by which the researcher uses the 8 closest neighbors in a semantic space in order to estimate the utility for untested elements.

b) Drive the tonality of the concept in the proper direction. Concept optimization, as we will see below, maximizes one criterion or several criteria simultaneously. By putting tonalities into the equation (via the dimensions) the marketer will push the concept towards several directions simultaneously, including persuasion (interesting the customer to frequent the restaurant) and tonality (communicating the right type of message to accord with the marketing strategy).

c) Segment consumers on the basis of each individual's utility Vs the semantic scale. This method of concept-response segmentation[8] is based upon the method of sensory preference segmentation for actual products[9]. The segmentation identifies subgroups of consumers in the population who share similar patterns of elements which they like versus which they dislike. Although segmentation may be done by using geo-demographics (age, gender, income, market), by purchase patterns (e.g., which restaurants the consumers frequent) or even by psychographics (profiles on personality or interest questionnaire), none of these segmentation methods focus in on the response of consumers to restaurants themselves. There is always the gap between the segmentation methods done for general purposes and the specific pattern of interest in restaurants. Concept response segmentation transcends this difficulty and makes the segmentation actionable by clustering consumers directly on the basis of their responses to concepts.

IdeaMap Step 5 - Create Experimental Designs For Each Consumer
Each panelist is exposed to a set of concepts comprising systematically varied elements. IdeaMap creates an individual model for each panelist which shows how the elements drive interest, and the other attributes rated by the consumer. The experimental design enables the researcher to identify the contribution of elements for each consumer in the following way. Each panelist evaluates 100 concepts, comprising 60 elements. Table 3 shows part of the schematic for the experimental design. The experimental design comprises four categories. Each category comprises three elements. Once the categories and elements are selected for an experimental design the design itself dictates what a concept will look like. Table 4 shows an example of a concept drawn from one experimental design.

Culinary Arts and Sciences 107

Concept	Category A	Category B	Category C	Category D
1	3	2	1	0
2	3	2	2	2
3	0	1	3	2
4	2	3	0	0

Table 3 - Schematic for the first four concepts of the 20 in the design. Each concept comprises four categories. Each number in the body of the table represents an element number from the category

PO10	Homestyle meals, prepared with card
LD4	Brightly lit dining room
SE13	You can phone in your orders up to a week in advance
PR1	Has set prices for various groups

Table 4 - Example of a concept comprising four elements from the set of 136, and created according to an experimental design

In a single session the consumer panelist rates 100 concepts created from five different experimental designs. Each experimental design comprises four different categories and three different elements from each category. Every consumer will therefore evaluate a unique set of 60 different elements and 100 different concepts, so that no consumer ever rates the same concepts. In this way the IdeaMap system is independent of the researcher's or marketer's state of knowledge. With a base size of 120 consumers, and with each consumer rating 100 concepts the O'Steers study generated 12,000 different concepts.

IdeaMap Step 6 - Run The Evaluation With Consumers Using The Computer To Present The Concepts And To Acquire The Ratings. Consumers find computer-based interviews fun to do because in many ways the interview resembles a computer game. The computer presents the systematically varied concepts to the consumer. Recall that 100 concepts were created for each consumer by creating five experimental designs comprising 20 concepts each. These 100 concepts are first randomized according to an internal computer program. Then the computer presents each concept to the consumer, who rates the concept on attributes. For this study the attributes were "interest in the restaurant", and "frequency of going to the restaurant". Both attributes were rated on 9 point scales, anchored at each end to reduce ambiguity.

The interview lasted approximately 35 minutes, and was conducted individually. Exit interviews with the panelists revealed that the panelists enjoyed the experience. Panelists ranged between the ages of 18 and 64. The study comprised panelists from three markets (East Coast, Midwest, West Coast) to ensure distribution and representativeness. Forty consumers from each marketer participated. The interviews were conducted in a room off a shopping mall. Each consumer panelist was recruited from the mall, qualified, and then invited to participate.

IdeaMap Step 7 - Build Individual Model And Estimate Utilities Of Untested Elements. As soon as the panelist finishes the interview the computer has sufficient information to create an individual model for that panelist showing how each element directly tested contributes to the ratings of "interest" and "frequency", respectively. Since the elements for each panelist were selected to fit an experimental design for that individual it is straightforward to relate the elements for the panelist to the ratings of interest and frequency. The ratings of interest were transformed from the original 1-9 scale (1 = not at all interested ... 9 = extremely interested) to a binary 0, 100 scale (0 replaces ratings of 1-6, 100 replaces ratings of 7-9). The transformation enables the researcher to change focus from the original rating of degree of interest to the odds that the consumer will change a vote from "not interested" (viz., 1-6) to "interested" (viz., 7-9). This transformation changes the research perspective from a psychological one (viz., interest on the individual's own mental processes) to a sociological one (viz., interest in the proportion of consumers who will exhibit a given behavior -- in this case the behavior being a statement of "I'm interested in this concept" because I have rated it 7-9 on the scale).

	Interest	Frequency
Winning Elements - Total Panel		
Wholesome and hearty meals, fresh from the grill	17	11
Sample our delicious breads right from the oven	16	12
Traditional dinners with a homey touch, featuring roast beef, meat loaf and burgers	15	8
Foods freshly prepared by trained chefs	15	9
A quiet environment where you can unwind and enjoy your meals	14	6
Losing Elements - Total Panel		
Highchairs and booster seats for young children available in the dining room	-3	-3
Harvest & Beef	-4	-3
Baby changing tables in the restrooms	-5	-7
Food portioned and served from a deli-type case	-5	-5
Sturdy plastic plates	-6	-7

Table 5 - Winning versus losing elements for total panel - interest utilities and frequency utilities

One of the key objectives of the research is to create a complete model for each individual showing how every element motivates. To this end the computer program imputes utilities for elements not tested, using the dimensionalization and routines explicated by Moskowitz[7]. Suffice it to say, within a few seconds the computer creates a model for the individual and imputes or interpolates utilities for those elements not directly tested.

Table 5 shows part of the final data aggregated across 120 consumers in the study. The aggregation is done by collecting the models from the full set of 120 consumers, and then averaging the utilities for corresponding elements. As Table 5 reveals, elements vary in their ability to interest consumers. The researcher can sort the utilities, either by interest or by frequency, in order to identify high scoring elements which promote concept acceptance, and also to weed out poor scoring elements. The negative values obtained by some elements indicate that some elements turn off consumers, and detract from the ultimate concept acceptance. The same type of analysis can be done on the frequency data to determine how often the consumer feels that he would frequent a restaurant. As we can see from the very little data in Table 5, the frequency of patronage is not necessarily the same as the interest in patronizing. [Similar disparities between interest and frequency also occur for foods, where top scoring foods on an interest scale would not necessarily be consumed most frequently].

IdeaMap Step 8 - Cluster Consumers On The Basis Of Their Individual Utilities. Individuals differ in what they find attractive in a concept, and in a restaurant. The question becomes how to identify naturally occurring clusters of individuals in a population who show similar preference patterns, without expending lots of money and time doing so. The resulting segmentation must also be immediately actionable for the food service operator, rather than being theoretical. It does little good to identify segments in the population based upon a general segmentation method only to discover later that these segments can't really be applied to the current food service problem. The stumbling block with most segmentation procedures is their generality rather than their immediate applicability, and the need to test hundreds of consumers in order to find the segments.

Moskowitz[8,9] presented a segmentation approach which allows the researcher to work with significantly smaller samples of consumers (around 100 or so), and which is immediately applicable to the product or service being offered. The underlying assumption is that the relation between an individual's liking (viz., the liking of a product, the utility of a concept element) in general follows an inverted U shaped curve, similar (at least generally) to that shown in Figure 1. Each individual generates curve similar that that shown in Figure 1, for every semantic differential scale. The specific parameters of the curve (viz., the shape, where the curve peaks, etc.) varies by individual and by underlying semantic scale. The relation between semantic scale value (X) and individual utility (Y) can be expressed by a quadratic function, so that the optimum may lie anywhere on the semantic scale. Each individual has a specific location on each semantic scale, respectively, where that individual's utilities peak. Thus each individual generates a vector of optimal levels on the same reference set of semantic scales. The researcher can cluster individuals based upon the set of optimum levels. Individuals with similar patterns of optimal levels on the semantic scales will fall into the same cluster.

The end result of the process is a set of clusters, with each cluster comprising consumers. The clustering and thus the segmentation method is independent of the sizes of the numbers used by the panelist because the sizes of the numbers used by consumers is a scaling artifact. We don't want to cluster together consumers who use high ratings simply because they use high ratings. We want individuals to lie in the same cluster if they show the same pattern of responses, which in this case consists of similar optimal levels on the semantic differential scales.

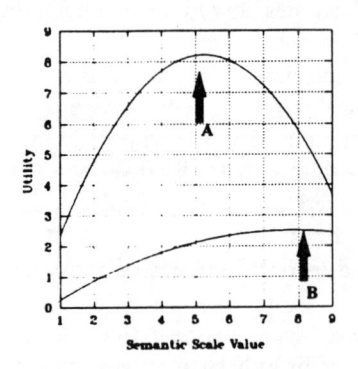

Figure 1 - Relation between semantic scale value (X axis) and utility (Y axis) for two individuals (A,B). The arrows show the optimal level for each panelist. The X value (semantic scale value) corresponding to the optimum utility is used as a basis to segment consumers.

	Total	Seg 1 - Trad	Seg 2 - Budget
Winning Elements - Total Panel			
Wholesome and hearty meals, fresh from the grill	17	25	17
Sample our delicious breads right from the oven	16	29	12
Traditional dinners with a homey touch, featuring roast beef, meat loaf and burgers	15	24	13
Foods freshly prepared by trained chefs	15	23	13
Winning Elements - Segment 1 - Traditionalists			
Sample our delicious breads right from the oven	16	29	12
Specializes in traditional foods that take a lot of time to prepare	14	29	4
Specializes in succulent, roasted meats and specialty casseroles	13	27	6
Fresh baked hams--like mama used to make	13	26	9
Winning Elements - Segment 2 - Convenience/Family			
Children's meals, including dessert and beverage, priced just right	8	1	24
Favorite kid foods like macaroni & cheese, burgers, corn and mashed potatoes	8	2	22
Less expensive than preparing a full meal at home	9	-1	21

Table 6 - Winning elements in the concept study - by total panel and the two concept response segments

This segmentation analysis generates individuals with radically different interest patterns, as Table 6 shows. Furthermore, the data suggest that there are two different segments -- a traditionalist food oriented group and a speed/convenience oriented group, and that these two groups are attracted by different messages. What turns on one group will not necessarily turn off the other group, but certainly will not attract them

IdeaMap Step 9 - Create New Concepts To Attract Customers. The final step consists of creating different restaurant concepts for O'Steers, by combining elements which score well, and which support the marketing objectives. Table 7A shows summary statistics for five concepts, and Table 7B shows the actual concepts, for five objectives:

    Objective 1 - Appeal to the total panel
    Objective 2 - Appeal to Segment 1 - traditional
    Objective 3 - Appeal to Segment 2 - budget
    Objective 4 - Appeal to the total panel, tonality of speed
    Objective 5 - Appeal to the total panel, tonality of relaxation

All of these optimizations are possible because the researcher has developed the algebra of the concept, in terms of knowing what elements turn on consumers (from the actual study), as well as how consumers feel about the "tonality" of the concept (from the dimensionalization). The researcher has to specify one or more objectives to satisfy, and then use the database and an optimization technique to identify those particular concept elements which go together, and which satisfy the objectives.

	Objective 1	Objective 2	Objective 3	Objective 4	Objective 5
	Total	Seg 1	Seg 2	Tot/Fast	Tot/Slow
Total	95	94	82	65	90
Seg 1	118	122	72	62	118
Seg 2	80	71	95	70	66
Relaxed/Fast	46	40	56	65	39

Table 7A - Expected ratings (utilities) for optimal concepts, obtained by summing the part-worth utilities of the components, and then adding in the additive constant from the regression model. The concepts all comprise the same four categories: positioning, food, ambiance and service.

Concept For Objective 1 - For Total Panel
Wholesome and hearty meals, fresh from the grill
Traditional dinners with a touch...featuring roast beef, meat loaf and burgers
A quiet environment where you can enjoy your meal
Foods freshly prepared by trained chefs

Concept For Objective 2 - For Segment 1 (Traditionalist)
Wholesome and hearty meals, fresh from the
Specializing in traditional foods that take a lot of time to prepare
A quiet environment where you can enjoy your meal
Foods freshly prepared by trained chefs

Concept For Objective 3 - For Segment 2 (Budget Conscious)
Wholesome and hearty meals, fresh from the
Real, wholesome foods for you and your
No need to dress up, our atmosphere is
The ultimate in restaurant convenience catering to your every need -- eat-in, drive-thru, pick-up and delivery

Concept For Objective 4 -Total Panel, With Tonality Move Concept Towards "Fast"
We do the cooking-when you don't have the
Traditional dinners with a touch...featuring roast beef, meat loaf and burgers
A lively and fun place to eat
You make the choices, we assemble the for you

Concept For Objective 5 - Total Panel, With Tonality Move Concept Towards "Relaxed"
Relax and enjoy--you know you're eating good food
Specializing in traditional foods that take a lot of time to prepare
A quiet environment where you can enjoy your meal
Foods freshly prepared by trained chefs

Table 7B - Five optimized concepts to appeal to consumers, yet convey a desired tonality

Accelerating The Process Even Further By Bringing Concept Development Into A Real-Time Mode. The foregoing description of the IdeaMap approach is designed for rapid (viz., weeks), but not immediate concept development (viz., hours). The IdeaMap method requires the research to do homework - viz., dimensionalize and restrict the elements, prior to the study. We can streamline the process even more by eliminating the dimensionalization and restriction steps, if the researcher is willing to live with fewer elements, and with no restrictions. This extension of IdeaMap has been implemented in the FastMap process.. The FastMap process follows the five steps listed below. These steps can be done in focus groups, with the same consumers acting both as creatives to develop the new ideas, and as evaluators who assess the concepts.

FastMap Step 1 - In focus groups identify select an experimental design. As before the design is nothing more than a schematic which comprises categories and elements. The design is a skeleton. The only difference now is that every category must comprise an equal number of elements, rather than being free to comprise as many elements as the researcher desired. There are many different experimental designs from which to choose, including designs with 3 categories, 4 categories, 5 categories, as well as designs wherein each category comprises 3 elements, 4 elements, 5 elements, etc. It is not vital to get the number of categories right at first, because the process will be iterative. It is more important to select an experimental design for each iteration which can accommodate the elements for that iteration. The marriage of spontaneity and discipline is what is important here.

FastMap Step 2 - Create elements for each category in order to fill the available slots for that category as dictated by the experimental design. In the focus group the moderator tells the consumer what the categories are for this particular design, or asks the consumers to suggest categories. Categories comprise related elements. The moderator then asks the consumers to provide elements for each category. The consumers at first are diffident in these creative groups, but quickly warm up. The problem is to winnow the large number of elements down to a workable number. Eventually, however, (usually within a few minutes) the moderator fills up the design. Sometimes the categories comprise pictures or videos, rather than words. Multi-media based programs can accommodate pictures, especially when these pictures are part of the library and can be selected by the "point and click" instructions available in most multi-media programs.

FastMap Step 3 - Create the completed experimental design, and prepare for evaluation. The elements are already selected and the design specifies which elements are to appear in what specific combinations. The only remaining setup step consists of selecting the appropriate rating attributes. Usually two attributes suffice (e.g., interest, frequency, as was done for the full IdeaMap setup), but one can choose more attributes if the interest focuses on other criteria besides interest and frequency, respectively..

FastMap Step 4 - Present the test concepts to the panelists in the focus group and acquire the ratings. This step usually takes about 10 minutes. The focus group consumers sit as a group in front of the computer monitor. As each concept is presented from the experimental design (viz., combinations of elements just selected) the consumers rate the concept. The average rating is entered into the computer for that concept. The number of concepts in a design is a function of the number of categories and number of elements within a category. It is important for the orderly progression of the project that the panelists give their first impression.

FastMap Step 5 - Obtain results from the panelists for the iteration, and create new elements if desired. The output from the evaluation is immediately available. The consumers can look at their results, discuss them, and then select new elements to replace old ones. The "debriefing" discussion may last a half hour or so, during which time the moderator can encourage the consumers to create new elements.

FastMap Step 6 - Repeat the procedure, starting with new elements, and a new design. Steps 1 - 6 can be repeated as many times as the interviewer wants, as long as each new iteration has its own identifying code. The multiple iterations yield a database comprising many elements, and the utilities of each element. Some elements will appear in only one iteration, whereas other elements will appear in many iterations.

FastMap Step 7 - Create an overall summary model, after the iterations have been completed. Step 7 summarizes the results from the entire study into one overall model, showing how every element developed by the focus group turns on or turns off consumers.

Discussion

As the risks in developing new restaurant concepts become ever greater and as the competition stiffens, restaurateurs are becoming increasingly interested in designing restaurant concepts with the aid of consumers. However, the pace of the food service industry is too rapid to accommodate the traditional, time-consuming tools of concept development, concept research and concept refinement. Opportunities arise and the food service entrepreneur or corporate manager must take advantage of those opportunities, even for a limited time.

As a consequence of the need for data consumer research is becoming more popular in food service. The popularity of "shoot from the hip" is beginning to wane, as research users are becoming more aware of the tools available to them, and as the researcher is increasingly able to muster the use of computers and technology in the service of concept design, data acquisition, modeling and optimization. Whereas a decade ago the idea of concept development was only relevant for the very large restaurant chains with management familiar with packaged goods, today concept development through consumers is increasingly popular.

The IdeaMap and FastMap technologies presented in this paper, and the case history provided for FastMap show the potential for incorporating consumer input at both the design and the evaluation stage. As time progresses one should expect to see far more integration of consumers into the development process, just as consumer feedback has been integrated into the measurement of customer satisfaction. We are just at the beginning of this integration process.

References

1. Box, G.E.P., Hunter, J, & Hunter, S. *Statistics For Experimenters*, New York, John Wiley, 1978.

2. Deming, S.M.. Experimental design: response surfaces, in: *Chemometrics, Mathematics And Statistics In Chemistry* (ed. B.R. Kowalski), pp. 251-266, D. Reidel, Dordrecht, 1983.

3. Green, P.E.. Hybrid models for conjoint analysis: An expository review. *Journal Of Marketing_Research*, 1984, 21, 155-169.

4. Cattin P., & Wittink, D.R.. Commercial use of conjoint analysis: A survey. *Journal Of Marketing*, 1982, 46, 44-53.

5. Moskowitz, H.R., & Martin, D. How computer aided design and presentation of concepts speeds up the product development process, in Proceedings Of The *46th E.S.O.M.A.R . Marketing Research Congress*, European Society Of Marketing Research, Copenhagen, pp. 404-424, 1993.

6 Moskowitz, H.R. *Consumer Evaluation And Testing Of Personal Care Products*, Marcel Dekker, New York, 1996.

7 Moskowitz, H.R. *Food Concepts And Products: Just In Time Development*, Food And Nutrition Press, Trumbull, 1994.

8 Rabino, S., Moskowitz, H.R., & Reider, I. Concept response segmentation and optimization: An international perspective, *In preparation*.

9 Moskowitz, H.R.. Sensory segmentation of fragrance preferences. *Journal Of The Society Of Cosmetic Chemistry*, 1986, 37, 233-247.

Catering systems revisited

W.G. Reeve, J.S.A. Edwards
*Worshipful Company of Cooks Centre for Culinary Research,
Bournemouth University, Poole. Dorset, BH12 5BB. UK.*

Abstract

Catering systems have been developed in order to reduce costs, increase overall efficiency and to satisfy consumer demand and expectations for variety and choice. Over the last two decades it has been assumed that the 'new' systems operate in such a way as to achieve these objectives. However, many of the weaknesses identified in traditional catering have not been corrected. In several cases, 'new' systems have been altered from their original concepts and in doing so faults re-introduced. This paper explores the development of the primary catering systems and asks whether some of the original assumptions made are still valid.

Introduction

Catering is defined in the Oxford English Dictionary as the purveying of food or other requisites, whilst others refer to it as the provision of refreshments in the form of food and drink. In the United States of America, the term 'food service' is favoured although this term has similar connotations, in that it is used to describe the provision of food and drink for people away from home[1]. The catering industry, both in the United Kingdom and overseas, comprises a diverse range of units and outlets operated by staff with mixed skills and levels of ability. It is also an industry which is highly fragmented, and whilst there are a number of large organisations operating, for example, chains of hotels, restaurants and other outlets, the bulk of the industry is comprised primarily of single outlets of various sizes and styles which are owner operated and managed.

In order to satisfy consumer demand and expectations for a varied choice, caterers often prepare and store food well in advance of requirements. It is perhaps, therefore, not surprising that catering systems have evolved and developed to meet the particular requirements of each outlet and its consumers. The purpose of this paper is to briefly consider the systems approach to catering, and highlight where it might be now.

The systems approach to catering is not entirely new. The partie system, developed and refined towards the end of the last century, sought to use a combination of labour, equipment, skills and other resources in order to achieve the most cost effective and efficient organisation. However, the application of the systems concept or systems approach to catering does not appear to have been taken up until the late 1950s or early 1960s but even here, its utilisation and development are generally attributed to food technologists[2]. More recently, developments and changes have been sought as, for example, labour has become more expensive, skilled staff more difficult to find, with the result that equipment has become more widely used.

In the catering industry, a number of subsystems can be readily identified. These include: food and beverage systems, service systems, and information systems. Jones and Lockwood[3] divide food service systems typically into three classifications:

> ***Integrated food service systems*** where both food production and food service are an integral function carried out as part of a single operation; found, for example, in a traditional restaurant.
> ***Food manufacturing systems*** where the production of food and meals is separated, that is decoupled, from the service of meals; found for example, in flight catering.
> ***Food delivery systems*** where little of no food production takes place and the operation focuses on assembly and regeneration of the meals.

The 'traditional' cook and serve system

It has been suggested[4] that catering ideally exists where food is cooked to order and consumed immediately, but this can normally only be achieved at home or in 'high class' establishments. At this level, the catering system involves three stages; inputs of the raw materials and other basic components; the processing stage during which those raw materials, for example foods are 'processed'; and their final outputs, that is the meals served. Although this basic concept is used as, and considered to be a 'traditional' system, in that the process is one continuous action, many variations have been introduced. These have incorporated 'newer' technologies and equipment in order to overcome staff shortfalls and to better satisfy consumer demand. Several catering systems have been developed but the primary systems are shown in Figure 1.

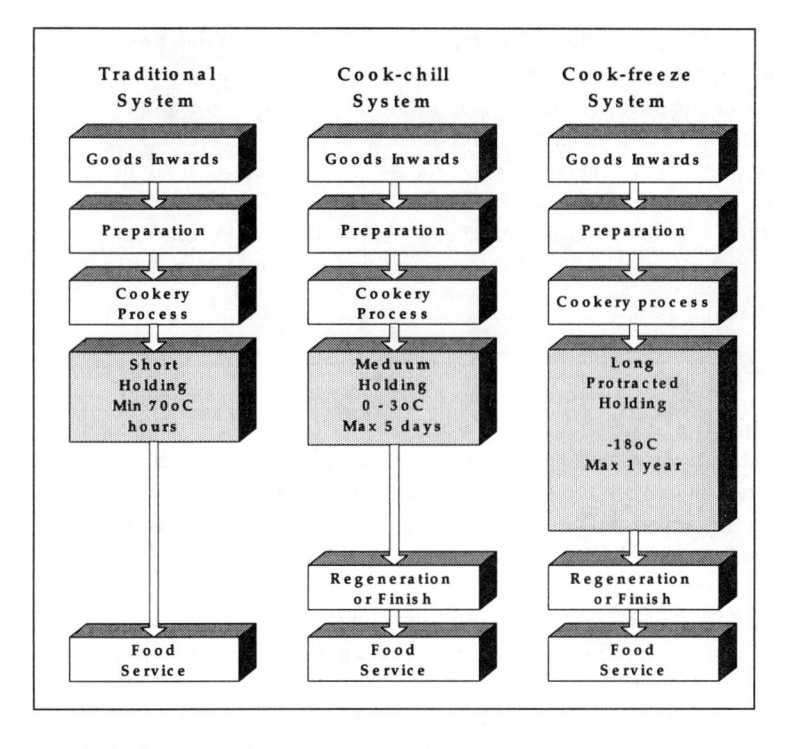

Figure 1. A Schematic Approach to Catering

Traditional System	Cook-chill System	Cook-freeze System
Goods Inwards	Goods Inwards	Goods Inwards
Preparation	Preparation	Preparation
Cookery Process	Cookery Process	Cookery process
Short Holding Min 70°C hours	Medium Holding 0 - 3°C Max 5 days	Long Protracted Holding -18°C Max 1 year
	Regeneration or Finish	Regeneration or Finish
Food Service	Food Service	Food Service

Centralised food production catering systems

Since the early 1960's systematised catering methods of cook-freeze and cook-chill have become more important where maximisation of quality, sensory, nutritional and bacteriological parameters, have been the main objectives.

Cook-freeze

The concept of freezing as a method of food preservation dates back to prehistoric times but the commercial use of freezing only really began in the 1940's with New Zealand lamb and garden peas[5]. Since then, freezing has become one of the most common and widely accepted methods of preservation for raw, ready prepared, and cooked food. This process was adopted by caterers in the UK in the 1960's and 1970's, but instead of relying solely on ready made frozen items purchased in, they produced their own in-house products which appealed to local tastes and met local requirements. The Hospital for Women in Leeds, for example, operated a cook-freeze system from 1967 where the advantages to other sectors, particularly school meals, soon became apparent. A study at the time calculated that overall costs were unlikely to be higher and, depending on the scale of production, could even be lower than those in traditional catering systems[6].

These early beginnings were important for the development of cook-freeze in the UK but the fuel crisis of the mid 1970's led to increased energy costs which reduced its cost effectiveness. Even today, high energy consumption, equipment costs and the need to adjust and modify recipes, preclude most small-scale caterers from producing cook-freeze meals. A considerable amount of research is available from the 1960's and 1970's on the application of cook-freeze and the subject has been covered in some depth, but little or no recent work is available, perhaps reflecting its decline for economic reasons, particularly in the institutional sector.

Cook-chill

Cook-chill is a system that uses a central production facility where trained and skilled staff prepare and prime cook food in bulk prior to it being rapidly chilled and stored at temperatures low enough to inhibit the growth of pathogenic and spoilage organisms. The storage life of cook-chill food is relatively short compared to cook-freeze, up to 5 days, so the linking of production to service has to be more rigidly controlled to avoid over or under production. The processed food is then distributed to finishing (satellite or end) kitchens which are smaller and staffed by semi-skilled or trained workers for final assembling, reheating and service.

Cook-chill methods were widely adopted in Europe from 1960 onwards[7] but in the UK, guidelines for its use were not issued until 1980, by which time technological advances had ensured that, provided it was properly controlled and operated, the system was practicable and safe. A study undertaken in 1986 identified at least 240 UK establishments using such a system[8] but from the literature it is not clear whether the operations were cook-chill units or conventional units equipped with chillers. What is apparent is that institutional caterers, particularly hospitals, education and the welfare establishments initially formed the largest body of users and that its value as a system to cater for large numbers has since been recognised in the industrial sector.

Prevalence of catering systems

The decision to change from one catering system to another, usually from a traditional system to cook-freeze or cook-chill is one faced by many organisations. Whether this decision is 'imposed', or arrived at following a detailed financial and cost benefit appraisal, is difficult to determine. Clearly, a number of organisations undertake feasibility studies into the viability of each system but whether the final decision is balanced and based on a thorough evaluation of all the factors is debatable. Although the issues are widely discussed, the decision in many cases is probably one of the least understood decisions made by catering managers.[9]

The prevalence of each system in the public sector in the UK today is difficult to determine as few studies have been undertaken recently. The most recent study of hospitals in the United Kingdom was undertaken between 1994 and 1995 and surveyed those hospitals with more than 100 beds[10]. This was part of a much larger survey, but from the 243 replies received, it was shown that 67.9% hospitals use a traditional catering system, 24.8% use a cook-chill system, 4.1% use a cook-freeze system and the remainder use a combination of the systems. In the USA, a similar survey, again using hospitals with more that 100 beds, was undertaken in 1993. Characteristics of these systems are shown in Table 1.

Table 1. Characteristics of Catering Systems in the USA

Hospital Size (No. of beds)	Conventional		Cook-chill		Cook-freeze		Other		Total
	n	%	n	%	n	%	n	%	n
<200	600	95.7	17	2.7	6	1.0	4	0.6	627
200-299	421	92.5	16	3.5	5	1.0	13	2.9	455
300-499	391	86.5	37	8.2	11	2.4	13	2.9	452
500 +	238	73.0	67	20.6	6	1.8	15	4.6	326
Total	1650	88.7	137	7.4	28	1.5	45	2.5	1860

As can be seen, although only 7.4% of those hospitals surveyed have adopted a cook-chill system, nearly one-fifth of them are the large hospitals with more than 500 beds. Few hospitals have adopted a cook-freeze system, (1.5%) and most hospitals (88.7%) continue with their conventional system. It is also noteworthy that the majority of those hospitals (86%) which have installed cook-freeze systems, did so prior to 1982, whereas the installation of cook-chill systems is spread over the entire period and shows little sign of slowing up.

The causes of systems failure

Recently, a number of Central Production Units (CPU) of large scale systems have failed to meet prior or planned expectations and in consequence have closed. One of the most recent cook-chill systems to close after two-years operation, was Orchard Food Services (OFS). This provides a useful example of the decision making processes involved in adopting a systems approach and also highlights some of the reasons for their demise. The reasons for failure can be broadly categorised under the following headings:

Historical: new systems are often required, perhaps as a matter of some urgency, to replace or refurbish existing catering provisions. The most pressing impetus in the late 1980's was loss of Crown Immunity by Government and Local Authority establishments and in consequence, rapid action was needed to meet existing and new food legislation.

Policy decisions need to be taken between refurbishing existing, or introducing alternative systems. It is at this stage that 'new' systems are expected to provide the answers to several problems[11]. In the case of OFS, the perceived problems were quality control, the logistics of feeding large numbers of patients in many hospital sites, and the practicalities of serving hot food at ward level. Often the apparent solutions offered by 'new' systems are viewed as a panacea with little consideration being given to an analysis of ways in which the current system could be improved.

Supply: once the decision has been taken on the 'new' system, other decisions follow including: whether to produce food in-house, or buy from outside suppliers. In the case of OFS the original intention was to buy-in, but no suppliers could be found which the Regional Health Authority (RHA) considered could combine the right quality and price with reliability and continuity[11]. As a result, the RHA built three CPU's of their own and produced in-house. This decision was regarded by some observers as the first and greatest mistake and should never have been made[12] (Rice 1995). ***Factors of change***: other factors compounded the demise of CPUs. Restructuring within the National Health Service (NHS) and replacement of Regional and Area management by local Hospital Trusts were not always known at the time original decisions were made. Neither were the rapid changes which led to hospital closures, increased patient turnaround, and the ability of Hospital Trusts to purchase on the open market. These affected the requirements for food supply and a CPU set up by OFS to provide 60,000 lbs of food weekly, never produced more than 40,000 lbs.

Although in this particular instance there are many factors influencing the final closure, in hindsight the question should be asked as to whether the choice of system was the most appropriate at the time? With prior knowledge of re-organisation, was it prudent to make a massive capital investment utilising a systems approach based on food manufacture? In this particular instance it may have been more feasible to refurbish and improve the existing catering facility and utilise other methods of food production.

An operations or systems deficiency?

However, in many instances, systems are successful and provide a range of high quality foods at an economic price. Others do not fulfil their full potential due often to poor practices and procedures. Problems perceived as being inherent in the traditional system are often re-introduced into the 'new' system which can be seen from a number of examples.

One of the primary advantages of the systems approach is that "...the production of food is separated from the immediate demands of service..."[13]. The systematised operation is an holistic approach which is summarised in Figure 2. The environmental boundary includes all production and service processes from receipt of raw ingredients to the service of the finished meal. Food production is not independent of service: both are part of the whole system and do not exist as independent entities.

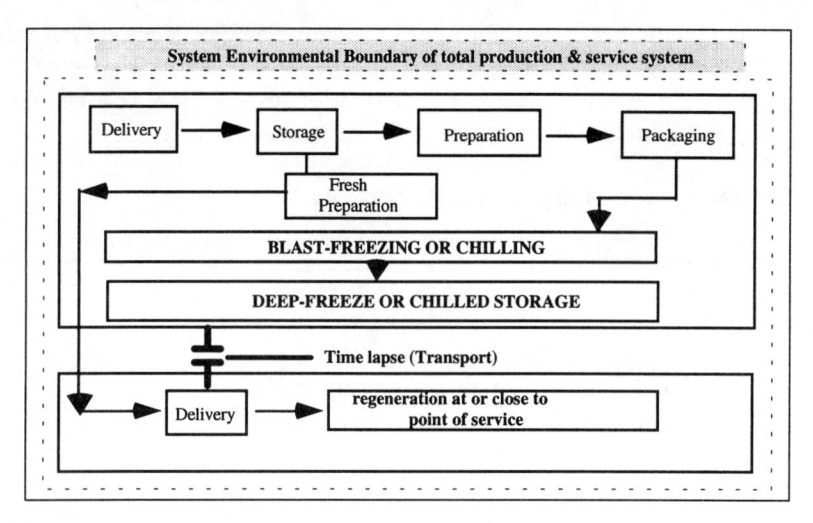

Figure 2. Systems Environment and Boundary

```
System Environmental Boundary of total production & service system

Delivery → Storage → Preparation → Packaging
              ↓
         Fresh Preparation
         
BLAST-FREEZING OR CHILLING
DEEP-FREEZE OR CHILLED STORAGE

Time lapse (Transport)

Delivery → regeneration at or close to point of service
```

Due to rapid freezing or chilling after cooking, and regeneration near to the point of consumption, the colour, taste, appearance and nutritional value of food does not deteriorate, and it can be served hot[14]. Any alterations or modifications to the 'whole' will effect the way in which the individual parts function. As Gestalt theory suggests ...the whole is greater than the sum of its parts...[15]. The synergy of a system may become unbalanced by the introduction of another operational procedure, that at face value, is thought to be a solution to another problem. The resulting outcome is that the equilibrium of the system becomes unbalanced and, as a consequence, does not achieve the required outcomes. One of the main criticisms made of the traditional system was the necessity to hold food for long periods before service[16]. The introduction, for example, of a heated 'holding trolley' to transport food to another location is a modification to the original system. Although the production process adheres to the systems approach, the service system destroys the ultimate objectives of the system. This situation is not new and was recognised as a major problem in the late 1980's where the concern was that people were 'taking short-cuts to save money'. It was considered that if cook-chill were to have any benefit it had to be undertaken to the highest standards with regeneration as close to the consumer as possible in order to minimise the problems arising from warm holding[17].

The human resource system

Moving the regeneration process closer to the consumer still does not guarantee a minimum holding time and associated loss of quality. If food is regenerated close to the point of service but not given the highest service priority, the same problem of warm holding remains. This is a re-occurring theme highlighted in a number of studies. The majority of hospitals, of all sizes, delegate responsibility for the final preparation and service of food to staff who are not employed specifically by the catering department. This effectively means the catering manager relinquishes control for some of the key elements of his system although he still remains accountable for the quality of the meal which the patient receives[10]. In an earlier study, the division of responsibility between caterers, nursing staff who served meals and dieticians was also criticised[18].

Other factors effecting food quality outputs

It is generally accepted that rapid cooling is able to retain the sensory and nutritional properties of food. However there are other factors that need to be considered:

1. Not all foods can be successfully frozen, chilled and/or re-heated, hence some products are of reduced sensory quality to conventionally cooked items. The most appropriate, as well as economical, main course dishes for regeneration are sauce (liquid) dependent leading to lack variety and reliance on a restricted range of cookery methods.
2. Some products are re-heated by inappropriate regeneration methods, for example, individual meals consisting of complex components cannot be regenerated independently according to their individual characteristics.
3. Production practices criticised in conventional systems, for example, the addition of bicarbonate of soda to vegetables, are used in the 'factory' production of meals. These processes in turn require other meal components to be fortified to counteract the loss of vitamins.

Are comparisons valid?

As with traditional systems, each of the 'new' systems have advantages and limitations, some of which are intrinsic to the system and others, due mainly to poor operational catering practices, extrinsic. In many instances, the issues surrounding a decision whether to adopt a different system is obscured by inappropriate observations. These include aspects such as long cooking times in traditional kitchens which are blamed on inefficient cooking equipment used for the task[15]. There is no place for inefficient equipment in any catering system. Furthermore, both traditional systems and CPUs employ, for example, high pressure steaming equipment which is capable of cooking vegetables from frozen in short periods of time. This sort of observation also questions the basis upon which comparisons between different systems are made. Is like being compared with like and are comparisons being made between old run-down premises, with old equipment inadequate controls, and 'new' systems with modern equipment with proper techniques of operation and control?

Efficiency and cost

When operational changes were made, most of the 'success stories' reported in literature involved hospitals switching from conventional to 'new' catering systems[19]. One of the main incentives for change was the promise of greater efficiency and lower costs. Many operators have reported improvements in operational and financial performance and several studies have concluded that changing from a traditional system to cook-chill or cook freeze result in lower labour costs. However, other studies question these findings and suggest, for example, that the labour savings in the production unit are forfeited by additional personnel required to assemble, plate, transport and control the food[18]. Other comparative studies have found that managers of traditional, cook-freeze and cook-chill systems employ similarly resources to achieve the same objectives. In areas such as full-time staff, turnover rates, absenteeism and salaries, caterers who expect to make savings in personnel costs by installing cook-chill and cook-freeze systems need to be certain that reductions in these areas can actually be achieved[20].

A food preservation process

It could be argued that most important factors to be considered in any catering operation are consumer acceptability and expectations. In traditional systems, food is prepared on the assumption that it will be served and eaten within a short period. However, most 'new' catering systems have introduced an additional stage between the cooking and service, a time buffer, during which food is preserved by chilling or freezing. It can be argued that the cook-chill and cook-freeze systems are firstly, cooked food preservation processes and secondly, food service systems. If this argument is used, it may change the way in which systems are perceived and what part they play in the food production and service. Unlike the operations concerned with canning and dehydration, all production and service can be performed by the caterer[16]. If this basic assumption is embraced, foods should be processed or preserved on the basis of which system best retains their quality. The assumption recognises that there are limitations to all systems and not all foods can be processed in the same way without changing their properties and characteristics. The caterer is not a food manufacture. He supplies a diverse range of foods, beverages and styles of services to the consumer. Most manufactures supply the caterer with a limited range of products, for example, frozen, chilled and dried products which are then assembled or processed by the caterer in order to meet a specific demand in their specific market.

Conclusion

In many instances, the failure of a catering system is attributed primarily to the characteristics of the system, whereas in reality, it may be due to policy and business decisions taken at the time. Nevertheless, some systems are ill conceived, poorly instigated and operated outside the parameters for which they were designed. Certain problems perceived as inherent in traditional systems have re-occurred and the promised advantages of labour savings and efficiency have not been achieved. This re-opens the debate on whether reliance on one system is desirable or preferable. Could it be that the various systems are seen as a panacea, whereas in reality, a multi-system approach utilising the most appropriate equipment for the task may be the best solution for large scale food provision? Perhaps we should step back into the past and re-examine some of the operational practices that were envisaged at the time. Cook-freeze is not a new process. Ready made frozen foods have been marketed for years and the equipment is available to enable more caterers to operate their own frozen food kitchens[21]. More and more conventional kitchens are purchasing blast-chilling equipment in order to meet the new requirements for cooling hot food[22]. Conceivably these operations are using technology for the benefit of their own particular systems and not being driven along the more specialised production systems that may not provide all the solutions to all the problems.

"...systems must be tailor made for the customer.."[21]

References

[1] Green, E.F., Drake, G.G. & Sweeney, F.J. *Profitable Food and Beverage Management: Planning.* Van Nostrand Reinhold, New York, 1991.

[2] Livingston, G.E. Development of the service approach to the design of food service operations. In Livingston, G.E and Chang, C.M. (eds). *Food Service Systems.* Academic Press, New York, pp 3 - 17, 1979.

[3] Jones, P. & Lockwood, A. Hospitality operating systems. *International Journal of Contemporary Hospitality Management.* **7(5)**, 17-20, 1995.

[4] Glew, G., Lawson, J. & Hunt, C. The effect of catering techniques on the nutritional value of food. In Cottrell, R. (ed). *Nutrition in Catering.* Parthenon Publishing Group, Carnforth, Lancs. pp 53 - 74, 1987.

[5] Sanders, T. & Bazalgette, P. *The Food Revolution.* Bantam Press, London, 1991.

[6] Millross, J., Speht, A., Holdsworth, K., & Glew,G. *The Utilisation of Cook-Freeze Catering System for School Meals.* University of Leeds. 1973.

[7] Tändler, K. In folienbehältnissen pasteurisierte fleischfertiggerichtee für die gemeinschaftsverpflegung. *Die Fleischwirtschaft.* **7,** 845-850, 1972.

[8] Walker, A.E. The update of cook-chill technology by the catering industry. In *The Future of Cook-chill Technology in Catering. A Workshop for Cook-chill Users.* Dorset Institute of Higher Education, Poole Dorset. 1986.

[9] Quoted in Nettles, M.F. and M.B. Gregoire. Operational characteristics of hospital foodservice departments with conventional, cook-chill and cook-freeze systems. *Journal of the American Dietetic Association.* **93(10)**, 1161-01163, 1993.

[10] Ervin, J. and J.S.A. Edwards. *Hospital Catering 1995.* A Worshipful Company of Cook Centre for Culinary Research Publication. Bournemouth University, 1995.

[11] Baker, J. Orchards bitter fruit. *Food Service Management*, July 1995.

12 Rice J. Quoted in Baker, J. Orchards bitter fruit. *Food Service Management.* July 1995.
13 Lawson, F. *Principles of Catering Design.* The Architectural Press, London, 1994.
14 Boltman B. *Cook-Freeze Catering Systems.* Applied Science Publishers Ltd, London, 1978.
15 Wertheimer, M. Laws of organisation in perceptual forms. In Ellis, W.D. *A Source Book of Gestalt.* Harcourt Brace, pp 71-88, 1938.
16 Glew G. Cook-freeze Catering. *An Introduction to its Technology.* Faber & Faber, London, 1973.
17 Lang T. What Future for Cook-Chill. *Catering and Hotel Keeper.* 30 July 1987.
18 Corsi, R.M. The ready foods system. *Consultant.* **17(1)**, 23, 1984.
19 Reed, R.M. Conventional system may be best. *Journal of the American Hospital Association.* **47(14)**, 161-162, 1973.
20 Greathouse K.R., Gregoire M.B., & Spears M.C. Comparison of conventional, cook-chill, and cook-freeze food service systems. *Journal of the American Dietetic Association.* **89(11)**, 1606-1611, 1989.
21 British Gas. *A Student's Guide to Gas Catering.* British Gas Marketing Division, Holborn, London, 1974.
22 Long S. Personal Communication, Food Services Manager, Foster Refrigerator (UK), Norfolk, 1995.

Catering systems in use in the British Army

S. Ridley, P. Jones
Department of Food and Hospitality Management, Bournemouth University, UK

Abstract

The overall purpose of the paper is to research the current food provision in barracks in the British Army, and to suggest options for a new food service system, leading to cost savings, improved customer satisfaction and employee motivation.

1 Introduction

The political and military environment in which the Army operates has changed since the end of the Cold War. The emphasis of the defence strategy has moved away from the strategic threat - the potentially hostile Soviet Union and Warsaw Pact - that has dominated the security concerns for many years towards an armed force that is now primarily concerned with peace keeping and creating stability in the international setting. There has been a major reduction in military commitments since the end of the Cold War, but the potential calls on the armed forces still remain significant, within NATO and the UN, since the military and political environment is uncertain and unpredictable.[1]

There has been increased political and economic pressure to make the Army, along with other government departments more cost efficient. This has led to a reduction in the size of the forces, but also an improvement in the quality of equipment, training and efficiency.[2] It has forced the Army to look at the other

Services to bench mark against them, to combine facilities where appropriate, to reduce duplication. The effects of the changes that the Army is being forced to make are having an impact on food services. It is being forced to reduce its manpower levels, to find alternative methods of feeding the soldiers, to determine the cost effectiveness of the service that is being provided and to make alternative suggestions for a more effective service. It is having to re-examine the role of its chefs, to determine new methods of making their trade a career again, and not just a job, to improve motivation, and to enable the continuation of recruitment.

2 Food Service Provision in the British Army

The Army expects a food service to be provided to meet their feeding needs under all conditions, in all locations, wherever the Army is called to deploy. It is the role of Army food services to ensure the provision of this service.

2.1 The role of food service

The functional aim for catering management in the Army is:
'To provide the most cost effective catering service for the Army, meeting operational requirements and prescribed standards.'
- The Functional Plan for Catering Management, 1987 [3]

Even though Army food services' primary objective is to feed the soldiers during war, the soldiers also need to be fed during peace on a daily basis, in barracks and on exercise. The soldiers need a good, nutritionally balanced diet, which is particularly important in the active roles they play, and is very important for their morale and well-being.

Thus, Army food services must provide meals of a prescribed standard, which are cost effective, on a daily basis under any military conditions.

2.2 The development of Army food services

The Army Catering Corps (ACC) was founded in 1941, with the primary aim of providing trained military chefs. During its 50 year history, it developed responsibility for maintaining standards, specialist training, the management of contracts and the provision of technical advice concerning building projects and catering equipment. The Army Catering Corps never had full operational or budgetary control of Army catering. Since its formation, time and effort have been focused on improving the quality of food prepared, with little attention paid to the logistics and management of food provision by the Catering Corps, which has always been left to the unit commanders.

The Directorate of the Army Catering Corps, DACC, 1992 [4] asserted that "The standards for food preparation, cookery and presentation provided for the British Army are probably the highest in the international military community, but, although this was the area which most required improvement at its inception, it is not clear why the Army Catering Corps' activity, authority or influence for most of its existence has been confined to an area most appropriate to an Army <u>Cookery</u> Corps". The Directorate continued with: "... the confining of the ACC's activity to cookery had more to do with good old-fashioned possessiveness and parochialism. Catering in its broader sense was not perceived to be a problem and, furthermore, was conducted by and managed by regimental personnel - and it is virtually a matter of principle in any organisation that manpower and authority are only to be surrendered under the strongest influences and pressures."

This situation resulted in positions for caterers in the upper echelons of power at policy level, but at unit level responsibility for the operational aspects was vested in commanding officers. "Regimental officers controlled the cooks' activities whilst their own Corps officers were restricted to providing consultative advice at a distance." DACC, 1992 [5]

In the 1970s, a unit based appointment of 'Specialist Catering Officer' (SCO) was created, to take responsibility for unit catering policy and procedure.

```
         Commanding Officer
                 |
           Quarter Master
                 |
   ┌─────────────────────────┐
   │    Catering Platoon     │
   │   Specialist Catering   │
   │         Officer         │
   │            |            │
   │       Master Chef       │
   │            |            │
   │          Chef           │
   └─────────────────────────┘
```

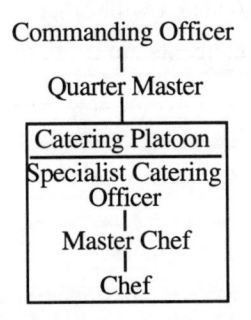

Figure 1: Position of Specialist Catering Officer in a unit

This position was replaced by the Area Catering Officer (ACO), in the 1980s. The ACO is responsible for the units within the brigades. The ACO works to the brigade commander, but within the Army Catering Service (ACS) establishment. The ACO can have great difficulty being accepted within the units/brigades as he is not a direct part of the unit. Catering officers often can feel unwanted - they are caught between trying to please the ACS and to integrate into the brigade.

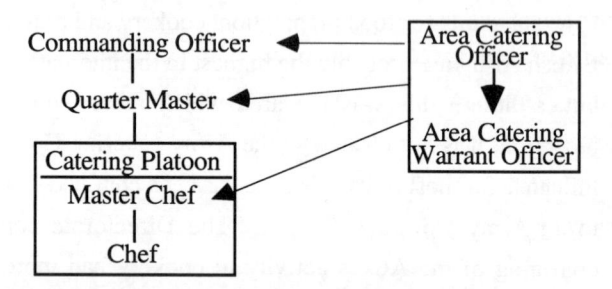

Figure 2: Position of Area Catering Officer in a unit

Since 1983, certain support services to the Army have been provided by civilian contracting organisations. A study of Army Catering concluded that savings could be made by contracting out or civilianising the work in static units.[6] Contractorisation has led to the reduction in the number of military chefs; a concentration of military chefs in the junior ranks' messes, with the sergeants' and officers' messes being catered for by a contractor.[7]

Since the end of the Cold War, several studies have been undertaken - Options for Change, the New Management Strategy and Front Line First.[8] Army food services has had to change in response to these studies. The Army Catering Corps has been amalgamated with other support Corps to form the Royal Logistics Corps, and all elements are continually seeking reductions in budgeted costs.

2.3 The current structure of Army food services

Army food services have to provide catering for the junior ranks, the senior non-commissioned officers and the commissioned officers in separate messes, within each regiment. Following the Army's restructuring and the reduction in manpower, new strategies and policies have been formed and implemented. Military chefs now run the junior ranks' mess and support that unit's training, and are also required to support training for minor units in the brigade, and undertake emergency tours. The Officers' and Sergeants' messes are now contracted out; and the shortfall of military chefs is made up with civilians.

Previously, the chefs were distributed throughout the Army, attached to units, under the command of the unit's Commanding Officer. There was sufficient chefs to man all static units and field forces during war. Now, there are only sufficient chefs available to be distributed to all field forces with no excess to man the rear units. The military chefs (15-19 per unit) are now attached to most Field Force Major units, under the command of the Commanding Officer of that unit, with the Area Catering Officer responsible for the deployment of the chefs, acting through the unit Master Chef.[9]

It is the responsibility of the Area Catering Officers (ACO), Area Catering Warrant Officers (ACWO), and on a daily basis the unit Master Chef to oversee the contractors. Although the Quartermasters of each unit, in their role of providing support services, have some involvement too. The contractors do not fit easily within the units' chain of command nor channel of communication, therefore new methods of communication are required to monitor and work with the contractors. The military chefs, in particular the Master Chefs, need to learn how to deal with the contractors, to understand their management and communication style, and adapt their own to be more flexible and consultative.
Thus, the situation exists where Army food services is a combination of contractors, civilians and military chefs. The Area Catering Officers and Master Chefs have the responsibility of maintaining morale and 'quality of life' for the military chefs; as well as working alongside and supervising the contractors.

3 Reasons for further change

Change has been forced onto Army food service in the past by the changes occurring in the Army as a whole. There are many problems with the present food provision system, possibly due to Army food services being unable to plan for change but having to adapt their old system to fit with the cost and manpower reductions imposed by the Ministry of Defence. If a survey had been undertaken to look at the present system - the flaws and strengths which exist in it, the satisfaction (or lack of) of the customers (the soldiers and officer) and the caterers' views of the system, possibly a better foundation would have been laid on which to create a new food provision system.

In preparing a case for further change, consideration should be given to customer satisfaction, chef morale and motivation, inefficiencies in the current system, the changing culture / background of the soldiers.

3.1 Customer Satisfaction

Discussions with soldiers and officers have revealed overall dissatisfaction with the system, which is organisationally, rather than customer oriented. There is no single reason for this, it is a combination of meal fatigue, through eating in the same establishment for three meals a day, seven days a week; the physical location of the dining hall in relation to the accommodation; the timing of meals, which are historically based rather than opportune; the nature of the soldiers' and officers' daily activities.

In general, it seems that the junior ranks are currently more satisfied with their food provision than the senior ranks or the officers. The reason that is often cited by the sergeants and officers for this is that the junior ranks' meals are provided by military chefs and the senior ranks and officers have their messes contracted out. Traditions, customs and practices have given rise to a level of service that is not manifest in the nature and terms of a contract. Some regiments are finding that their current expectations are not being met by the contractors and they feel that their traditions and customs are being thwarted through the use of contractors.

3.2 Chef Morale and Motivation

Since the military chefs now only work in the dining halls and not the officers' and sergeants' messes, there is less opportunity to develop and practice varied catering skills, leading to a reduction in career prospects. The reduced quality of life for the military chefs due to an increased number of exercises and chefs no longer being attached to one regiment but deployed as and when necessary, may lead to lack of job satisfaction and decreased motivation and ultimately to the chef leaving the Army. There may be recruitment difficulties if the prospective trainee chefs cannot see a career future for themselves within the Army.

3.3 Inefficiencies in the system

The costs of the food service are suggested as being high, through having to provide meals to several different outlets, and in the case of the sergeants and officers to only a few people in each mess. Manpower demands, building and energy costs are high, with service in several different kitchens and messes.

Payment for meals is by the day and is deducted from pay at source and assumes that the individual will consume three meals a day. The mechanism the Master Chef uses for budgeting for the meals is totally unrelated to the food charges. This is through a Daily Messing Rate (DMR) - a costed daily amount based on the concept of a standard scale of food sufficient for one man for one day. The DMR provides sufficient money for the chef to meet the soldiers' expectations of quality, variety and choice only if some soldiers do not turn up for meals, e.g. at the weekend or at breakfast. If all those entitled to meals ate every day, there would be insufficient money available to supply them all with meals to the expected standards.

It could be argued that the use of only one supplier (NAAFI), and thus no competition, creates higher costs and a mediocre distribution and delivery system.

It appears that the specifications for the contracts for the provision of food and accommodation in the messes have been badly drawn up, resulting in a system which is unacceptable to the sergeants and officers in their messes.

3.4 The changing lifestyle of the soldiers

Historically, the majority of soldiers lived in barracks, and the Army assumed the responsibility for providing them with a balanced diet. Now many more are married and live in quarters, or rent their own accommodation, their food intake is thus determined more by their family rather than the Army. Does the Army therefore have to ensure those that do live in barracks have a balanced diet when they cannot influence the food intake of the soldiers who live out of the barracks?

Society is changing, eating habits are changing, many more children are growing up in families with both parents working, they make their own meals, and look after themselves. The soldiers, coming from similar backgrounds, therefore should be more capable of catering for themselves, and their attitudes should be more independent, more inclined to providing for themselves.

4 Options for a new food service system

Three options for food provision are contemplated in this paper, taking into consideration the reasons for further change. These options are:

4.1 Remove the provision of food and accommodation in barracks

Is it necessary for the Army to provide accommodation of the current type and standard, since today, it is not necessarily the norm to live in barracks? Viable alternatives, currently being considered, may be local shared rented accommodation, hotel type provision or Army provided self-catering facilities. The provision of a mid-day meal, similar to employee feeding in a large commercial organisation, would be the only catering provision necessary.

4.2 Amalgamate the distinct food and accommodation facilities

Each regiment has its own junior ranks' dining hall, sergeants' mess and officers' mess. Within a camp, such as Tidworth, there could be 10 regiments within a limited radius. These 30 messes could be amalgamated so there was one central junior ranks' dining hall, one sergeants' mess and one officers' mess. This would considerably reduce the cost, requiring the use of less manpower, less equipment and heating, and increase the efficiency of food provision, reducing wastage, whilst probably increasing the choice and quality of meals. Such an amalgamation would fit with option 4.1, since there would be no accommodation provision or only self catering provision in barracks, hence no need for catering provision, by Army food services, at each accommodation site.

4.3 Central Production Unit (CPU)

A possible option which has evolved from options 1 and 2 is to reduce the number of production areas in a camp through the use of a CPU, providing hot meals to one location for immediate consumption, and then chilled or frozen meals to satellite positions within the camp, for immediate consumption or consumption at a later date. It could be operated along the line of a hospital or a large industrial plant site, managed by military chefs or contract caterers.

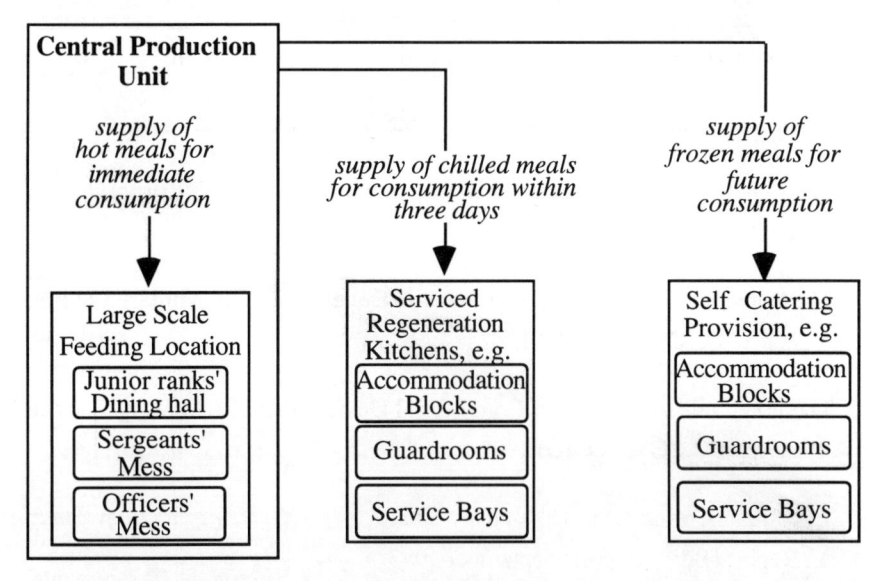

```
┌─────────────────────┐
│ Central Production  │────────────┬──────────────────┐
│        Unit         │            │                  │
└─────────────────────┘            │                  │
    supply of                                      supply of
   hot meals for         supply of chilled meals  frozen meals for
    immediate            for consumption within      future
   consumption                 three days          consumption
        ▼                          ▼                   ▼
┌─────────────────┐      ┌─────────────────┐  ┌─────────────────┐
│   Large Scale   │      │    Serviced     │  │  Self Catering  │
│ Feeding Location│      │  Regeneration   │  │ Provision, e.g. │
│ ┌─────────────┐ │      │ Kitchens, e.g.  │  │ ┌─────────────┐ │
│ │Junior ranks'│ │      │ ┌─────────────┐ │  │ │Accommodation│ │
│ │ Dining hall │ │      │ │Accommodation│ │  │ │   Blocks    │ │
│ └─────────────┘ │      │ │   Blocks    │ │  │ └─────────────┘ │
│ ┌─────────────┐ │      │ └─────────────┘ │  │ ┌─────────────┐ │
│ │  Sergeants' │ │      │ ┌─────────────┐ │  │ │ Guardrooms  │ │
│ │    Mess     │ │      │ │ Guardrooms  │ │  │ └─────────────┘ │
│ └─────────────┘ │      │ └─────────────┘ │  │ ┌─────────────┐ │
│ ┌─────────────┐ │      │ ┌─────────────┐ │  │ │Service Bays │ │
│ │  Officers'  │ │      │ │Service Bays │ │  │ └─────────────┘ │
│ │    Mess     │ │      │ └─────────────┘ │  │                 │
│ └─────────────┘ │      │                 │  │                 │
└─────────────────┘      └─────────────────┘  └─────────────────┘
```

Figure 3: A model of a Central Production Unit feeding system in barracks

The perceived benefits of this system would be reduced manpower, reduced operating costs, increased choice in the sergeants' and officers' messes, and flexibility in the provision and consumption of meals.

An adaptation to this system could be to have a commercial food manufacturing company supplying appropriate food products to the camp(s), which are then incorporated into meals and regenerated in suitable locations.

This option, the use of a CPU, is put forward for discussion, its viability - the practicalities, the costs and efficiencies, to be examined in future research.

5 Conclusion

Hypotheses can be developed for discussion about new food systems, about removing the provision of all but the mid-day meal, of amalgamating messes, but at the end of the day the constraints imposed by the Army - the culture and traditions of the regiments, the officers and soldiers, the economic and functional conditions, will determine future provision.

References

1. *Statement on the Defence Estimates*, HMSO, London, 1992.

2. *Statement on the Defence Estimates*, HMSO, London, 1992.

3. The Functional Plan for Catering Management in *The QMG's Corporate Plan*, MOD, London, 1987, p. 5-1.

4. The Potential Application of Commercial Developments to Military Catering (A Paper by DACC for QMG) 1992.

5. The Potential Application of Commercial Developments to Military Catering (A Paper by DACC for QMG) 1992.

6. Executive Committee of the Army Board, *Army Catering Strategy Study*, Final Report (Berry Report) September 1993

7. Executive Committee of the Army Board, *Army Catering Strategy Study*, Final Report (Berry Report) September 1993

8. *Britain's Army for the 90s*, HMSO, London, 1991

9. *Statement on the Defence Estimates,* HMSO, London, 1991

10. *Front Line First: The Defence Cost Study*, HMSO, London, 1994

11. Executive Committee of the Army Board, *Army Catering Strategy Study*, Final Report (Berry Report) September 1993

The evaluation of trolleys used in hospital meal delivery systems

A. West

The Hotel and Catering Research Centre, Department of Food, Nutrition and Hospitality Management, University of Huddersfield, Queensgate, Huddersfield HD1 3DH, UK

1. Introduction

The objective of hospital catering is to provide a food service to patients that is economic, efficient and effective. The food should be enjoyed by patients, provide a good source of nutrition and be safe to eat. The distance meals have to travel once prepared can make this objective difficult to achieve.

1.1 Preparation Methods

The type of delivery system required will depend upon the preparation methods employed. Most hospitals utilise the traditional cook serve system, nearly one third use cook chill and a minority use cook freeze and the meals assembly technique. All the systems utilised require a method of transportation to deliver meals to patients on wards. This is usually achieved by using a trolley specifically designed for a particular method of preparation.

In the conventional cook serve system, food is fully cooked, portioned and transported to wards. The food during transportation is kept hot or cold depending on the type of meal. The problem with this method of service is that the food can be held for long periods prior to being served, spoiling both the sensory and nutritional qualities and therefore leading to wastage.

The use of cook freeze and cook chill systems can avoid this problem by transporting frozen or chilled meals to wards and reheating immediately prior to service. The meals assembly technique utilises chilled and frozen meal items purchased from food manufacturers. As required frozen items are thawed and held at chill temperatures. The chilled meal items are then assembled into meals and transported to the wards for reheating, immediately prior to service.

1.2 Types of Trolleys Available

An important consideration in hospital catering is the selection of trolley required to transport the meals to patients. For cook serve, hotline trolleys are used, designed to maintain the temperature of hot items and keep cold items at chilled or ambient temperatures. Regeneration trolleys are used for the delivery of cook chill, cook freeze and meals assembly meals and are designed to regenerate foods to be served hot and keep cold items at chilled temperatures. Both types of trolleys are able to take either bulk meals for service at the wards or plated meals, for the tray service method.

1.3 Trolley Design.

Hotline Trolleys
Heating Methods - Natural convection is used to heat and cool the compartments. Insulation is relied on to maintain the meals hot or cold during transportation to wards.

Regeneration Trolleys
Heating Methods Forced air convection (FAC) - one or more fans circulate hot air around the compartment or conductive plates - solid cast aluminium plates with a heating element either embedded in the plate or fitted underneath the plate in specially designed grooves or loose fitting aluminium or stainless steel plates which slide over the elements exposed within the compartment.

Cooling Methods Mechanical refrigeration, incorporating a condenser unit to cool the compartment by natural convection currents or eutectic plates containing refrigerant liquid. These require freezing at -18°C prior to loading into the trolley.

The hotline and regeneration trolleys currently available are not ideal. Patients therefore can receive meals which are too hot or too cold and unacceptable in terms of sensory and nutritional qualities. A common problem with hotline trolleys is that once switched off they do not maintain adequate temperatures for the service of meals. The problem with regeneration trolleys is that the food can be unevenly reheated and too hot for patients to eat. However patients can still receive meals which are too cold due to poor catering practices and delays in the wards. The National Audit Office report[1] on the standard of hospital catering in England noted that 16% of patients surveyed thought that hot meals were too cold.

In 1992 the National Health Service (NHS) Supplies Authority commissioned the Hotel and Catering Research Centre, at the University of Huddersfield to develop a specification for both hotline and regeneration trolleys and provide an independent testing facility for trolley evaluations

2. Aims

The primary aim was to :
achieve an acceptable specification for trolleys in terms of performance and construction with special reference to materials and hygienic design.With the long term aim of securing improvements in trolley performance and design and therefore providing patients with better quality meals.

Over a period of three years 38 trolleys have been tested and the results disseminated into the NHS for the benefit of those making purchasing decisions. This paper describes the performance tests carried out against the specification and discusses the results to highlight the differences in performance and to make recommendations for improvements in trolley design. The performance criteria are listed in Table 1 for hotline and regeneration trolleys, for both bulk and plated meals. The full specification and the results for the trolleys tested are contained in the Meal Delivery Trolleys : Evaluation and Guidance Reports[2].

3. Methodology

Tests were devised to evaluate each trolley against the performance specification (Table 1). A standard test meal was selected, typical of the food and portions provided in hospitals. Plated trolleys were tested using standard NHS plates and bowls. Bulk trolleys were tested using 250 x 250 mm aluminium foil containers. Lids were used during all tests and all trolleys were fully loaded. All tests were carried out at a room temperature of 22 ± 2°C. Food temperatures were recorded at geometric centres using a Grant Squirrel Data Logger with T type thermocouples, with an accuracy of ± 0.5°C.

Table 1.   Performance Criteria for Hotline and Regeneration Trolleys

Hotline
Hot Compartment - air temperatures
1. achieving 85°C throughout the compartment within 60 minutes.
2. maintaining a maximum of 100°C throughout the compartment during the 60 minutes heating cycle.
3. maintaining a maximum differential of 14°C throughout the compartment during the 60 minutes heating cycle.
4. maintaining 65°C for 30 minutes throughout the compartment, once the heating cycle is switched off.

Cold compartment - air temperatures
5. achieving 8°C throughout the compartment within 60 minutes.
6. maintaining 10°C for a further 30 minutes once switched off.

Cold compartment - salad temperatures
7. maintaining food loaded at 5°C at a maximum of 10°C during the 30 minutes holding time.

Regeneration
Regeneration compartment - food temperatures
1. maintaining meals loaded at 0-3°C at a maximum of 5°C for 30 minutes, prior to regeneration, once the chilled sycle is switched off.
2. regenerating its load of normally presented food to an internal core temperature of 75°C within 50 minutes.
3. maintaining food temperatures below 95°C during the regeneration cycle.
4. heating the same foods to a maximum temperature differential of 5°C on completion of regeneration.

Cold compartment - salad temperatures
5. Maintaining food loaded at 0-3°C at a maximum of 10°C throughout the initial 30 minutes, regeneration cycle and 15 minute holding time.

4. Results and Discussion

This section reports on the significant findings over the testing period during which 38 trolleys were evaluated. The results have been considered in relation to the maintenance of food quality and to illustrate features of note in the trolleys. Thus, the section reports the performance of trolleys generally in meeting the performance criteria and in particular on the reheating characteristics of regeneration trolleys.

4.1 Hotline Trolleys

Hot Performance
All the trolleys heated up to 85°C well within the 60 minute heating period, most achieving this after 20 minutes. Some trolleys overheated and reached temperatures above 100°C at times, within the heating period, 103°C was the highest temperature recorded. In addition, during the 60 minute heating period there were wide variations in temperatures, outside the 14°C allowed in the specification. As expected after switching off, the differences narrowed as the temperature fell towards 65°C. However, the fall was not even and in some cases there were temperature differences as large as 22-34°C in different parts of trolleys. Trolleys with better thermostatic control maintained a narrower difference e.g. 9-10°C.

These results are to be expected as the majority of hotline trolleys utilise the natural convection method of heating. The excessively high temperatures recorded in some trolleys and the temperature variations will have a damaging effect on food quality during holding periods. Most trolleys maintained a minimum temperature of 65°C for 30 minutes after switching off, the range recorded being 62-75°C. In reality this period could be extended due to delays at the ward and further temperature drops occur as meals are removed for service to patients.

Cold Performance
All trolley cold compartments cooled to 8°C within 60 minutes but some could not maintain a temperature below 10°C for a 30 minutes after switching off. In tests using food, all salads, loaded at temperatures below 5°C, maintained temperatures below 10°C for the 30 minute holding time. This test was stringent to meet the temperature requirements in force at the time.

4.2 Regeneration Trolleys

Regeneration Performance
Maintain 5°C for 30 minutes
The ability of individual trolleys to maintain chilled temperatures prior to reheating in the regeneration compartment was greatly influenced by the type of shelving in the compartment. There were 3 types of shelving used :
    loose fitting shelving made of lightweight conductive plates
    fixed shelving made of solid, heavyweight conductive plates
    open wire shelving, used in FAC regeneration compartments.
Trolleys with fixed conductive plates did not maintain chilled temperatures, after only 8 minutes holding time, one such trolley recorded food temperatures above

5°C. Trolleys with loose fitting conductive plates maintained lower temperatures than the solid conductive plates, recording temperatures up to 6°C after 30 minutes. Trolleys with open wire shelving maintained the lowest temperatures, rising to no higher than 5°C after 30 minutes holding. Of all the trolleys tested, only those with wire shelving were capable of maintaining 5°C for the full 30 minute and therefore keeping food temperatures below 10°C.

Regenerate to 75°C within 50 minutes

All the bulk trolleys, with both conductive plates and FAC regenerated a full load to 75°C within the 50 minutes. The average bulk regeneration time was 40 minutes. (Figure 1). Plated trolleys required a longer regeneration cycle to achieve 75°C, the average time being 50-55 minutes. (Figure 2). Where a trolley failed to achieve 75°C at the end of the cycle, temperatures continued to rise and 75°C was reached within a few minutes, in the case of Trolley B, in 52 minutes.

Figure 1 Temperatures in Bulk Regeneration Trolleys

Regeneration Compartment

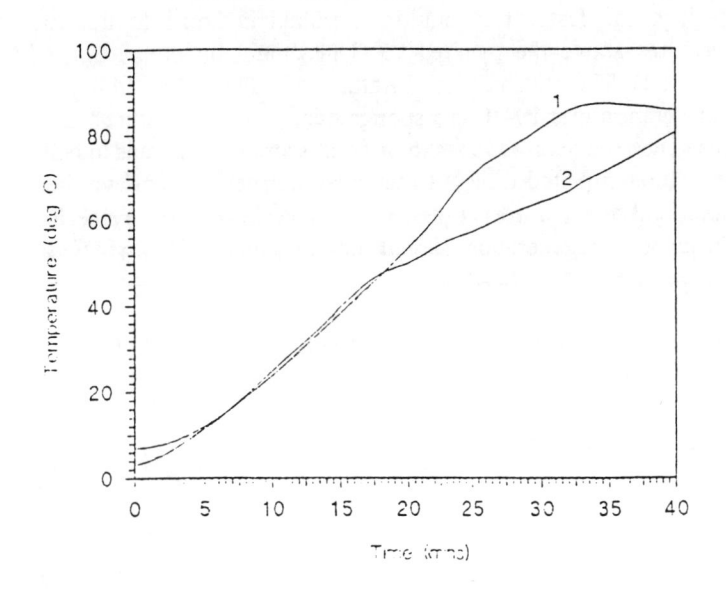

1-Conductive plates
2-FAC

Figure 2 Temperatures in Plate Regeneration Trolleys

Regeneration Compartment

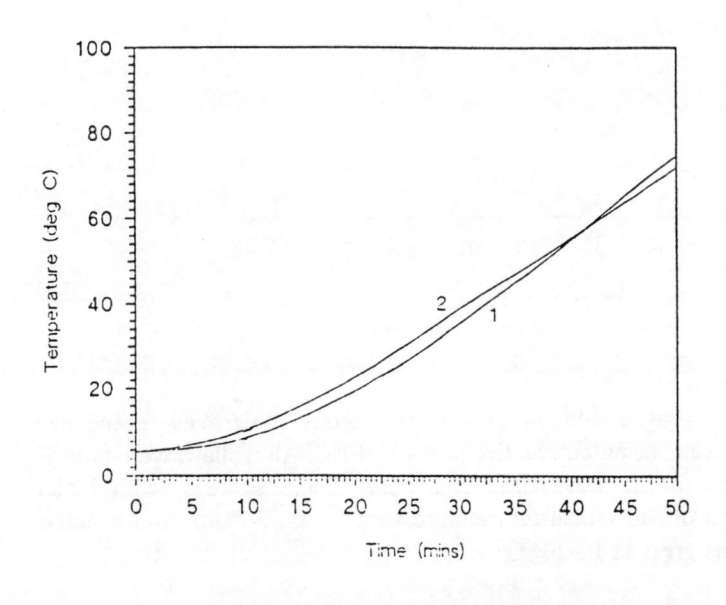

1-Conductive plates (A)
2-Conductive plates (B)

In the tests both FAC and conductive plate heating gave similar results in the reheating of chilled meals, both in bulk and plated trolleys.

Maintain food temperatures below 95°C during regeneration
In most bulk trolleys the faster regeneration resulted in some food items recording temperatures above the required 95°C maximum on completion of regeneration. (Figure 3). The maximum temperature recorded by any trolley on completion of regeneration was 104°C in a sponge item (air temperature). The high temperatures generated caused foods to dry out, scorch and liquids to boil. The slower regeneration in plated trolleys to temperatures below 95°C resulted in more acceptable food temperatures at the end of the cycle.(Figure 2) Ideally, temperatures throughout regeneration should be no higher than 85°C to preserve food acceptability.

Figure 3 Temperatures in Bulk Regeneration Trolleys at the End of Regeneration

Regeneration Compartment

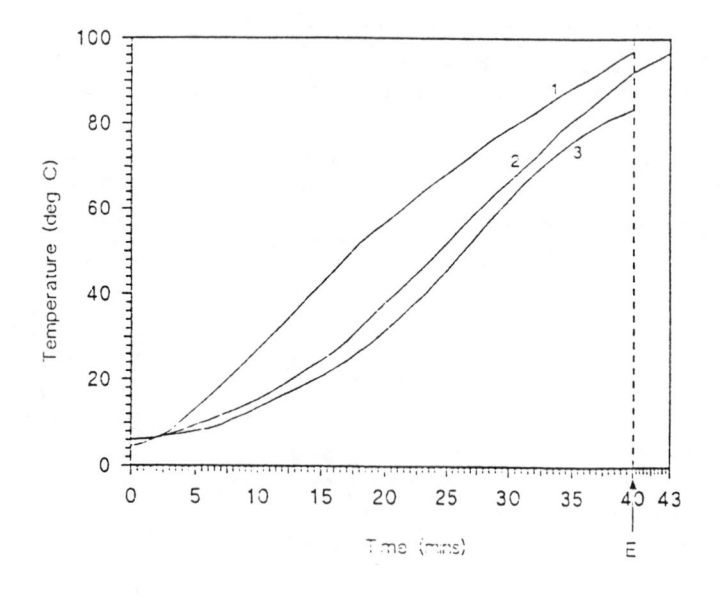

1-FAC
2-Conductive plates
3-Conductive plates

Maximum difference of 5°C
This requirement proved difficult to achieve whether the trolley regenerated bulk or plated meals, using conductive plates or FAC. All trolleys were outside this requirement for at least one food item. Some trolleys, with better thermostatic control had small temperature variations, (6-7°C) whilst others had differences as great as 10-20°C.

This was a difficult requirement of the specification to meet for all trolleys due to poor temperature control and the different thermal characteristics of foods. However, it is important to narrow the gap as far as possible because food served at differing temperatures can have lowered acceptability to patients.

Cold Compartment
Maintain a maximum of 10°C
The tests showed that mechanical refrigeration has no advantage over eutectic plates both maintain chill temperatures in cold compartments and keep salads below 10°C.

5. Conclusions

The main problems in both hotline and regeneration trolleys are over heating and large variations in temperature distribution. The over heating and uneven heating in regeneration trolleys causes liquids to boil and food to scorch and stick onto containers. This could be alleviated by using forced air convection heating in hotline trolleys and better thermostatic control.

There is no link between power rating and performance, an important practical consideration in some hospitals. Both forced air convection and conductive plates can give satisfactory results in reheating. A significant factor affecting heat distribution being the type of shelving used in the regeneration compartment. The use of mechanical refrigeration over eutectic plates has no advantages. In fact the use of these plates can have benefits for trolley design as it avoids the need to accommodate a refrigeration unit. Again in the chilled compartments the type of shelving influenced the temperature distribution. Heat creep is a problem in some trolleys during regeneration due to inadequate insulation between the chilled and regeneration compartments. There is a similar problem in hotline trolleys with bain maries set in the top, causing the chilled cabinets to warm up during holding times.

The main design features influencing performance in both types of trolleys are the siting and sensitivity of thermostats, insulation properties and the design and construction of shelving.

In general both types of trolley, hotline and regeneration meet the legal requirement of maintaining meals at a minimum temperature of 63°C for hot food and 8°C for cold food. However, in meeting this requirement some trolleys over heat the meals to temperatures in excess of 85°C, and some to above 95°C. This excessive heating of meals causes severe losses in sensory properties resulting in lowered acceptability. Nutrient losses also occur, particularly in heat labile vitamins such as vitamin C, Hill et al[3], Lachance and Fisher[4] showed that holding foods at temperatures above 80°C for short periods resulted in significant losses of vitamin C. This needs consideration especially for long stay patients and in the light of the Health of the Nation guidelines[5] for hospital catering.

References

1. National Audit Office, *National Health Service : Hospital Catering in England*, HMSO, 1994.

2. NHS Supplies Authority, *Meal Evaluation Trolleys Evaluation and Guidance Reports*, The Hotel and Catering Research Centre, The University of Huddersfield, 1994.

3. Hill, M. Baron, M. Skent, J. Glew, G. The effect of hot storage after reheating on the flavour and ascorbic acid of pre-cooked frozen food, Chapter 22, *Catering, Equipment and Systems Design*, ed G. Glew, pp331-339, Applied Science Publishers Ltd. London, 1979.

4. Lachance, P. & Fisher, B. Effect of food preparation procedures on nutrients in food service practices, Chapter 16, *Nutritional Evaluation of Food Processing*, ed R S Harris & E. Karmas, 3rd edition, pp 463-528, The AVI Publishing Company, Inc. 1988.

5. Health of the Nation. *Nutritional Guidelines for Hospital Catering*, Department of Health, 1995.

Section 1.3: Education and Training

Greening the food and beverage curriculum
J. A. Wade
Sheffield Hallam University, UK

Abstract

Over the last few years sections of the hospitality trade press, professional bodies and industry have taken proactive steps to support and introduce responsible environmental practice. The aim of this paper is to establish a snapshot of how environmentalism is currently included in the hospitality higher education (HHE) curriculum, with particular reference to food and beverage studies, and look briefly at some of the curriculum development implications. A brief literature review gives some background into the development of environmental education and describes the perspectives of the major stakeholders in HHE, professional bodies, government, industry, and education. A survey of HHE courses reveals the level of environmental teaching and its perceived importance, the perceived expertise of staff, the environmental topics currently included in the curriculum, and the subject areas where teaching staff think environmental subject matter should be integrated

The investigation found that many courses offer little or no environmental education despite the overwhelming support for such education by the stakeholders. The likely implications for the food and beverage curriculum are discussed should environmentalism be integrated. These include the need for staff development, resourcing, defining appropriate subject content.

Introduction

Research into the greening of the curriculum across sport, leisure, hospitality and tourism was undertaken by the Council for Environmental Education's "Education and Training for Industry and the Environment" project. The hospitality industry's position is well documented in Roberts[1]. This paper focuses on the impact upon the food and beverage curriculum.

Methodology

A brief literature review offers some insights into the perspectives of the various stakeholders in HHE. This is followed by an analysis of an investigation into how environmental issues are currently being taught in UK HHE courses. The definition used here for hospitality higher education is full time hospitality management courses including higher national diplomas (HND), undergraduate degrees and postgraduate courses. Some hospitality management courses (HNCs and professional diplomas) were not included, because of the time available for the project. A thorough analysis would have to include these courses and in the future hopefully this can be achieved.

Questionnaires were sent to all course leaders listed in the HCIMA's*"Careers 95" publications, which lists all the HHE courses available in the UK. Some 64 were returned out of 137, a 47% response rate. The analysis describes the environmental topics currently covered in the curriculum, how environmental issues are integrated into the curriculum, the perceived importance of environmental issues, the expertise of the respondents and which areas of the curriculum should integrate environmental issues.

The questionnaire does not define "environmentalism", this was done purposefully to allow respondents to make the judgement themselves about whether they thought they were dealing with environmental aspects in their curricula. "Environmental" and " green issues" are popular terms and in the general context have no widely accepted definition, whereas terms such as sustainability do have clearer definitions because of their use in official policy documents (DoE[5]).Where questions were asked about specific topics these are either widely accepted terms, for instance "waste management" and "marketing", or are again deliberately broad to allow some interpretation by the respondent, such as "general green issues." The questionnaire results give an impression of the current position rather than a precise definition. They are intended to inform a discussion about curriculum development.The analysis is followed by a discussion of some of the curriculum implications.

The Background

Throughout the 1960s and early 1970s concern about the environment grew rapidly, culminating in the Stockholm UN Conference of the Human Environment in 1972. The International Environmental Education Initiative (IEEI) was also launched as a result of the conference(Tilbury[2]). Environmental education was now a permanent feature of UN environmental initiatives. In 1975

*Hotel, Catering and International Management Association

64 countries attended an environmental education workshop in Belgrade, from which emerged the Belgrade charter. This outlined environmental education objectives, these included the need to encourage environmental awareness, knowledge, attitudes, and skills and the ability to critically evaluate environmental issues. The Belgrade Charter was *"the first international statement to outline the objectives, concepts and guiding principles for environmental educational"* (Tilbury[2]) . The 1992 Rio Earth Summit strongly supported the role of education. These comments from Chapter 36 of the United Nations Agenda 21 presented at the Rio Earth Summit are typical " *"Education is critical for promoting sustainable development and improving the capacity of people to address environment and development issues"* and *" "To be effective, environmental and development education should deal with the dynamics of both the physical biological and socio economic environment and human development, and should be integrated into all disciplines .."* Environmental education is supported by the European Union (EU), *"..it has been recognised by the ministers of the EC as an integral and essential part of every European citizen's upbringing and furthermore a key instrument in the successful implementation of environmental policy at EC and member state level"* (Sterling[3]). In 1990 the UK government published a white paper on the environment which amongst many other points made clear that *"Although the Government has to be in the lead, responsibility for our environment is shared by all of us ..businesses, central and local government, schools, voluntary bodies and individuals must all work together to take good care of our common inheritance. That is a job for all of us"* (DoE[4]). The UK Strategy on Sustainable Development was a response to the 1992 Earth Summit in Rio de Janeiro. The principles of sustainable development (DoE[5]) do not preclude economic growth or a halt to any change in the environment but that decisions throughout society should take their impact on the environment into account. This has been taken on board by stakeholders such as customers, employees, banks and insurers. They are asking questions such as *"what is your environmental policy "* and *"what are you doing towards environmental improvement?"* (CBI[6]). Employers recognise the need for staff with "green" expertise (DFE[7] and Ecotech[8]). A hospitality management course can combine many disciplines and it would be difficult to turn the hospitality undergraduate into a technical specialist . However what is necessary is to ensure the hospitality student understands the environmental implications of their industry. In-depth technical expertise will probably remain the province of specialists such as manufacturing chemists, engineers, architects, energy consultants, and agriculturists.

The hospitality trade press has published articles on many aspects of the environment including green marketing, energy conservation, recycling and waste management. The International Hotels Environmental Initiative's (IHEI) members (such as Marriott, Forte, Accor, Hilton) cover the globe. The IHEI published manuals, videos and workbooks to assist hotels in greening their operations, and they have recently launched a new publication "Green Hotelier".

Tourism facilities including hotels can be accredited with the "Green Globe" mark (Editor[9]). The institutional sector is also responding, Aramark, an international contract catering company, has gained Green Globe accreditation on behalf of one of its clients and seeks to save energy and food (Anon[16]). The Hotel and Catering International Management Association (HCIMA) published a technical brief no.13 "Environmental Issues", and publications on "Living in Harmony with the Environment [10]" and "Training and Education in Harmony with the Environment[11]". HCIMA is also involved in the piloting of BS 7750 (Environmental Quality Management Systems) for the hospitality industry. The professional bodies are very proactive and can be seen as "environmental evangelists". The obvious importance placed on environmental good practice by these bodies is crucial in raising awareness generally across the hospitality industry. These bodies can support curriculum developments (Roberts[1]), and HHE can support the professional bodies' initiatives to influence the management thinking and practice of future managers, by ensuring students are thoroughly prepared for their professional environmental responsibilities.

Khan[12] comments that environmental education is one of the National Curriculum Council's five cross curricular themes and in the future, students will enter higher education with some environmental knowledge and awareness. Tilbury [2] goes on to conclude that from a historical review of environmental education its main goals and characteristics, in summary, are;

a) to produce citizens with the requisite knowledge, skills and attitude to be critically aware of environmental issues and their own responsibilities
b) to link environmental education with everyday life (be relevant!)
c) to integrate poverty and development , relating socio economic growth and environmental issues
d) to be concerned with problems of the contemporary world, including poverty and development, and focusing on current and future trends
e) to encourage active participation
f) to teach environmental issues at all levels
g) to encourage an interdisciplinary and holistic approach
h) to encourage the consideration physical and spiritual needs
i) to encourage co operation

At first glance these should form the aims and objectives of hospitality environmental education, especially *a,b,e,f,g*, and *i*. Objectives *c,d*, and *h* have socio-cultural/ political dimensions which require a more thoughtful analysis to produce any realistic learning outcomes. These will provide interesting points to debate! HHE has the potential to play an important role in preparing future hospitality management in this key area. *"Everyone has some scope for doing his or her job in a more environmentally responsible way"* (DFE[7]), and by doing so is supporting government policy, meeting employers needs, supporting professional bodies' initiatives, and continuing a theme started in the national curriculum.

Current Practice

The investigation found that some 19% of HNDs, 10% of undergraduate degrees and 17% of postgraduate courses do not include any environmental education in their formal curricula, while others only offer this topic in options. On all these courses students may not receive any environmental education unless they choose certain options. In other courses the subject is included in core (and therefore compulsory) units and/or embedded in all or most units.

fig 1: How environmental topics are delivered in the curriculum.

Course level	Not included	In optional units	Included in core units	Included in most units
HND	19%	26%	51%	33%
Undergraduate	10%	30%	60%	43%
Postgraduate	17%	22%	33%	11%

In some instances respondents indicated that the subject was delivered both through core units and included in all or most units. Over half of the HND (51%) and undergraduate courses (60%) have compulsory environmental education. At post graduate level the exposure is significantly lower (33%). As a cross curricular theme environmentalism is less well advanced although 43% of undergraduate courses appear to have assimilated the topic across all or most units. The investigation found that all the respondents thought environmental issues are, at least, an important topic for hospitality students to study. There seems to be a majority opinion that environmentalism should be integrated into all aspects of the hospitality curriculum. Many of the respondents were aware of their lack of expertise. Most thought they were "adequate" at best, and this in itself could deter curriculum development. The need for staff development is urgent. Tutors must feel competent to integrate environmental education within their particular subject area. If we accept that environmental education should encourage environmental responsibility both personally and professionally, teaching strategies must inform and challenge the individual.

Environmental Topics

From a list of given topics (see fig.2) it became clear that environmental teaching is concentrated in the applied aspects of the curriculum. Excluding

fig 2 : Course Comparison of Environmental teaching by topic

topic	hnd	degree	postgrad
food aspects	60%	60%	33%
impact studies	23%	43%	22%
green ethics	42%	57%	22%
green marketing	47%	70%	34%
environmental auditing	12%	30%	22%
general green issues	63%%	73%	39%
BS 7750	37%	70%	22%
finance aspects	40%	47%	28%
hazardous materials	93%	80%	28%
environmental services mgt.	70%	67%	22%
waste management	67%	67%	17%
energy management	91%	97%	33%

postgraduate courses, those areas covered extensively are often the subject of current legislation. The management of hazardous materials is controlled by the Control of Substances Hazardous to Health Regulations and waste management covered by the Environmental Protection Act 1990 .Other popular topics are proven cost saving strategies, such as energy conservation. How the respondents interpreted "food studies" has not been analysed, however, 60%, of undergraduate and diploma courses, do appear to have integrated some environmental aspects into their food curricula. Respondents were asked to suggest where in the curriculum integration should be made, even if this is not undertaken at present (see fig. 3). The food and beverage area comes high on the list alongside other applied areas such as facilities and accommodation management. Environmental matters are seen as less relevant in the business and management subject areas. This must be reconsidered because positive environmental thinking impacts on all aspects of a business not just in operational matters.

fig.3: Curriculum sectors in which environmentalism should be included

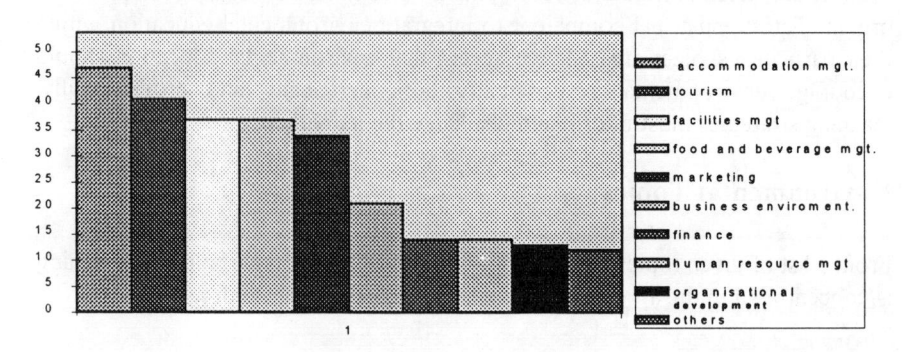

Evidence has been gathered that there are students who do not receive adequate environmental education within their courses and thus their education could be said to be incomplete. It seems foolish to ignore the value and importance of environmental education within the formal hospitality management curriculum given the value placed on it by the different stakeholders. In the long term a planned approach is needed so that the learning outcomes of hospitality environmental education are clear to both students and teaching staff.. The Toyne Report (DFE[7]) recommends that all further and higher education institutions implement environment education and adopt good environmental practice in all aspects of their operations. White and Senior[13] comment that *"Educating the future decision makers at an early age will also, it is hoped, develop more long term ripples".*

Greening the Curriculum

A green curriculum is a curriculum with an environmental dimension which helps students to understand the way their chosen subjects connect with the environment, whether this be the natural environment or manmade environments of agriculture, towns and cities. (CPD [14])

Managers can adopt different environmental roles (DFE[7]), those of environmental practitioners (experts), co ordinators (managers with some environmental responsibilities) and responsible individuals (those whose work can contribute to good environmental practice). Many hospitality managers will be co ordinators and responsible individuals. Within the broad subject range a hospitality curriculum embraces, it seems unlikely that producing expert practitioners will be a feasible objective. Learning outcomes should focus on ensuring graduates are effective co ordinators and responsible individuals. The objectives identified by Tilbury [2] are those which could form the basis of learning outcomes. For instance (these not definitive and only given as examples) :

- students will have the requisite knowledge, skills and attitudes to be critically aware of how environmental issues will impact on the food and beverage system
- students will be able to make a personal contribution to effective environmental performance
- students will be aware of how environmental issues impact upon the day to day operations of a food and beverage operation
- students will be able to identify how to actively involve staff in environmental management

Hospitality managers must be able to make informed decisions by being aware of environmental issues and the likely implications for all aspects of the hospitality operation. At the same time the need to make a profit cannot be ignored just because a manager wants to adopt environmentally friendly policies. As

Dorweider and Yakhou[15] conclude " *the influence of environmental requirements and societal demands are to compel consideration of profit relationships to environmental objectives*". Students must be able to appreciate that effective environmental management must be delivered within a competitive commercial environment. Roome[17] also identifies that managers need to understand not just environmental principles but also the tools and techniques that underpin effective management. These include environmental auditing, product life cycle analysis, environmental statements and reviews. Roberts[1] comments that *"..it is important to build a basic understanding of environmental principles appropriate to the level and background knowledge of the student group..discussion may then move forward to a more focused view of associated with specific leisure areas"*. Roberts found that hospitality professional bodies and employers identified the following as being particularly important.

definitions of sustainability	waste management	consumption of resources
water conservation	transportation policies	environment and health
environmental economics and impact assessment	policy making	land use planning.auditing
	risk management	societal & cultural effects
environmental communication	energy conservation	environmental regulations
operations of physical systems	pollution from chemicals	

These topics were assessed for their importance across sports, tourism, leisure and hospitality curricula. The particular vocational aspects of the hospitality curriculum were not made explicit. If a curriculum is to integrate environmental education effectively, it must decide whether just to raise awareness by introducing students to the issues or in addition to teach underlying principles and techniques so that students can use these in problem solving and management practice. Roberts describes principles and techniques which a food and beverage curriculum development would have to encompass. A useful approach to incorporating these into the curriculum is to look at the food and beverage operation, see fig.4.

fig. 4: A model of a food and beverage operation

THE FOOD PRODUCTION AND SERVICE SYSTEM

INPUTS	staff		OUTPUTS
labour			tired staff
capital			profit
food and beverages	food delivery	service system	customer satisfaction
utilities and power	food storage	reception	
	food processing	beverage service	waste food
equipment &plant	food presentation	entertainment	non food waste
expertise	ood service	general waste	emmissions
	food waste	heating and lighting	obsolete equipment
	menu planning	administration systems	
	customers	use of IT	

If the model is taken as a simple representation of the food and beverage system, the various elements represent parts of the food and beverage curriculum. The following table (fig.5) suggests how some principles and techniques could be applied.

Fig.5.:Integration of elements into the Food and Beverage Curriculum

ELEMENT	PRINCIPLE	TECHNIQUE	APPLICATION
food delivery	reducing pollution	food miles concept transportation policy	local produce
food storage	sustainability- use of energy	cost comparison & evaluation	compare taste of fresh & frozen foods
food processing	sustainability of finite resources,	cost evaluation	compare packaging waste
food presentation	sustainability of finite resources	cost & quality evaluation environmental regulations (hygiene(	compare service systems.
food waste	sustainability of finite resources	pollution control risk management	minimisation of waste
menu planning	"	cost comparison	seasonal menus
service system	sustainability (energy.water,)	market research society effects	customer preferences
reception	sustainability of forests	cost evaluation	reduce paper
purchasing	sustainability of finite resources	cost comparison product life cycle evaluation	purchasing of eco friendly products
beverage mgt.	"	market research	organic wines
general waste mgt	"	pollution control environment and health land use planning	recycling possibilities
buildings and facilities mgt	"	life cycling costing environmental auditing land use planning energy conservation	ensuring effective energy use
staff	personal responsibility environmental communication	training	staff work methods e.g. sorting glass and waste

Conclusions

There is a need to thoroughly re-evaluate the subject content and philosophy of the food and beverage curriculum; in particular, to identify opportunities for incorporating environmental principles, techniques and applications. The level must be appropriate to the course. The appropriate subject content for the food and beverage curriculum must be defined and the preferred learning outcomes determined.

The resources of an institution will also be influential in how quickly and in what depth the introduction of environmentalism can be achieved. Resources are needed for staff development, new teaching materials, curriculum development, and research. Courses which refuse to integrate environmental topics may risk recruitment difficulties, or dissatisfaction with curriculum content, given the expectations of students used to a national school curriculum that acknowledges the environment as an underpinning theme (Khan[12]).

Environmentalism in the food and beverage curriculum offers students and staff opportunities to explore innovative management strategies and new operational techniques. The student should realise both personal and professional development, to become an effective hospitality manager and a responsible citizen.

References
1. Roberts, C.(1995)*The Environmental Agenda,Taking Responsibility, promoting Sustainable Practice through HE Curricula - Sports Leisure,Hospitality and Tourism*. Pluto Press.
2. Tilbury, D. (1994)"The International Development of Environmental Education : a basis for Teacher Education Model?", *Environmental Education and Information*, Vol 13,No.1,pp.1-20.
3. Sterling,.(1995)"Towards a Sustainable Europe"*Environmental Education*,Spring 1995,p6-7.
4. Department of the Environment (Dec 1990) *Our Common Inheritance - a summary of the White Paper on the Environment*, HMSO.
5. Department of the Environment (DoE) (May 1994) *Sustainable Development - the UK Strategy, An Outline*, DoE.
6. CBI (1994) "What's your green policy?", *CBI News*, October 1994, p.15.
7. Department for Education (DFE), (1993) *Environmental Responsibility - an agenda for further and higher education*, H.M.S.O.
8. Ecotec,(1993)*Educating &Training Personnel concerned with Environmental Issues relating to Industry* European Foundation for Improvement in Living & Working Conditions.
9. Editor,(1995)"News and Views -What is Green Globe", *Cornell Hotel and Restaurant Administration Quarterly*, February 1995, p.15.
10. Hotel and Catering International Management Association (1994) *Living in Harmony with the Environment*, HCIMA,London
11. Hotel and Catering International Management Association (1991*) Training and Education in Harmony with the Environment*, HCIMA ,London.
12. Khan, S. (1992) *Colleges going Green*, FEU and Council for Environmental Education.
13. White, S. and Senior, J. (1994) "The Ripple Effect - a New Approach to Environmental Education for Adults", *Adults Learning*, April 1994,pp.214-215.
14. Committee of Directors of Polytechnics(1991)*Greening the Curriculum Working document*, May 1991, CDP
15. Dotweider V.P. and Yakhou,M. (1994) "Environmental Management and Environmental Education - New Dimensions in Business Education", *Environmental Education and Information*, Vol. 13, No. 2, pp.131-136.
16. Anon. (1995)"Huge savings from Green Project",*Cost Sector Catering*, August 1995.
17. Roome N.(1994*)The EnvironmentalAgenda, Taking Responsibility, promoting Sustainable Practice through Higher Education Curricula - Management and Business*.Pluto Press.

Gastronomy - towards vocational reality
J. M. Schafheitle, P.A. Jones
Worshipful Company of Cooks Centre for Culinary Research,
Bournemouth University, Poole, Dorset, BH12 5BB, UK

Abstract

This paper aims to explore the rationale, objectives and outcomes for the study of gastronomy as an integral element of the final year programme of studies for students intent on following a career in food and hospitality management. Gastronomy provides the opportunity for a critical evaluation of the gastronomic influences of consumers including cultural and religious conventions and the internationalisation of food behaviours. These theoretical underpinnings need to be well grounded in reality of a business framework if the study of gastronomy is to be seen as having practical value for future managers. The aim of the study of gastronomy is to provide the students with several major goals, that culminate in a gastronomic event, managed and carried out by a student group in collaboration with a chef and establishment of international repute. This model of sequenced development in learning will be evaluated and considered including the importance of structured readings and workbooks as a means of facilitating student centred learning. The practical implications of business enterprise are explored by the students to ensure that the gastronomic events are not only efficiently and cost effectively organised but also meet commercial realities.

Introduction

Gastronomy has been broadly defined (Cracknell and Nobis, 1985[1]) as "a love and true appreciation of good food and wine, two of the pleasures of life which when supported by good service and jovial company help to create a truly great meal experience". The great 18th century philosopher of eating, Jean Anthelme Brillat-Savarin, ascribes the study of gastronomy as 'the intelligent knowledge of whatever concerns man's nourishment" (Physiology of Taste, translated by M F K Fisher, 1972[2]).

Professor John Fuller (1972[3]) stated that "today many people in the western world devote more time, thought and conversation to dining than to religion or politics" and opines that caterers as well as educational establishments and training courses should recognise this phenomenon and reflect upon the needs of customers to extend knowledge of cultural differences and the pleasures of kitchen and table.

Despite recent economically lean times, this situation has not diminished, in fact it has increased. More people are looking at the amount of food and beverage related issues being portrayed in the popular media and the increasing trends of people eating away from the home, at least 1.78 meal experiences per person per week (National Food Survey, 1994[4]). Customers have become tremendously sophisticated with varied meal expectations and experiences (Cracknell et al 1987[5]). This challenges modern caterers to evaluate and meet the complex and special needs, drives and motivation of customers in relation to food, drink and the environment. A professional caterer must therefore be in a position to assess the satisfaction level required, and provide a service equal to the clients' expectations.

Establishments responsible for the education of future caterers in whatever sector or level therefore have an onus not only to develop student skills in extrinsic commercial activities, but to develop awareness of the intrinsic reasons for man's need for nourishment.

The subject of Gastronomy has been recognised and developed in order to bridge the gap between skills competence in food and beverage service and emotive attitudes necessary to not only meet, but anticipate consumer needs.

The Academic/Educational Context

Gastronomy at Bournemouth University is one of five units taught to 3rd year HND students in the Department of Food and Hospitality (within the School of Services Industries). The overall global curriculum design of the course but the method of delivery - the how, when, why - of educational experiences is ongoing and changing. This process has been considerably influenced by five main elements, set out in Table 1.

Table 1 Global curriculum design of the course

Influencing Element	Effect
Strategic Departmental Development Plan	The overall philosophy, ideals, and direction in which the teaching/learning process continues to evolve and to provide integrative problem based solving/learning experiences for the students. Realistic working environment and practices within a collaborative based project combined with a proactive approach to management style, aim to capitalise on supervisory management skills.
BTec HND	This has direct consequences for the overall course aims of the BTec HND Programme of Studies in Food and Hospitality Management which have been designed to enable students to a develop a positive understanding of human behaviour in relation to workers and customers b plan, organise and manage realistic food and hospitality operations c develop an awareness of the business variables and environmental influences surrounding the food and hospitality industry. The total effect is the provision of a vocational emphasis to the course, providing students with a sound basis from which they will be able to progress towards their future careers.
Industry	Responses from Industry have been elicited through encouraging consultation with representatives of the catering industry. The Department has assessed not only what employers expect of Diplomates, but also if any gaps exist between those variables, how they can be bridged to benefit both interested parties. In trying to ascertain, with particular reference to Gastronomy, what aspects should be emphasised, it was primarily found that the secondment of educators into the industry was considered very important. Employers required students to have a good attitude to work and the business environment. Intelligence, an ability to communicate and hands on experience were all highlighted as basic qualities employers expected of students. Awareness of profit orientation was an underlying factor acknowledged as being crucial.

Table 1 Global curriculum design of the course (Continued)

Influencing Element	Effect
Subject Development	Past delivery of Gastronomy as a subject was very much influenced by the early development of a HND for the hospitality industry at Bournemouth & Poole College at The Lansdowne and Dorset Institute of Higher Education. Shades of influence came from Oxford College via Harry Cracknell and Ron Kaufmann in the late 1960's and early 70's. Distinct emphasis on topics such as Brillat Savarin's food philosophies were mixed with regional/national food differences and religious adherence and customs. Lessons took shape in a four hour weekly timetabled teaching activity underpinned with student seminars, wine and food tasting activities which excelled in finding intrinsic enjoyment factors, followed at the end of the academic year by final examinations. This was, however, an optional subject and it had to compete for students with other options.
Student Feedback	Throughout the academic year, a structure and formal mechanism exists where student experiences can be expressed in order to create an appropriate environment in which they can excel in their studies and where problems can be solved or further examined. Within the last five academic years, specifically for the Gastronomy unit, a questionnaire has been used to evaluate student's satisfaction levels on the usefulness and relevance of the unit to them and inviting opinions as to where changes could be made to improve the topic.

Implications of the Course Unit - the learning experiences

As a result of the five major forces outlined in Table 1 above, the following topic aims, objectives and "method of delivery" have been developed:-

i   critical study of the theoretical and practical concepts of gastronomy, its origins, literature, values of culture and the artistic aspects of dining;
ii   a keen personal and consumer appreciation of food and beverages;
iii   collaborate with an establishment/celebrity chef of high professional standing and manage with them a realistic food/beverage function.
iv   provide the students with a guided task to pursue a special interest research project (a portfolio) within this wide and varied subject area, with possible future application to his/her career.

These aims are interpreted to make the learning experiences varied and interesting with a mix of :

cognitive skills, to be found in reading, writing up researched material; evaluating the importance of the affective skills in the positive attitude of a group/team skills; and
customer skills in the expression of personality, behaviour at work and dealing with members of the public (e.g. being pleasant, helpful, tactful).

The students existing knowledge and dexterity, developed through industrial experience in their second year of study, is underpinned to improve competence and ability to perform skills related tasks. Interpretation of a whole range of skills is necessary, including those of small business management, financial management, and product development, as gastronomy can never practised in isolation. Gastronomy provides an integrated and holistic approach which incorporates a study of influences and origins of food and culture complemented by the development of cognitive, intellectual and performance skills. This is underpinned by a comprehensive understanding of the consumer attitudes and beliefs focused into a formal business plan that is monitored and evaluated against pre-set and agreed performance criteria with the aim of providing a total gastronomic experience and ensuring that the students and the customers enjoy the service of food and wine.

Design of Formal/Informal Teaching and Learning Experiences

The Department of Food and Hospitality Management operates a system of block timetabling, the effect of which is that students have one week in five allocated to the Gastronomy unit with two x three hour formal sessions per block. Such timetabling has benefits as well as drawbacks, e.g. students can use the whole week for informal work relating to set assignments through student centred and group work. One drawback is that students have no formal contact with the subject lecturer for five weeks. This may be overcome by an open access policy by lecturers to encourage students to keep in touch.

Methodology

To realise the perceived benefits of block timetabling, students have to be well motivated in order to undertake self centred study for which there is ample opportunity to undertake research for set assignments. The onus is on the Lecturer to ensure that such informal activities need to be well designed and sequential with the formal class contact. This needs to be instilled from the onset of the academic year so that students are aware of content, assignment schedules and assessment criteria, can make full use of their input and are aware of the commitment required. The sequence and emphasis of each formal timetabled block session is:-

Block 1　　Introduction to the Academic Year

Students are provided with an overview of the academic year's work and organisation, i.e. Seminars, Gastronomic Visit, Gastronomic Evening Function (Practical). They are made aware of the teaching/learning experience and equation and the assessment strategy and criteria. The roles of the Lecturer and students are discussed and commitment to excellence emphasised.

Formal lectures are used to introduce the definition of Gastronomy, its relationship to other subjects, e.g. history, physiology, psychology, sociology; organisation of students work assignments for the academic year; time schedule - areas of responsibility to be taken up by individual students or as part of a group;. Supporting material in the form of a Reader for this subject is distributed to students.

Block 2　　Student Seminars

Student seminar presentation to a group of first year students by a team of two 3rd year students. The subject of the presentation is the food and wine connection. For the presentation students are given a menu course, e.g. starter, fish, main course or dessert for which they must select an appropriate wine to complement the dish, and a second wine (or beverage) as a contrast which is inappropriate.

Students are required to undertake research, prepare presentation material and serve the food and beverages at a formally arranged dining table for tasting.

First year students acting as tasters, listeners and questioners are invited to make an assessment of each seminar presentation based upon previously informed and guided advice on 'how to make an objective assessment'. A marking scheme has been developed by the Lecturer in the form of percentage marks with associated remarks and comments in relation to what constitutes a successful seminar presentation from the top of the scale, downwards through the levels of success to a limited, poor presentation.

At the end of the seminar the 1st years' assessment marks are averaged and their remarks recorded. Once their seminar presentation has been given, gastronomy students are required to submit a written report summarising key facts on their wine research, seminar material used in the presentation, and a conclusion. This report is marked by the lecturer using the following criteria
i.e.,　　communication skills/presentation
　　　　introduction to the topic
　　　　width/depth/answer/interpretation - how research material was summarised;
　　　　conclusions on wine selected and seminar presentation skill;
　　　　the use of references, acknowledged in text and listed.

The marks obtained from the report and the 1st year marks are combined and averaged and the total mark represents 20% of the marks for the academic year.

One addition in the current year has been the inclusion of a short tutorial with the lecturer following completion of the seminar and report in order to record the student group's own views of how they feel their seminar was received; assessment of the positive outcome and where and how the students felt improvements could be made in content, organisation and presentation skills. These comments, combined with those of the 1st year students and lecturer combine to make an objective overview which provides students with an indication of their good and weak areas. In addition to displaying knowledge of the food and wine connection comments may include remarks on the style of dress, communications skills and use of presentation aids.

Block 3   Gastronomy Visit
The theme of this block is a gastronomic visit. The purpose of the visit is to enable students, as paying customers, to discover and understand the benefits of observing and experiencing the operation of a successful (in high gastronomic terms) establishment. Learning and evaluating through observation how customer expectations are perceived and met. Students use a questionnaire to record their comments on the environment, the table (layout), menu card, menu selection to be tasted, selection of appropriately matched beverages, presentation of the food, the eating sensation, the service and value for money aspects.

Following the visit students will evaluate each individual aspect and after group discussion each student is invited to mark the set criteria from 1 (excellent) to 6 (poor) in relation to their perceived experiences the marks are then averaged out for the group as a whole.

Obvious benefits are derived from visiting a selected hotel/restaurant, particularly in relation to the ambience created by the surroundings and the style of a specially created 6 or 7 course meal epitomising the best in food, wine and service for which such establishments are associated.

The serious intent of the visit should, however, be highlighted. The establishment to which the visit is made is also the one with which that particular student group will be collaborating on a gastronomic themed event. Having themselves experienced the quality aspects of the establishment's style of dining, the students are then able to appreciate the quality aspects which they must then replicate.

Block 4   Gastronomy Function
This block is seen as the pinnacle of the whole of the students gastronomic academic year. Students have prepared themselves to stage a gastronomic event - a charity evening function. This function is designed to provide the students with realistic management/work experiences working with and alongside professional caterers. The main aim and underlying reason for these collaborative functions is to provide the students involved with intensive practical involvement and to extend and re-examine food and beverage knowledge and perceptions in the light of current and new evaluation of gastronomic values and norms. Students previous industrial experiences are

168 Culinary Arts and Sciences

linked and measured with the collaborative establishment in terms of skills, attitudes and business awareness to create profit.

The evening function provides an opportunity to develop management skills, through practical delegation and decision making. The exercise helps to gel the theoretical understanding of gastronomy with real-life practical considerations required to organise, implement, control and meet the customer expectations of paying guests.

In recent years the following celebrity chefs and establishments have participated and contributed their specific philosophies and expertise in management and competence as indicated in Table 2.

Table 2 Chefs and Establishments Participating in Collaborative Gastronomic Events

Chef	Establishment	Year(s)
Albert Roux, Visiting Fellow and Professor	Le Gavroche	1984 - 95 (incl)
Anton Edelmann	The Savoy	1991 - 96 (incl)
Anton Mosimann	Mosimann's	1993, 95
Jean-Christophe Novelli	Le Provence & The Four Seasons The Dorchester	1993, 95
Willi Elsener	Le Manoir aux Quat' Saisons	1993
Raymond Blanc	Gidleigh Park	1994
Shaun Hill	Chewton Glen	1994
Pierre Chevillard	The Greenhouse	1994, 95
Gary Rhodes	The Ritz	1995
David Nicholls	The Lanesborough	1996
Paul Gayler		1996

At the end of the exercise students are required to evaluate and contribute to a group written report an account of their input in terms of personal and group effectiveness and competence, the level of success achieved and what key points were important, also those areas where improvements could be made. The collaborating establishment contributes to the delivery, guidance and direction of the event and the student group has to make objective judgements on their own personal input and that of their group

The specific objectives are set out in the subject Reader at the beginning of the academic year with suggestions made for planning before/on the day/night, at the conclusion of the event. The division of labour and areas of responsibility are decided by the students themselves but cover the traditional management roles.

At the conclusion of the evening function Block 4 will be assessed by diners who have been provided with a questionnaire (covering the same criteria used for the gastronomic visit). Comments are also invited which are used in the feedback session.

Marking of this exercise is based upon peer group, lecturer and chef personality (mainly comments) assessment with marks averaged to produce a 60% weighting. Students who have demonstrated outstanding commitment in the practical and managerial implementation of the exercise will receive a personal letter of achievement from the lecturer emphasising their proactive approach. Second year students working as operatives who have demonstrated positive commitment to the event will be given a letter of acknowledgement signed by the chef personality and the lecturer in charge. Distribution of the letters takes place at the end of the event following feedback with the respective student managers under whom they worked.

Within the many cumulative multi-tasks and skill performance areas which such an exercise provides for students, one additional factor needs to be explained and highlighted that is the 'business venture' aspect. Whether real or imagined the ultimate goal is that at the end of a business transaction a profit must have been generated. Within the policies of the Department, once the costs of the exercise have been deducted from the income, the surplus money is donated to the nominated charity. In the last academic year £6,400 was raised from profits, raffles and auctions during six collaborative evening functions.

Following completion of the exercise and submission of the written report, a feedback session is held, findings and observations from which are noted for future reference. The written report is placed in the University library and acts as a form of reference for students in future academic years, highlighting the pitfalls and input required.

Block 5   Summary/Feedback Session
This block is divided into three different aspects:
 i) feedback from the practical exercise;
 ii) short seminar presentation highlighting individual research activity and presented in the form of a portfolio. This portfolio represents 20% of the overall mark allocation;
 iii) providing students with their combined grades and inviting them to complete (anonymously) a questionnaire evaluating their perception of the content and relevance of the Gastronomy unit and their recommendations/conclusions on the gastronomic year. The results from last year's questionnaires will be discussed later

The Handbook/Workbook/Reader

Whatever name is used for providing students with written information right from the onset of their academic year, the purpose is of vital importance and should be acknowledged as being one of the main tools in support of the teaching/learning process. The Gastronomy Reader provides a template of

intent of the students work activities. It has been designed to help them to organise, manage and complete specific tasks and comprises:- .
an overall perspective of this subject;
an outline of the main areas of research study and exercises **"TO DO"** for inclusion in the Gastronomy Portfolio;
the criteria for assessment;
"Codes of Practice" for best practice in relation to seminar presentations, gastronomic visit, the charity evening function and portfolio research;
lecture notes in the form of position papers covering the topics:-
definition of gastronomy; Origins/History; Aesthetics of Eating/Dining; Physiology; Psychology; Sociology; Regional, National and International Cuisines (East meets West); Food Culture/Habits; Chinese, French, British Food Characteristics; Food and Wine Connection; Styles and Trends of Cooking; Beverages; The Creation of Special Gastronomic Occasions; Feast and Famine;
lesson notes are explained in the form of covered aims, objectives, direction, reading list and further research activities for use in the portfolio;
information where further help may be forthcoming;
a feedback questionnaire giving students a chance to assess and give comments on where and how changes to the set up, design and delivery of this topic could be made in support of their learning.

Student Feedback Questionnaire

This consists of six different questions asking students to assess and comment upon

i   timetable arrangements
ii   which of the blocks they enjoyed most and least
iii the support and input received from the Lecturer/Reader/Facilities and Organisation of the Department;
iv   where any changes could be made to improve the educational and operational aims and objectives;
v   asking about themselves - were they diligent students?
vi   any additional comments.

Summarising the results of the last academic year, 90% of the gastronomy students returned their questionnaires. Almost all of the students' assessments of this topic were positive:

i   the gastronomic visit, most enjoyable
ii   functions (each group managed two) most challenging
iii the Reader was highlighted as being a supporting aid throughout the academic year and students stated that they found it very useful
iv   the Lecturer was, in their opinion, a positive influence

One result from this year was that each management group was in charge of organising and managing two functions and the students voiced their views that this was one too many, considering the amount of work involved and in addition to their other four units. They opined that involvement in one function provided sufficient "experiences" for gastronomy. This particular issue was acknowledged as a 'case in point' by the HND Year Course Team and, as a result, the unit was changed to one function involvement in the current academic year.

In Conclusion

This paper has highlighted some of the main features and characteristics of the Gastronomy unit, how it has been designed, developed and delivered. Students are provided with a rich variety of educational experiences, including the addition of real-life learning projects in collaboration with industrial partners, with the emphasis on innovative, challenging enterprise activities. The student is deliberately placed at the centre of teaching/learning activities, taking joint responsibility with the Lecturer facilitating progression of understanding.

The main focus of the unit is the enabling of students to be proactive in business orientation, professional skills competence, quality delivery of services and development of positive values, aspects which modern consumers expect and demand from catering professionals. Students have the opportunity to emulate with enthusiasm and realism the 'Good Catering Practices' demonstrated by the collaborative establishments in their interpretation and execution of producing a quality product and overall excellence.

The role of the lecturer is a supporting one, strengthening, encouraging and enhancing the students input within an organised frame-work, the programme for which is documented in the Reader. Initiation of links between education and industry brings the two closer together and enables the two to assess how the stated goals have been achieved by the parties involved, where and how improvements can be made to make the subject of Gastronomy relevant, enjoyable, worthwhile and inspirational for the future labour force of the catering and hospitality industry.

References

1. Cracknell, H.L. and Nobis, G. *Practical Professional Gastronomy,* 1985. Macmillan Education Ltd.
2. Fisher, M.F.K. (Translation of Brillat-Savarin) *The Physiology of Taste,* 1972 Alfred A Knopf Inc.
3. Fuller, J. "Gastronomy in Catering - How relevant is Gastronomy to Professional Caterers?" *HCIMA Journal* Feb. 1980, 7 - 9.
4. MAFF *National Food Survey 1994*. 1995, HMSO.
5. Cracknell, H.L., Kaufman, R. and Nobis, G. *Practical Professional Catering.* 1987. Macmillan. Education Ltd.

Eating with your eyes: a teaching programme for Food and the Media

P. Hann, A. Colquhoun

School of Food and Accommodation Management, University of Dundee, 13 Perth Road, Dundee DD1 4HT, Scotland

Abstract

From still photography to TV and video, food imagery can be used to entertain, inform, educate and sell. Today's consumers see food images and photography almost daily - from the pack on a supermarket shelf; the Sunday colour supplement and the vast array of cookery books to the recent innovation of recipe ideas on television advertisements. Like any visual medium, it is constantly subject to the dictates of style, fashion and marketing trends.

Traditionally, Home Economist training offered few opportunities for developing awareness of the specialist nature of this form of food preparation/presentation. The Home Economist or Food Stylist can be part of a team which includes a marketing manager, an art director, the food editor and the photographer. Collectively they create the photographic illustration for the product, production or feature.

Responding to the demands of industry, we have included introductory food photography and demonstrations in the second year of our undergraduate programme and augmented this with a specialist *Food and the Media* Senior/Junior Honours option in food photography, television and advanced food demonstration techniques. Here students have the chance to work on `live' projects' for the food industry. The organisation and presentation skills required for food photography are extended with work for food on video or television. The paper outlines stages in the conception, planning and development of a specialist *Food and the Media* course geared to the needs of the food and catering industries.

Introduction

Everyday in the media we encounter photographs, film, video and illustrations of food. Food photography and related media sell food, images of food and even lifestyles. For example the *Recipe for Love* TV advertisements by the Meat & Livestock Commission, suggest a view of relationships congruent with the final years of the 20th century.

Food photography is a predominantly static medium; the exception being TV or video portrayal. Further, as we can neither smell, taste nor touch the food in a photograph, the picture must impart to its audience all the aforementioned characteristics. It not only informs the customer about the recipe, ingredients or product but, through its styling, can convey a particular lifestyle.

The paper follows the development of a food photography and demonstration course and the subsequent development of a specialist *Food and the Media* course at honours level within a consumer studies degree programme and shows how the disciplines of design and food studies can be combined to forge a new and exciting partnership focusing on food within the media.

Historical Development

Food photography, as an integral part of the University of Dundee MA (Hons) in Food and Welfare Studies programme was a logical progression from the close links already developed between food and design on the former Diploma in Home Economics. The phasing out of the Dress and Textiles option on the Diploma, offered the opportunity for design staff involvement in the emergent area of teaching - food photography and demonstration. An understanding of colour, pattern, texture, shape, form and composition -ie design principles - is just as crucial to the food stylist. Only the medium differs.

Second Year Food Themes

Food Photography
From experience on the Diploma course, it was decided to further develop food photography for inclusion in the *Food and Nutrition* course on the second year of the new degree programme. Revisions to both organisation and teaching methods were necessary giving a more structured approach to the management of workshops.

Food photography was designed as part of a course, undertaken by all second year students, which introduced students to all stages of food preparation, presentation and styling. Students are introduced to the theoretical aspects of food photography and, thereafter, to techniques for creation of the food image for photography. Skills acquired are used in the interpretation/implementation of a food brief selected from a number of food topics which require the student to relate her/his investigation to the needs of a particular client, commissioning agent or potential market. Every opportunity is allowed for discussion and teamwork at this stage of the project. Furthermore, a photograph is never simply viewed as an end in itself but rather as a focus for a detailed study involving selection, interpretation, research, testing, planning, organisation, execution and presentation.

Food Demonstration Techniques
Linked to the food photography course is an introduction to food demonstration techniques. Referring to the same project brief, students plan, prepare, script and present a food demonstration to their peers. This experience forms the basis of advanced demonstrations in Year 3/4 *Food and the Media*. The Year 2 demonstrations also feed effectively into the TV and video section of the *Food and the Media* syllabus.

Throughout the course, efforts are made to show links with other parts of the degree programme - in particular, design studies, marketing, communications, health promotion and food product development - and to show how the range of knowledge and experience acquired can be applied to honours options.

The success of the course has given rise to a number of developments. It has helped to forge positive relationships with local food companies with whom we have co-operated on several `live' projects and opened the door for placements at local and national level. Furthermore, food photography and demonstration techniques at this level serve as a platform for honours options on the programme - *Food Product Development* and *Food and the Media*.

Third/Fourth Year Options

The concept of *Food and the Media* as an option in the MA(Hons) Food and Welfare Studies degree programme was conceived in 1992 and implemented for September 1994. The course has developed from the second year food themes course outlined above but was developed with the intent of furthering the creative talents of students allowing them to build on their knowledge and technical skills with regard to food, its preparation and presentation.

From its initial concept, the idea of the course development was to bring together the knowledge and technical expertise within the two Schools of Television & Imaging and Food & Accommodation Management. This fusion of the creative talents of academic and technical staff and students of the two Schools created a synergy which resulted in the formation of a unique multidisciplinary team equipped to focus on new teaching initiatives in the areas of food photography, demonstration, TV and video.

`Live'Project Work

Industrial links have been forged by the School over a long period of time. These links help foster an atmosphere in which students learn to hone and develop their skills in food studies by gaining first-hand experience working to a `live' client brief. Students are thus given added impetus in their studies and confidence to put theoretical and practical knowledge into a real-life context. A number of `live' projects have been undertaken to date by staff and students.

- Clients have included local food producers and retailers, a local publishing organisation and the National Trust for Scotland. For the latter, we have researched and produced a range of food historical texts, teachers' information packs and childrens' activity packs which were illustrated by associated food photography shot on location at National Trust properties (see Appendix 2).

- Two short videos for a Fife-based food retailer and restaurant - Scotland's Larder - focusing on the selection and promotion of quality Scottish fresh seasonal produce. This project was conducted in conjunction with staff and students from the School of Television and Imaging (see Appendix 3).

- A number of high profile demonstrations including one arranged in association with Dundee Heritage Trust and featuring a guest celebrity demonstrator. Other demonstrations this session have included a major presentation for Perth Food Festival and a series of smaller demonstrations for a `demonstration day' at the aforementioned Scotland's Larder.

Most demonstrations have been sponsored by local food companies and food retailers. Both the video and demonstration projects have formed a key part of an industrial placement period for students based with Scotland's Larder and the Perth Food Festival. Placement experiences of year three students reflect the growing demand for student expertise in this area of work. To date, students have had the opportunity to further develop skills and abilities within organisations as diverse as public relations companies, major food retailers and producers, freelance operators and publishing companies - both locally and nationally.

Conclusion

Including a *Food and the Media* option as part of the degree programme can be fraught with problems. The development of `live' projects and the balancing of a number of skills, abilities and client intervention creates an intensive yet highly exciting medium in which to work. The passion shared by both staff and students for `food' shows through in every aspect of the creative process. The whole process has had a number of valuable dimensions. It has created the opportunity for cross-fertilisation between staff research/consultancy and teaching. In addition, a placement period in Year 3 of the degree programme gives students the opportunity to apply skills in the `real world' thus gaining experience which we hope will help them achieve that ultimate goal - employment within the industry.

Appendix 1: Resources and Equipment for *Food and the Media*

At the outset we had no equipment, studio or `props' (ie the range of tableware and backgrounds required to `style' the food). The problem of camera equipment was initially solved by our liaison with the School of Design but the needs of Graphics students in relation to their own curriculum and timetable necessitated our own purchases. Over a two-year period we made a capital investment of approximately £3000 helped by advice from the School of Television & Imaging, and training for our graphic designer AV technician. Purchases included a Mamiya RB 67 SD, 3.5 127mm lens medium format studio camera with a polaroid back, two Elinchrome studio flash lights 1500s and a Minolta auto meter IVF. Subsequent investment in photographic equipment includes a Bowen's Hi-Glide System; three Manfrotto Expansion Sets; a Bowen's Prolite 120 and a Manfrotto Background Set.

Over a number of years, we have steadily built up our `props' store buying some new equipment and scouring our local flea market for others - from a Victorian butter churn to complete period dinner services. This session we have been fortunate to win a capital bid for the installation of a new £20,000 food photography, TV and video studio. It comprises a kitchen space capable of a visual metamorphosis by the use of interchangeable clip-on kitchen doors. These allow us to change from a country kitchen to a slick contemporary kitchen in minutes. As the kitchen units are mobile, the space can also be quickly transformed into a TV kitchen for demonstration to camera. In addition to flexibility of use in the kitchen, we have also built `flats' containing a window and fireplace, in order to help re-create a variety of interior shots.

The School has a fully-equipped 50 seat food demonstration theatre with a demonstration bench containing an electric halogen cooker. A monitor linked to a ceiling video camera allows close-ups of the bench and the demonstration.

Appendix 2:
Example of a Food Photography Brief for *Food and the Media*

CLIENT: The National Trust for Scotland
Barry Mill, near Carnoustie, is an early to mid 19th Century oatmeal mill. It was acquired by the Trust in 1988 and is now a fully restored working (oat) meal mill. As an educational resource, it is often visited by parties of school pupils. Once they have viewed the process of meal milling from the grain to the meal (flour), they see it ultimately in its final bagged form. At this stage the miller is often asked what the meal is subsequently used for. Children of the deep-freeze/ convenience food/supermarket age may come from a background where little or no baking is done and find it difficult to equate this light brown floury substance in a hessian sack with something which can be turned into food.

Working in two groups of two, research, select, plan, write, organise and execute one of the following:-

(a)   a teacher's information pack for P4 to P6 (one group of two)

(b)   a children's activity pack   for P7 to S2 (one group of two)

These packs are to be used as part of the Environmental Studies Curriculum - Understanding People in the Past: The Age of Revolution (1700-1900). The material should cover the story of the grain from the grading to it's uses in recipes and it's place in the diet 150-200 years ago. You will undertake one photograph each which should show a minimum of four traditional oatmeal based dishes. The photography will be undertaken on location at Glamis Folk Museum. You may support your food photography with other photographs taken from the National Trust Photo Library. You may also take your own photographs at Barry Mill (prior permission must be negotiated with the resident Manager) but they must be of good enough quality to use in your pack.

Indicative Reading List:

Beeton, I., *Mrs Beeton's - All About Cookery*, Ward, Lock & Bowden: London, 1893.
Brown, C., *Broths to Bannocks: Cooking in Scotland 1690 to the Present Day*, John Murray: London, 1990.
Brown, C., *Scottish Regional Recipes*, Chambers: London, 1992.
Davis, J., *The Victorian Kitchen*, BBC Books: London, 1989.
Davis, J., *The Victorian Kitchen Garden*, BBC Books: London, 1991.
Geddes, O.,*The Laird's Kitchen: Four Hundred Years of Food in Scotland*, HMSO: London, 1995.
Macleod, I.,*Mrs McLintock's Recipes for Cookery and Pastry Work*, Aberdeen University Press: Aberdeen, 1986.
Paston-Williams, S.,*The Art of Dining: a History of Cooking and Eating*, National Trust: London, 1993.
Seymour, J., *Forgotten Household Crafts*, Dorling Kindersley: London, 1987.
Wilson, A.,*Traditional Country House Cooking*, Weidenfeld and Nicolson: London, 1993.
Wilson, A., *Luncheon and Other Meals: Eating with the Victorians*, Alan Sulton Publishing: London, 1994.

Appendix 3:
Example of a Video and TV Brief for *Food and the Media*

CLIENT: Scotland's Larder, Fife
Referring to the information pack supplied, select, plan, organise, storyboard, write and create a 10-15 minutes video for Scotland's Larder. This is a high quality retail and restaurant outlet which aims to promote the very best of Scottish fresh seasonal produce.

Working in a group of nine, your topic will be either Spring/Summer or Autumn/Winter. In this group, you must meet and agree sub-groupings of responsibility to work on particular aspects of the video.

Sub group (of three) - Storyboard and script
Sub group (of three) - Design, styling and food presentation
Sub group (of three) - Presenter/s and food preparation

However, it is essential that as a group of nine you agree the storyboard idea and the main concept - *the look and style of the whole video*. Teamwork, dialogue and compromise will be the key to success.

180   Culinary Arts and Sciences

From the information pack you will see that it would be impossible to cover all the material included. Within your group of nine you must select, discuss and agree an <u>emphasis</u> or <u>theme</u>. The following are only suggestions, feel free to use one of these or devise your own.

1) From plot to plate.
2) Entertaining with Scotland's best.
3) Quality produce to the fore.
4) The pick of the crop.
5) Sea and shore.
6) Meating or meeting the quality challenge?

This project is taught primarily by Steve Flack, a free-lance TV director who teaches in the School of Television & Electronic Imaging (TVI). The video will be planned and filmed in conjunction with post-graduate students from TVI. They will also work in a team ie. director, editor, camera-person, sound-person, lighting person. You should realise that the Director in particular, and the production team generally will also have a view about the overall `look' of the video. Again constructive dialogue and discussion are crucial to success. Ultimately the Director holds final editorial control.

APPENDIX 4:
Example of a Demonstration Brief for *Food and the Media*

CLIENT: Discovery Point Visitor Centre, Dundee

Research, plan, script, organise, present, select and test recipes for a major food demonstration to be mounted at Discovery Point for an audience of approximately 150, entitled a *Taste of Tayside: Traditional and Contemporary*. For this project you will, develop on previously acquired demonstration skills, recipe development techniques and presentational/communication skills.

The guest demonstrator will be Moyra Fraser (Cookery Editor at the Good Housekeeping Institute) presenting six dishes and both recipes and the event will feature in the *Good Housekeeping* magazine. Students will assist Ms Fraser with her demonstrations.

Each student will demonstrate one recipe based on `Traditional Foods from Tayside'. This can be interpreted in the widest sense.

(a)   Traditional recipes brought up to date eg. substitute ingredients (see Moira's work, Term 1),

(b)   Use traditional local produce in a new way (eg. work for Frank Yorke and

Howgate).

It may be that you decide to keep the traditional recipes for the buffet afterwards and demonstrate contemporary recipes with traditional references.

You will work in teams of two to style three large tables for the presentation of the demonstrated food and finger buffet. As with all other styling consideration must be given to the overall theme, the interpretation, the appearance and colour of the food and the fact that the table must have visual impact from a distance -unlike a photograph.

As well as the researched information about the recipe and its ingredients you may wish to refer to other factual historical/humorous/anecdotal information. With this in mind we have prepared a `fact-file' of mainly historical information to which you may refer. The event will be fully sponsored by local food and drink producers and retailers.

Appendix 2: Examples from a range of approximately 40 second year food themes offered

TOPIC OR BRIEF	FORMAT	COMMISSIONING AGENT OR CLIENT	MARKET	DEMONSTRATION AUDIENCE	DEGREE THEME LINKAGE
Healthy Eating over 60 Whether you've retired recently or have been a pensioner for some time, eating a good diet, taking regular exercise, keeping interested and involved will all help you to remain active and enjoy your retirement years.	Booklet	Help the Aged	For the over 60's	University Keep Fit/Leisure Group	Health Theme
The Glorious Fruits of Summer Exploit the variety and abundance of summer fruits which can contribute to exciting picnic hampers and attractive buffets.	Magazine	Country Living Magazine	Up-market magazine readers	Highland Show	Food Theme
Cheap and Cheerful How to feed 2 adults and 2 children, 3 times a day, on tasty, healthy food for under £5.50 a day. Good food that everyone will enjoy - and your purse and your health will benefit too.	Booklet	DHSS	Low income families	Family Centre	Welfare Theme

Posters

Barriers to skills transfer of hotel management students and graduates in Hong Kong

J. Kivela
*Department of Hotel and Tourism Management,
The Hong Kong Polytechnic University, Hung Hom, Kowloon,
Hong Kong*

What stops students and graduates from transferring skills from classroom to job? The real test of any hotel management programme is whether or not graduates can successfully transfer and apply knowledge and skills learnt during the course of studies to their jobs, without having to be trained in the same skills once on the job. The advantages and benefits to the hotel organisation in students' (during industrial placements) and graduates' being able to fully or near-fully transfer their college-learnt skills to their new jobs, are obvious.

It is suggested that environmental and organisational differences between the learning and work environment, often inhibits knowledge and skills transfer, which results in negative transfer. The findings of a recent survey about the barriers to students'/graduates' skills transfer, reveal some specific environmental and organisational attributes that reduce the potential for skills transfer in Hong Kong hotels. The paper gives succinct recommendations how to minimise negative skills transfer.

Hotel management education in Hong Kong at crossroads

J. Kivela

Department of Hotel and Tourism Management, The Hong Kong Polytechnic University, Hung Hom, Kowloon, Hong Kong

Is it time for industry/educator partnerships and for the re-engineering of hotel management curriculum?
In Hong Kong, the quality of hotel management education outcomes, may not be meeting the needs of graduates and the hotel industry in the future. It is suggested that there is a need for a systematic investigation of hotel managers', hotel management educators', and hotel management graduates' perceptions, about how relevant are some of the hotel management skills to graduates' jobs, and about graduates' competencies in applying these skills to their jobs.

This paper is a result of a survey of 220 hotel managers, Hong Kong Polytechnic University (HKPU) faculty staff, and HKPU hotel management graduates. The paper examines the respondents' perceptions about the relevance of some of the hotel management skills, and graduates' competencies in applying these skills to their jobs. The findings reveal that some major gaps in "relevancy" and "competency" perceptions between the respondent groups does exist. The paper offers a set of recommendations that can be applied to narrow these gaps.

Quality improvement in foodservice management

L.R. Ferrone, P. Jones

School of Service Industries, Bournemouth University, Bournemouth BH12 5BB, UK

The concept of 'quality' has differing meanings to different people at different times in different places. However, the British Standards Institute defines quality as: 'The totality of features and characteristics of a product or service that bear on its ability to satisfy a given need'. The hospitality industry has always revered quality and has the same concerns as that of other industries when it comes to improving quality. Whilst the hospitality industry is diverse and all-embracing, it is hard to imagine any of its services not including some form of foodservice, even if it may only involve the provision of a beverage. This poster looks at the roles of those directly involved with foodservice, namely chefs and waiters, and seeks to identify ways of improving the internal market product in order to establish a more appropriate model for quality improvement. The model is based on the concept of an interfacing social 'tool kit' acting as the medium for greater flexibility, and faster adoption of quality management initiatives as used in other industries.

SECTION 2: FOOD SCIENCE AND TECHNOLOGY

Section 2.1: Microbiology

The development and implementation of a hazard analysis system in a welfare feeding organisation

D. Worsfold
Cardiff Business School, University of Wales, Park Place, Cardiff, Wales, CF1 3AT

Abstract

This paper describes the development of a food safety hazard analysis system for a large voluntary organisation. A research methodology previously employed to investigate the hygiene of domestic food preparation practices was adapted for use by the organisation. A representative sample of food projects were subjected to hazard analysis using direct observation and food temperature measurements. The data was used to compile a set of check-lists (THE SAFE FOOD SYSTEM) for food project organisers which enabled them to identify and control activities critical to the safety of the food. The system may have application to other institutional food businesses.

Introduction

The voluntary organisation works independently and in co-operation with the Health and Local Authorities to help people needing support to preserve their independence and improve their quality of life. Elderly and disabled people are aided by the provision of 'community meals', shopping and library services for the house-bound and assistance in residential homes and day centres. 'Community meals' include:
> meals delivered to people's homes
> meals provided for those living in sheltered accommodation, and
> meals served in lunch clubs.

It is currently responsible for the annual delivery of 15 million Meals on Wheels (MoW) which is more than half of the meals delivered nationally and also provides 3 million lunch club meals each year. The organisation is staffed by 70,000 workers, 99.5% of whom are unpaid. The volunteers are drawn from all walks of life and include many housewives and retired people. They work in local food projects, each run by a project organiser. District and county food managers supervise the work of local projects and report to six divisional food managers and a director of the food service.

Hot meals for daily delivery are produced from fresh ingredients cooked on the premises, or are regenerated from frozen food. Frozen meals for weekly delivery to the clients are bought-in or produced by the organisation in its own kitchens. Meals for lunch clubs may be produced by prime cooking or may be regenerated from frozen or chilled food provided by the local authority or food manufacturers. Some lunch clubs simply serve hot food supplied by a local authority kitchen. If the lunch club kitchen is very small or has very restricted facilities some of the food may be prepared and cooked at the homes of the volunteers, then transported to the club and re-heated.

New food legislation

The Food Safety (General Food Hygiene) Regulations 1995 came into effect in September 1995. They aim to ensure common food hygiene across the European Community as set out in the Food Hygiene Directive[1]. Proprietors are required to carry out a hazard analysis of their food business. This involves making an assessment of activities to identify steps which are critical to food safety and ensuring that adequate safety procedures are identified, implemented and reviewed. Food manufacturers may have already adopted the principles and detail of formal Hazard Analysis Critical Control Point (HACCP) within their general quality assurance systems and procedures. A formal HACCP system is not, however, a legal requirement and would be inappropriate for lunch club or MoW operations.

A hazard analysis system for the organisation

When developing a hazard analysis system suitable for use in its food projects the organisation gave consideration to the following factors:

1. The vulnerability of the consumers to food poisoning. Elderly people who are the recipients of MoW and luncheon club meals are more at risk than the general public
2. The wide diversity of catering operations in the organisation. All the projects handle a variety of open high-risk foods. Most cook and serve or deliver it immediately which poses fewer hazards than where food production is divorced from consumption
3. The size of the work force which would have to be supervised and trained/ instructed to control and monitor food safety hazards. A busy urban MoW scheme will be staffed by approximately 400 volunteers
4. The level of food hygiene training of the volunteers. Leaders of food projects hold a Basic Hygiene Certificate and have trained the members

of their teams with an in-house hygiene awareness package[2] which won Department of Employment Regional and National Training Awards in 1994. Some, however, might find the new hazard analysis approach initially challenging and would need a user-friendly hazard analysis system which could be easily implemented

5   The existence of documented procedures and systems. An internal hygiene audit system, FOODCHECK[3], designed to measure the compliance of the projects with the prescribed standard of hygienic operation was already in use and project leaders and managers were familiar with the check-list format.

A food safety consultant was used to design and help implement a hazard analysis system for the organisation. An initial risk assessment of a representative sample of food projects was undertaken. It was established that the priority for hazard analysis were projects which cooked on a daily basis, used fresh ingredients and delivered or served meals to a large number of elderly clients. The basis of the system would be in accord with the guidance given in Assured Safe Catering[4] and S.A.F.E (Systematic Assessment of Food Environment)[5].

To comply with the new Regulations, project leaders together with their catering staff needed to examine their catering process; decide where potential food safety hazards could occur; determine the actions (preventative measures) which would control the hazards; decide how to check that these measures were being used and regularly review the system.

Designing the system

To enable the project leaders to make and document their hazard analysis system, simple check-lists were developed and piloted in a number of projects. Three check-lists were designed:
1   a list of common food hazards with risk ratings
2   a comprehensive list of the main catering operations with the main generic hazards identified. The hazards relevant to the particular operation could then be selected for attention
3   a list of appropriate generic preventative or control measures and target values

The first check-list was produced after analysing menus from a representative sample of projects. The foods were assessed by reference to the literature, for the likely presence of pathogens or their toxins and the severity of their outcome and risks of occurrence. In a representative sample of projects, each step in the food production process was examined by observation and temperature measurements, to establish sources and specific points of contamination and the potential for micro-organisms to survive any process and to multiply. A check-list which had been used originally to gather information in the domestic environment[6] was modified for use with the projects. The equipment, premises, and methods of food production in some lunch club projects often bears more similarity to the domestic environment than the commercial catering sector.

Food projects have to demonstrate that they have determined the control measures necessary to eliminate or reduce the food safety hazards to a safe level. Since some steps in the production of food are more critical to the safety of the food than others, it is especially important to employ preventative measures at these critical control points. As the projects produce cooked food, the critical control points will include cooking and any processes that take place after cooking, such as cooling, holding, re-heating etc. until the food is eaten by the clients. A control target was set for every critical control point that had been identified. In the domestic research, realistic generic control measures for each identified hazard were determined. These were used as the basis of a check-list of appropriate control measures that was piloted in a representative sample of projects. The check-list requires the leader to confirm which control measures are in place in the project and indicates appropriate monitoring methods.

Using the system

The following example illustrates how the system works. Using the first check-list, a project would identify the cooked ham it uses for salad as a high risk food that is easily contaminated. Using the second list, they would identify the relevant handling operations. They would be advised that the potential hazards involved in purchasing this product are damaged packaging, out-of-date stock and unreliable and unhygienic suppliers. These hazards might lead to the food becoming contaminated or to an increase in the level of contamination.

The potential hazards associated with delivery of this product would be identified, namely: delivery temperature is too high, the delivery containers/vehicle is dirty and/or the delivery temperature of the food is too high. These are conditions which might all lead to an increase in contamination. The potential hazards during storage of this product before use are identified as follows:

    it remains at room temperature for too long
    it is stored at the wrong temperature
    no stock rotation is observed so the food is out-of-date
    raw and cooked foods are stored together
    its packaging is unclean and/ or inadequate

These conditions might lead to an increase in contamination and present opportunities for cross-contamination. Potential hazards and their likely consequences would be identified for all the other food handling operations the product might be subjected to until it is given to the client.

Project leaders and their catering teams are then required to confirm that they are using the recommended control measures, using the final check-list. In this example, when purchasing and receiving deliveries of cooked ham they would only use reliable and hygienic suppliers. They would reject food with damaged packaging, check date stamps and delivery temperatures. When storing the ham, they would ensure that it is stored quickly and at the correct temperature. They would rotate stock and would check the shelf life of the product. Raw and cooked foods would be stored separately and foods would be covered. Appropriate control measures are given for all the other handling steps the product might be subjected to. Guidance is provided on the use of monitoring procedures to ensure that the identified critical control points are effectively controlled. Whilst the Regulations do not require written records of monitoring, records will be valuable if due diligence has to be established. Four pro-formas have been designed to enable records to be easily maintained.

The SAFE FOOD SYSTEM

The hazard analysis system, which was developed for the organisation was called the SAFE FOOD SYSTEM. The food managers have attended half-day workshops where they were introduced to the principles of hazard analysis and were taken through the stages of risk-based decision making using a number of appropriate examples. They were given the check-lists and advised on suitable monitoring methods such as visual checks, temperature measurements and time checks. There was discussion on the necessity for clear work instructions which identify who is responsible for checking that a preventative measure has been carried out correctly and how and when checks must be undertaken. The requirements for staff to be trained to carry out the necessary checks and to be informed about the appropriate action in the event of a control measure not being carried out correctly were emphasised.

The food managers have subsequently introduced the SAFE FOOD SYSTEM to the project leaders of the highest risk projects in their areas. They are now in the process of assisting them complete their hazard analyses. The documentation is customised for each unit and then one copy is returned to the food manager for checking and filing. The Food Safety Regulations do not demand a documented hazard analysis system but a written explanation of the system would help to demonstrate compliance with the Regulation.

Food managers do not report any great difficulty in getting projects to comply with the hazard analysis requirement of the new legislation They do, however, think that the maintenance of monitoring records in some projects will take some time to become routine. The projects were provided with four pro-formas for monitoring critical control points that related to the steps of delivery, food storage, cooking and cleaning. In some projects these have been introduced with no alteration and are being maintained by the staff. In other projects, the leaders or food managers have simplified the monitoring sheets or prefer to use record sheets previously in use prior to the implementation of the system. In a minority of projects, the food managers report a reluctance on the part of the staff to complete additional documentation.

Discussion

Evaluation of the hazard analysis workshops has shown that most of the managers found them useful. The provision of a user-friendly system that could be tailored for each project was welcomed. An early and incomplete review of the hazard analyses undertaken by some of the projects has been made. The range of foods that were identified in the pilot study have proved to be sufficiently comprehensive. In the workshops. food managers were found to be well informed about the dangers of high risk foods especially eggs and poultry. Projects have occasionally identified additional foods in regular use, but these have all been items which present low food safety risks.

Many of the projects do not routinely prepare and cook well in advance of delivery or service. They are not, therefore, involved with the potentially hazardous practices of cooling cooked food, storing this food until required and subsequent re-heating procedures. In those projects where food is cooked in advance, it is usually frozen, stored for less than one month and subsequently re-heated. The potentially hazardous processes of hot holding and transport of heated food are obviously a feature of projects providing a daily hot service of food.

Appropriate simple effective control measures are in routine use. For example, regular checks are made on the condition and shelf-life of delivered food and the temperature of frozen and chilled foods. Most projects use suitable control measures during food storage and preparation. Temperature records of refrigerators and freezers are maintained by most MoW projects. The traditional cooking techniques used would appear to ensure that the food reaches an internal temperature of $75^{\circ}C$. The routine recording of the final temperature of high risk food is routine in MoW projects which regenerate frozen food, but has not been practised by most lunch club projects. Guidance on the purchase and use of digital temperate probes has been provided. The minority of projects that need to cool cooked food for later use, do not appear to be using the most effective methods for speeding the process. Since a blast chiller would be too expensive an acquisition for these units, it has been recommended that water or ice-baths are used. Guidance has been provided on the protection of cooling food.

Recommended cleaning practices are in use and volunteers follow a regular cleaning rota. However, in the many projects a cleaning schedule is not displayed. The cleaning schedule should indicate the frequency and method of cleaning and/or disinfection to be used for different surfaces and items of equipment and who should carry out the tasks. It should be monitored by the supervisor. The organisation has since invested in laminated display schedules, which provide the essential information.

It will be necessary for the food managers to check at regular intervals that the food safety management system in each of their projects is up to date. Suppliers, recipes, equipment and staff may change and this may call for a revision of the hazard analysis system

Conclusion

The voluntary organisation has over 2000 food service projects, each of which must complete a hazard analysis. This will obviously take some time, however, it has made an early start on the task and is building on a firm foundation of food hygiene training for all food handlers and established documented practices and procedures. The SAFE FOOD SYSTEM was developed before the Catering Industry Guide to good hygienic practice[7] was published. It does, however, comply closely with the guidance given in this document. The SAFE FOOD SYSTEM uses the HACCP principles in a way that can be easily applied to different sized catering operations. It has been developed by surveying representative units and using currently available information on hazards and risks. The generic model which has been developed can be customised by individual units to provide their own hygiene management system. The system saves time for the project leaders and it provides confidence that they have focused on the activities that are critical to the safety of food. Nether the less, the organisation recognises that if the system is to be successfully implemented all the volunteers will have to demonstrate the same level of commitment they have previously shown in becoming trained in food hygiene.

The introduction of the SAFE FOOD SYSTEM is viewed by the organisation as an opportunity to reinforce the use of existing adequate control measures and to explain the rationale for their use. It will be used as a training vehicle to ensure that staff are instructed in control and monitoring procedures. The system may have application to other institutional food businesses.

References

1. Department of Health, *Council Directive 93/43/EEC Hygiene of Foodstuffs*, Department of Health, 1993.

2. Worsfold, D., Training help for Food Handlers, *Network Wales*, Issue 95, 1992.

3. Worsfold, D., Food Check, *Nutrition and Food Science*, No. 6, 1992, pp 25-27.

4. Department of Health, *Assured Safe Catering*, HMSO, 1993.

5. British Hospitality Association, *S.A.F.E (Systematic Assessment of Food Environment*, BHA, 1995.

6. Worsfold D. An Evaluation of Domestic Food Hygiene and Food Preparation Practices, *Ph. D Thesis*, Open University, 1994.

7. Joint Hospitality Industry Congress, *Industry Guide to Good Hygiene Practice: Catering Guide*, HMSO, 1995

Microbiological safety of recipe mayonnaises

H.I. Dodson, A.S. Edmondson, M.A. Sheard
Food Research Group, Leeds Metropolitan University, Calverley Street, Leeds LS1 3HE, UK

Abstract

Ten easily available catering and domestic recipes for home-made mayonnaise were investigated for their microbiological safety. The mayonnaises were prepared using accurately measured ingredients. Safety was assessed by measuring pH and percentage of acid, using the CIMSCEE formula (IFST), and following the survival of inoculated *Salmonella enteritidis* PT4 either by direct plating on XLD or by using RABIT (indirect impedance measurements) following enrichment.

Two of the recipes produced a microbiologically unsafe product by all the criteria. One catering recipe (DHSS) stood out as being very safe by all the criteria. This recipe was used to investigate the effects of recipe abuse, in which the proportion of vinegar was reduced.. Reduction of the weight of vinegar added, to 75% of that in the original recipe, gave a visually acceptable and microbiologically safe product. However, if only 50% of the vinegar was added the product was visually unacceptable and unsafe. This recipe would thus be valuable for use in a busy kitchen and would ensure a level of control of accidentally introduced organisms.

Introduction

There have been several reported outbreaks of food poisoning, caused by salmonella, associated with the consumption of mayonnaise (Ortego-Benito & Langridge[1], Radford & Board [2], Irwin et al.[3]). Recipe mayonnaise made from fresh shell eggs can be contaminated by organisms present on the shell of the egg, and, in the case of *Salmonella enteritidis* phage type 4 (PT4), by organisms present in the yolk of the egg. To prevent this problem caterers were advised to use pasteurised egg in uncooked dishes (HMSO[4]). However, the use of pasteurised egg in recipe mayonnaise is an inconvenient restriction for caterers and does not necessarily ensure product safety. Poor kitchen hygiene could still allow cross contamination of the product. Subsequent storage of the product at room temperature, for example on a buffet table, could allow the development of high levels of salmonella and the potential for an outbreak of food poisoning.

Previous work on mayonnaise safety has tended to concentrate on the commercial formulations. Smittle[5], reviewing previous work, suggested that a pH of 4.1 with acidity over 0.25% using acetic acid was needed for safety. Manufacturers of sauces containing acetic acid use the CIMSCEE formula (IFST[6]) to calculate safety from pathogens. It is based on the work of Tuynenburg Muys[7] who used *E. coli* to represent the growth of salmonella in oil-in-water emulsions and determined that the fate of microorganisms was dependent on the chemical composition of the water phase of the emulsion. The code definition of intrinsic safety is that there should be a three log reduction in numbers of viable *E. coli* in less than 72 hours at 20°C.

Perales & Garcia[8] used graphs of the fall in viable count to follow the effect of alteration in pH and acidulant on *Salmonella enteritidis* PT 4 in home-made mayonnaise. A five log fall in viable count was achieved when vinegar was used to produce a pH of 3.6 and incubation was at either 24°C or 35°C. When lemon juice was the acidulant to produce a pH of 3.6 incubation at 35°C produced little reduction in viable count during the first 24 hours. This was followed by a five log fall from 24 to 72 hours. At 4°C there was little fall in viable count with the exception of mayonnaise prepared using vinegar to give a pH of 3.6 where a fall between 24 and 48 hours was not continued over the next two days. The protective effect of low temperature (4°C) was also shown by Lock and Board[9] in their work with commercial mayonnaise. Radford and Board[2] in their review, also note the protective effect of refrigeration, and recommend a holding time of 24 hours at 18°C to 22°C before refrigeration for safety of home-made mayonnaise. Radford et al[10] investigated the toxicity of different oils when used to prepare mayonnaise.

The standard method for demonstrating the absence of salmonella is enrichment of a 25g sample followed by detection using selective media (Roberts et al[11]). Impedance methods have also been shown to give good detection of salmonella (Donaghy and Madden[12]).

A wide variety of recipes are available for the preparation of home-made mayonnaise. The purpose of this work was to ascertain the safety of some of these recipes when contaminated and stored at room temperature. Ten recipes, from both catering and domestic texts and using a variety of oils and acidulants, were tested for microbiological safety by both theoretical and practical methods. Also, because the proportion of ingredients used in practical catering production varies, the safest recipe was then used to check the effect of recipe abuse.

Materials and Methods

The selected mayonnaise recipes contained a variety of oils and acidulants as shown in Table 1. Predictive methods used for safety determination were pH, titratable acidity and the CIMSCEE formula. Microbiological analysis was by viable counts and the presence/absence test.

Mayonnaise

Table 1 Mayonnaise recipes

Recipe Number	Source	Oil	Acidulant
1	Martland & Welsby[13]	Vegetable	Wine vinegar & lemon juice
2	Finch[14]	Vegetable	Malt vinegar
3	DHSS[15]	Vegetable	Wine vinegar
4	Ceserani & Kinton[16]	Olive	Malt vinegar
5	Reay[17]	Olive	Wine vinegar
6	Stevenson[18]	Olive	Malt vinegar & lemon juice
7	Hanbury-Tenison[19]	Sunflower	Wine vinegar
8	Good Housekeeping Institute[20]	Sunflower	Wine vinegar
9	Miller[21]	Olive	Lemon juice
10	Readers Digest Assoc.[22]	Sunflower	Wine vinegar

Recipes 1 - 6 are from texts aimed at caterers whilst recipes 7 - 10 are for domestic use. Mayonnaises were prepared from preweighed ingredients according to the published recipes.

Recipe abuse
Recipe 3 was prepared using reduced amounts of vinegar. Batches were made with 100%, 75% and 50% of the weight of vinegar stated in the standard recipe. The weights of all other ingredients were as in the published recipe.

pH and acidity measurements
pH was measured directly using a Phillips PW9420 (Pye Unicam, Cambridge, UK) both immediately after preparation and after overnight storage at 10°C and return to room temperature.

Titratable acidity was measured by preparing a mixture of 10g food in 100ml water. 25ml aliquots were titrated against 0.1M sodium hydroxide using phenolphthalein as indicator (Egan et al.[23]).

CIMSCEE formula (IFST[6])

The formula utilises the percentage of acetic acid, salt, hexose and disaccharide and the pH to give a numerical value for microbiological safety:

$$15.75(1-\alpha)(\text{total acetic acid\%}) + 3.08(\text{salt\%}) + (\text{hexose\%}) + 0.5(\text{disaccharide\%}) + 40(4.0\text{-pH}) = \Sigma_s$$

where $(1-\alpha)$ is the proportion of the total acetic acid which is undissociated and is related to the pH of the product and the pK of acetic acid where $pH = pK + \log(\alpha/1-\alpha)$ and the pK for acetic acid is 4.757.
For microbiological safety $\Sigma_s \geq 63$.
The formula is only suitable for products between pH 3.0 - 4.5.

Microorganism

Salmonella enteritidis PT 4 was obtained from the Central Public Health Laboratories, Colindale. The strain had been isolated following an outbreak of food poisoning attributed to mayonnaise. It was maintained on Nutrient Agar slopes (Oxoid), incubated overnight at 37°C and stored at 22°C.

Contamination of the food with salmonella

A culture of the salmonella was prepared by inoculating 0.1% peptone (Oxoid) containing 10% egg yolk emulsion (Oxoid) from the stock slope and incubating at 37°C for 18 hours. This culture was added to the mayonnaise at the rate of 1ml per 100g, to give approximately 10^6 organisms per g, and blended in a Colworth 400 stomacher for 2 mins. The contaminated samples were stored at 22°C.

Microbiological analysis of the inoculated food samples

Viable Counts: Samples were taken at hourly intervals for the first 7 hours and then at 24 hours after contamination. At each time interval 10g contaminated food was blended with 90ml Buffered Peptone Water (BPW)(Oxoid). Serial tenfold dilutions were prepared using 0.1% peptone as diluent and plated out using the surface drop method (Roberts et al. [11]) on XLD agar (Oxoid). Incubation was at 37°C for at least 18 hours.

Presence/Absence test: Samples were taken at 7 and 24 hours after contamination. A 25g sample of food was weighed into a sterile container and 225ml BPW added, swirled to mix and incubated at 37°C. After 24 hours 100µl were added to preprepared indirect RABIT tubes (Rapid Automated Impedance Technique, Don Whitley Scientific, Shipley, U.K.) following the method of Donaghy and Madden [12].

Results and Discussion

The mayonnaises prepared according to the ten recipes differed greatly in colour and texture. Recipe 8, which used egg white rather than egg yolk in order to reduce cholesterol intake, produced an unusually white mayonnaise. The use of malt vinegar tended to produce a brown colour which may not be considered desirable. Recipe 9, containing lemon juice, was rather too bright yellow in colour and too stiff for use. The other factor affecting appearance was the level of mustard, larger amounts producing a deeper yellow colour.

The values determined for the safety criteria of the ten mayonnaise recipes are given in Table 2. It can be seen that the pH values of seven of the mayonnaises immediately after preparation were below 4.1, considered by Smittle[5] to be the level for safety. However, it was found that the pH value increased slightly on storage of the mayonnaise, resulting in three of these having an unsafe value when used. Recipes 3,6,9 and 10 had a pH below 4.1 when used. Because facilities for pH measurement are not available in kitchens, a recipe which consistently produces a product with a pH well below 4.1 is desirable.

Table 2: Values for safety criteria of mayonnaises

Recipe number	pH made	pH used	Titratable Acid (%)	CIMSCEE value	Log decrease in viable count after 7 hours	Presence of salmonella after 24 hours
1	4.05	4.12	0.35	27	1.53	+
2	4.15	4.22	0.30	31	1.22	+
3	3.79	3.88	0.50	85	>6	-
4	4.45	4.50	0.18	44	0.17	+
5	4.80	4.90	0.13	-[a]	0	+
6	3.92	4.06	0.32	32	1.53	+
7	4.07	4.15	0.43	60	1.08	+
8	4.03	4.11	0.44	22	1.55	-
9	3.94	4.03	0.00	-[b]	1.28	+/-[c]
10	3.94	4.03	0.43	70	3.27	+
Safe limit	*< 4.1*		*> 0.25*	*> 63*	*>3.00*	-

Key: [a] CIMSCEE formula not valid as pH > 4.5. [b] CIMSCEE formula not valid due to absence of acetic acid. [c] result variable from different batches

It can be seen that only two recipes, 4 and 5, resulted in a titratable acidity produced by acetic acid below the value of 0.25% considered as the minimum for safety (Smittle[5]). Recipe 9 contained no acetic acid and, therefore, could not be included. It has lemon juice as the acidulant, and in spite of using a bottle of lemon juice with preservative to eliminate the variation of individual lemons, it was not possible to obtain a consistent product.

The CIMSCEE formula is not easy to handle and it requires chemical analysis of the product. In this study calculations were performed on the basis of the recipe formulations and showed good correlation with the other safety criteria values. However, the level set for safety in the formula, a three log reduction of *E. coli* in 72 hours, would probably not be adequate to prevent food poisoning from cross-contaminated mayonnaise. In calculating the formula values it was noted that the salt concentration has a large effect on the final value when the pH is over 4. This could be a problem in domestic or catering preparation when it is unlikely that small amounts of salt would be weighed accurately. Besides recipe 3, only recipe 10 gave a safe value. However, repeated preparation of recipe 10 showed that occasionally the pH would be above 4.1 and the safety compromised.

Considering the fall in viable counts, mayonnaise recipes 3 and 10 were again the only ones to achieve the safe limit of greater than three log decrease in seven hours.

The presence/absence test after 24 hours gives a good indication of the safety of the product provided that it is stored at room temperature. Two recipes, 3 and 8, were safe in this test but, because of its unusual appearance, recipe 8 was not investigated further.

Curves for fall in viable count are presented in Figures 1a and 1b. These support the acidity data in that for recipe 5 there is some evidence of growth of the inoculum, whilst for recipe 4 the inoculum decays very slowly. Both these would be very unsafe products. Conversely, the rapid fall in viable count for recipe 3 again indicates a very safe product. The most noticeable feature of these graphs is the different shape of the curve for recipe 9. This shape indicates an unsafe product because, although there was initially a good reduction in viable count, the rate of fall decreased with time with the possibility of viable organisms being recovered after several days. Comparison of the curves for recipes 7 and 10, where the differences are that recipe 10 contains more vinegar and mustard plus some sugar, suggest the importance of acidity plus a possible effect of the other ingredients (Lock and Board[9]). Radford et al[10] found that, among the oils they tested, sunflower oil was the least toxic to salmonella. No evidence was found in this study to indicate any difference between the toxicity of vegetable oil, sunflower oil and blended olive oil.

It was concluded that many of the readily available domestic and catering recipes for mayonnaise do not result in a product which is intrinsically safe if salmonella organisms are introduced via eggs or by cross contamination. Of the ten recipes tested only recipe 3 produced a mayonnaise which was safe by all the predictive and microbiological criteria. In small scale preparation of mayonnaises in a kitchen the proportion of ingredients used is not accurately controlled. Therefore, recipe 3 was further investigated to determine the effect of recipe abuse on product safety.

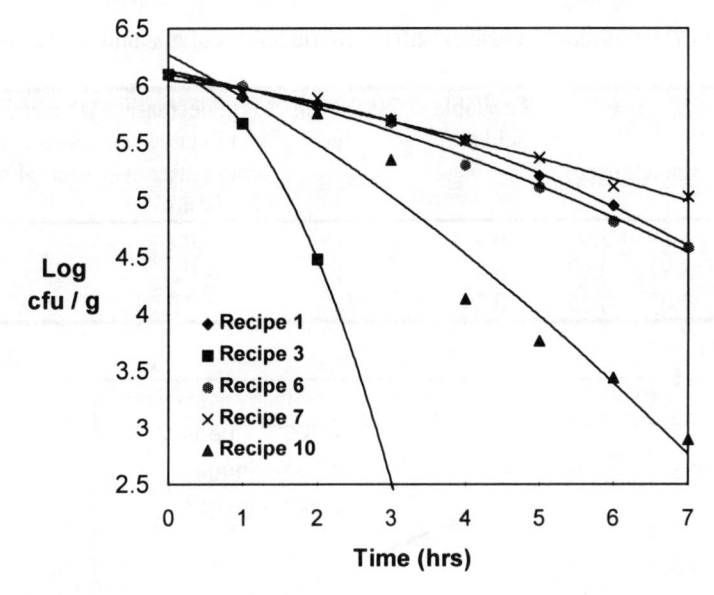

Figure 1a: Survival of *Salmonella enteritidis* PT4 in mayonnaises 1,3,6,7 and 10

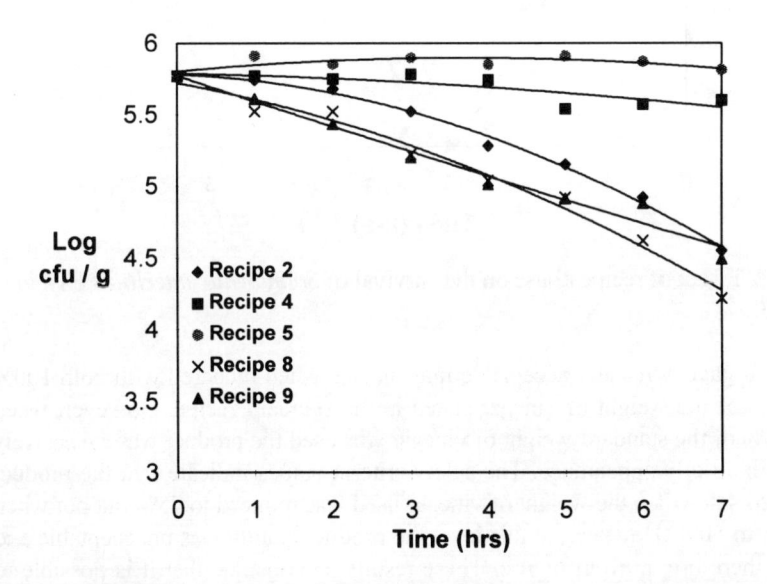

Figure 1b: Survival of *Salmonella enteritidis* PT4 in mayonnaises 2,4,5, 8 and 9.

The effects of recipe abuse on the safety criteria values of mayonnaise recipe 3 are shown in Table 3 and the effects on the fall in viable count are shown in Figure 2.

Table 3: Effect of recipe abuse on safety criteria values of mayonnaise recipe 3

Amount of vinegar	pH made	pH used	Titratable acid (%)	CIMSCEE value	Log decrease in viable count after 3 hours	Presence of salmonella after 24 hours
100%	3.89	3.92	0.51	79	5.00	-
75%	3.91	3.97	0.41	72	4.02	-
50%	4.09	4.15	0.29	61	1.23	+

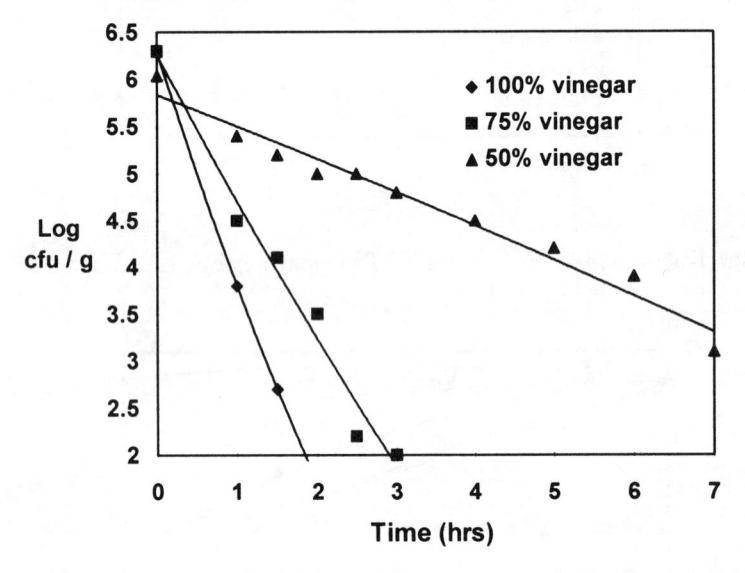

Figure 2 Effect of recipe abuse on the survival of *Salmonella enteritidis* PT4 in recipe 3

Recipe 3 gave a visually acceptable mayonnaise when prepared with both 100% and 75% of the weight of vinegar stated in the standard recipe. However, when only 50% of the standard weight of vinegar was used the product was excessively stiff with an oily appearance. The safety criteria values indicate that the product remained safe when the weight of vinegar used was reduced to 75% but not when reduced to 50%. However, at this level the product quality was unacceptable and would, therefore, prevent its use. These results demonstrate that it is possible to have a microbiologically safe recipe for mayonnaise where safety is correlated to the appearance of the product so that it can be assessed in practical use.

Acknowledgements

Thanks to Helen Khan and Rosemary Jones for preparation of the mayonnaises and performing the pH and acidity measurements.
Thanks to UKEPA for financial assistance.

References

1. Ortega-Benito, J.M. & Langridge, P. Outbreak of food poisoning due to *Salmonella typhimurium* DT4 in mayonnaise, *Public Health*, 1992, **106**, 203-208.

2. Radford, S.A. & Board, R.G. Review: Fate of pathogens in home-made mayonnaise and related products, *Food Microbiology*, 1993, **10**, 269-278

3. Irwin, D.J., Rao, M., Barham, D.W., Pencheon, D.C., Lofts, P., Jones, P.H.,O'Manony, M.,Soltanpoor, N., Ward, L.R. & Threlfall, E.J. An outbreak of infection with *Salmonella enteritidis* phage type 4 associated with the use of raw shell eggs, *Communicable Disease Report*, 1993, **3**, R179-R183.

4. Advisory Committee on the Microbiological Safety of Food. *Report on Salmonella in Eggs,* HMSO, London, 1993.

5. Smittle, R.B. Microbiology of Mayonnaise and Salad Dressing: A Review, *Journal of Food Protection*, 1977, **40**, 415-422.

6. Institute of Food Science and Technology. *Shelf Life of Foods - Guidelines for its determination and Prediction*, IFST, London, 1993.

7. Tuynenburg Muys, Ir. G. Microbial Safety in Emulsions, *Process Biochemistry*, 1971, **6**, 25-28.

8. Perales, I., & Garcia, M.I. The influence of pH and temperature on the behaviour of *Salmonella enteritidis* phage type 4 in home made mayonnaise, *Letters in Applied Microbiology*, 1990, **10**, 19-22.

9. Lock, J.L., & Board, R.G. The fate of *Salmonella enteritidis* PT4 in deliberately infected commercial mayonnaise, *Food Microbiology*, 1994, **11**, 499-504.

10. Radford, S.A., Tassou, C.C., Nychas, G.J.E. & Board, R.G. The influence of different oils on the death rate of *Salmonella enteritidis* in homemade mayonnaise, *Letters in Applied Microbiology*, 1991, **12**, 125-128.

11. Roberts, D., Hooper, W. & Greenwood, M. *Practical Food Microbiology*, 2nd Edn., Public Health Laboratory Service, London, 1995.

12. Donaghy, J.A. & Madden, R.H. Detection of *Salmonella* in animal protein by Rappaport-Vassiliadis broth using indirect impediometry, *Int. Journal of Food Microbiology*, 1993, **17**, 281-288.

13. Martland, R.E. & Welsby, D.A. *Basic cookery: Fundamental Recipes and Variations*, 2nd edn., Heinemann, London, 1988.

14. Finch, C.F. *Food Preparation*, Pitman, London, 1977.

15. Department of Health and Social Security. *Manual of Health Service Catering Vol III: Standard Recipes*, HMSO, London.

16. Ceserani, V. & Kinton, R. *Practical Cookery, 5th edn.*, Edward Arnold, London, 1981.

17. Reay, J. *Cooking for Large Numbers: A Simple Guide to Quantity Catering*, Northwood Books, London, 1980.

18. Stevenson, D.R. *Professional Cookery - the Process Approach*, Hutchinson, London, 1985.

19. Hanbury-Tenison, M. *The Sunday Telegraph Cookbook*, Granada Publishers, London, 1980.

20. Good Housekeeping Institute. *Good Housekeeping - Healthy Eating, Healthy Heart*, Ebury Press, London, 1979.

21. Miller, N. (ed). *The Food Processor Cookbook*, Octopus Books, London, 1979.

22. Readers Digest Association. *The Cookery Year*, Readers Digest Association, London, 1973.

23. Egan, H., Kirk, R.S. & Sawyer, R. (eds). *Pearson's Chemical Analysis of Foods*, 8th edn., p 375, Churchill Livingstone London, 1981.

Section 2.2: Technology

Application of fuzzy mathematics in *sous vide* cook-chill food process development

G. Xie, H. Keeling, M.A. Sheard
Food Research Group, Leeds Metropolitan University, Leeds, UK

Abstract

Assessment of food quality is a key factor in food process development. It still mainly depends on sensory analysis, particularly for eating quality and overall acceptability. One of the major problems with this method is the large associated variations. In this study a fuzzy model, consisting of three key elements (a factor set, an evaluation set and a fuzzy transformation), was applied in the process development for *sous vide* cook-chill foods. Four alternative process conditions were compared. Three food samples (carrot, chicken and salmon) and four quality attributes (texture, juiciness, colour and flavour) were used in sensory analysis. The model was modified and results were compared to those from the original one. Fuzzy mathematics was found to be an easy tool for handling the uncertainty and multiple factors associated with the sensory analysis and overall quality assessment of food.

1 Introduction

Fuzzy sets theory was first introduced by Zadeh[1] as a mathematical tool to handle the uncertainty and vagueness inherent in the real life environment. Computer programs based on fuzzy mathematics and neural network have been used to control a variety of industrial processes, schedule trains, and select operating conditions for washing machines, etc. Similar techniques have also been applied to control food processes[2,3,4,5,6].

The decision-making power that has made fuzzy mathematics valuable for process control can also be of value in process development. Zhang and Litchfield[7] demonstrated its application in the development and comparison of products where both objective and subjective factors were included. The power of fuzzy mathematics lies in its ability to (a) utilise inexact or inconsistent information; (b) consider multiple factors; and (c) evaluate both subjective and objective information. In food process development the most important quality attributes of the products must be identified and assessed. Food quality attributes such as texture, colour, flavour and juiciness and overall acceptability are still assessed mainly by sensory analysis. In practice, there are always large variations and uncertainty associated with this kind of analysis. It can also be a very difficult task to identify the single major contributing quality attribute to the overall quality because information from the analysis may be ambiguous and/or conflicting. Zhang and Litchfield[7] claimed fuzzy mathematics to be a powerful tool to handle these kinds of problem. The objective of this study was to apply a comprehensive fuzzy evaluation model, developed by Chen et al.[8], in process development for *sous vide* cook-chill foods.

2 Materials and method

2.1 Materials

Three food samples, carrot, chicken and salmon, were selected for this study. All raw materials were purchased fresh, on the day required for production, from Morrisons supermarket. After preparation the average portion weights for carrot, chicken and salmon were 69±3 g, 140±3 g and 232±3 g respectively. Prepared materials were then packaged in 140 mm × 200 mm Cryovac NOV 710 pouches (W. R. Grace Ltd., London, UK). Thermocouples were inserted through pouches into the geometric centre of the food and sealed with the use of Ellab GTK-21009-C00 polyoxymethylen packing glands (Ellab Ltd., King's Lynn, UK) and waterproof silicone sealant (Henkel Home Improvement and Adhesive Products, Winsford, UK). The pouches were finally vacuum sealed in a Gastrovac A300/42 packer (Multivac UK Ltd., Swindon, UK). After each run pouches were inspected for vacuum integrity and location of thermocouples.

2.2 Heat treatment

Pouches were heated in a Rational CC6 six-grid combination oven (Rational, London, UK) on vario-steaming mode and rapidly chilled to 3°C in a WCC-100 conduction chiller (Williams Refrigeration, King's Lynn, UK). Process conditions and their assigned labels are shown in Table 1.

Table 1 Food core temperatures and holding times for heat treatment of *sous vide* cook-chill pouches[9]

Temperature (°C)	Holding time (min)	Label
80	26	P1
85	11	P2
90	4.5	P3
95	2	P4

Temperatures were measured using T-type Copper/Constantan mineral insulated thermocouple assemblies covered with a stainless steel sheath (Labfacility Ltd., Teddington, UK) and recorded on a Squirrel 1205 data logger (Grant Instruments Ltd., Cambridge, UK) at time intervals of 0.5 min.

2.3 Sensory analysis

The panel members were carefully selected on the basis of their willingness, availability, general health, i.e., colour blindness, and personal habits, i.e. smoking, eating habits and food likes and dislikes[10]. 12 panel members were employed for each food and duplicate tests were conducted for each process condition. An eight-point, bipolar graphic scale was used to record the taste panels' opinions[11]. Four individual quality attributes: colour, texture, juiciness and flavour, and also overall product acceptability were assessed.

3 Fuzzy model

An evaluation that includes all the key factors is considered comprehensive. If one or more of the factors is ambiguous, the evaluation is fuzzy. A comprehensive fuzzy evaluation model developed by Chen et al. (1983) consists of three key elements: a factor set **U**, an evaluation set **V**, and a fuzzy transformation Γ. The factor set **U** includes all of the key attributes, both objective and subjective, and the evaluation set **V** includes a range of possible values for each factor. Each factor can be adjusted numerically in the weight set **X**, which is a subset of the factor set **U**.

The factor set is transformed onto the evaluation set. This transformation means that for any element of the matrix **U** there corresponds an element in the matrix **V**, and the elements are related by a "fuzzy membership". These fuzzy memberships can be thought of as the "extent or likelihood of belonging", and can be represented in a matrix form $R_{m \times n}$, where there are m investigating factors in the factor set **U** and n rankings in the evaluation set **V**. Therefore, $R_{m \times n}$ is a mathematical representation of the transformation Γ from **U** to **V**. This set of three matrices (**U**, **V**, $R_{m \times n}$) forms the model for a comprehensive fuzzy evaluation.

The rankings are determined by the maximum-minimum composition of X_m and $R_{m \times n}$ and are reported in a ranking subset **Y**:

$$Y = X_m \circ R_{m \times n} \qquad (1)$$

where the symbol o represents the operation of determining the maximum among minima. Therefore, the ranks y_j in the ranking subset **Y** are determined as:

$$y_j = \bigcup_{i=1}^{m}(x_i \cap r_{ij}) = \max_m \left\{ \min_n (x_i, r_{ij}) \right\} \qquad (2)$$

where x_i represents the weight of an investigated factor in the factor set, and r_{ij} represents the fuzzy membership of an investigated factor i contributing to the quality rank j. The symbols $\cup$ and $\cap$ represent fuzzy operations of determining the maximum value and the minimum value, respectively.

The crucial step in building a fuzzy comprehensive evaluation model is determining the fuzzy memberships. For subjective factors the fuzzy membership μ is calculated by simple ratio:

$$\mu = \frac{number\ of\ u_o \in A}{total\ number\ of\ u_o} \qquad (3)$$

where μ represents the fuzzy membership of the u_o tests contributing to subset **A**. The notation $u_o \in A$ means that u_o is a member of **A**.

4 Results and discussion

4.1 Comprehensive fuzzy model

The numbers of votes for each quality rating were converted into fuzzy memberships using eqn (3). The weight for each quality rating was assigned subjectively. The membership for each quality attribute μ_q was calculated using eqn (4).

$$\mu_q = \sum_{i=1}^{8} w_i \cdot \mu_i \qquad (4)$$

where μ_i represents the fuzzy membership of quality rating i, and w_i represents the weight of quality rating i in a quality attribute. In this study a total of eight quality ratings were applied.

To perform an overall quality analysis the fuzzy memberships for each quality attribute were converted into fuzzy memberships for overall quality by normalization, i.e. each membership was divided by the sum of the fuzzy memberships for a process condition. The normalized fuzzy membership matrix $R_{4\times4}$ used in this study is shown in eqn (5).

$$R_{4\times 4} = \begin{bmatrix} r_{1\times 1} & r_{1\times 2} & r_{1\times 3} & r_{1\times 4} \\ r_{2\times 1} & r_{2\times 2} & r_{2\times 3} & r_{2\times 4} \\ r_{3\times 1} & r_{3\times 2} & r_{3\times 3} & r_{3\times 4} \\ r_{4\times 1} & r_{4\times 2} & r_{4\times 3} & r_{4\times 4} \end{bmatrix} \begin{matrix} Colour \\ Texture \\ Juiciness \\ Flavour \end{matrix} \qquad (5)$$
$$\quad\ \ P1\ \ \ \ P2\ \ \ \ P3\ \ \ \ P4$$

Weights for the contribution of quality attributes to overall quality are based on consumers' or experts' opinions. The sum of weights should be 1.0 in all cases (eqn (6)).

$$X_4 = \left\{ \frac{x_1}{Colour} \quad \frac{x_2}{Texture} \quad \frac{x_3}{Juiciness} \quad \frac{x_4}{Flavour} \right\} \quad (6)$$

By substituting eqns (5) and (6) into the comprehensive fuzzy model (eqn (1)), the ranking subset **Y** can be obtained (eqn (7)). The elements in the subset **Y** are the likelihood of each process condition, P1 through P4, yielding the best overall quality. The larger the membership, the higher the likelihood.

$$Y_4 = \{\mu_{P1} \quad \mu_{P2} \quad \mu_{P3} \quad \mu_{P4}\} \quad (7)$$

where μ_{P1} through μ_{P4} represent the overall quality memberships of P1 to P4.

As an example, the taste panels' votes, memberships and weights for salmon colour are summarised in Table 2. Memberships are given in brackets.

Table 2 Taste panel response, weights and fuzzy memberships for salmon colour

Rating	P1	P2	P3	P4	w_i
Extremely pale			1 (0.043)		0.1
Very pale	1 (0.042)	4 (0.174)	2 (0.087)	6 (0.250)	0.3
Moderately pale	5 (0.208)	9 (0.391)	4 (0.174)	4 (0.167)	0.7
Slightly pale	6 (0.250)	5 (0.217)	4 (0.174)	6 (0.250)	0.9
Slightly deep	5 (0.208)	2 (0.087)	7 (0.304)	3 (0.125)	0.9
Moderately deep	4 (0.167)	3 (0.130)	3 (0.130)	4 (0.167)	0.7
	3 (0.125)		2 (0.087)	1 (0.042)	0.3
Very deep					0.1
Extremely deep					
Total votes for colour	24 (0.725)	23 (0.691)	23 (0.702)	24 (0.659)	

For salmon the memberships for the four quality attributes and four process conditions are given in eqn (8):

$$R_{4\times 4} = \begin{bmatrix} 0.725 & 0.691 & 0.702 & 0.659 \\ 0.583 & 0.628 & 0.608 & 0.687 \\ 0.534 & 0.508 & 0.491 & 0.475 \\ 0.555 & 0.661 & 0.477 & 0.563 \end{bmatrix} \begin{matrix} Colour \\ Flakiness \\ Juiciness \\ Flavour \end{matrix} \quad (8)$$

$$\phantom{R_{4\times 4} = }\ P1 \quad\ P2 \quad\ P3 \quad\ P4$$

The normalized memberships are given in eqn (9) and weight set $X_{1\times 4}$ is in eqn (10)

$$Rc_{4\times 4} = \begin{bmatrix} 0.302 & 0.279 & 0.308 & 0.276 \\ 0.243 & 0.249 & 0.267 & 0.288 \\ 0.223 & 0.205 & 0.216 & 0.199 \\ 0.232 & 0.267 & 0.209 & 0.236 \end{bmatrix} \begin{matrix} Colour \\ Flakiness \\ Juiciness \\ Flavour \end{matrix} \quad (9)$$
$$\phantom{Rc_{4\times 4} = }\ P1 \quad\ P2 \quad P3 \quad\ P4$$

$$X_4 = \left\{ \frac{0.210}{Colour}\ \frac{0.250}{Flakiness}\ \frac{0.270}{Juiciness}\ \frac{0.270}{Flavour} \right\} \quad (10)$$

The results from the comprehensive fuzzy evaluation model are shown in eqns (11) and (12).

$$X_4 \circ Rc_{4\times 4} = \begin{bmatrix} 0.210 & 0.210 & 0.210 & 0.210 \\ 0.243 & 0.249 & 0.250 & 0.250 \\ 0.223 & 0.205 & 0.216 & 0.199 \\ 0.232 & 0.267 & 0.209 & 0.236 \end{bmatrix} \begin{matrix} Colour \\ Flakiness \\ Juiciness \\ Flavour \end{matrix} \quad (11)$$
$$\phantom{X_4 \circ Rc_{4\times 4} = }\ P1 \quad\ P2 \quad P3 \quad\ P4$$

$$Y_4 = \left\{ \frac{0.243}{\substack{Flakiness \\ P1}}\ \frac{0.267}{\substack{Flavour \\ P2}}\ \frac{0.250}{\substack{Flakiness \\ P3}}\ \frac{0.250}{\substack{Flakiness \\ P4}} \right\} \quad (12)$$

From eqn (12) it is found that P2(0.267) has the highest likelihood of yielding the best overall product quality, and the major contributing attribute is flavour. P3 and P4(0.250) and P1(0.243) are ranked sequentially lower because of their flakiness. Eqn (13) shows the result from the taste panels' assessment of overall acceptability. The membership has also been normalized, i.e. the membership for a process condition was divided by the sum of memberships for the four process conditions. It is found that P2(0.265) has the highest value, followed by P4(0.250), P1(0.249) and P3(0.236) sequentially. The values for P1 and P4 are very similar. These two results agree quite well.

$$Overall\ acceptability = \left\{ \frac{0.249}{P1}\ \frac{0.265}{P2}\ \frac{0.236}{P3}\ \frac{0.250}{P4} \right\} \quad (13)$$

For carrot, P4(0.265) has the highest likelihood of yielding the best overall product quality, and the major contributing attribute is crunchiness. P3(0.253) and P1 and P2(0.250) are ranked sequentially lower because of their crunchiness and juiciness (eqn (14)). For overall acceptability the taste panels gave the highest rank to P2(0.266) and the lowest one to P4(0.227) (eqn (15)). The reasons for the disagreement between these two results may be: (i) the assigned weights in subset **X** do not match the taste panel response, and (ii) for the overall quality more individual quality attributes may need to be included, and (iii) the fuzzy model has a drawback.

$$Y_4 = \left\{ \frac{0.250}{\underset{P1}{Juiciness}} \quad \frac{0.250}{\underset{P2}{Juiciness}} \quad \frac{0.253}{\underset{P3}{Crunchiness}} \quad \frac{0.265}{\underset{P4}{Crunchiness}} \right\} \quad (14)$$

$$Overall\ acceptability = \left\{ \frac{0.241}{P1} \quad \frac{0.266}{P2} \quad \frac{0.266}{P3} \quad \frac{0.227}{P4} \right\} \quad (15)$$

For chicken, P1 and P4(0.280) have the highest likelihood of yielding the best overall product quality, and the major contributing attribute is chewiness. P2(0.259) and P3(0.257) are ranked sequentially lower because of their chewiness (eqn (16)). Again, this result differs from that derived from the taste panel assessment of overall acceptability (eqn (17)). Similar reasons for the disagreement may apply.

$$Y_4 = \left\{ \frac{0.280}{\underset{P1}{Chewiness}} \quad \frac{0.259}{\underset{P2}{Chewiness}} \quad \frac{0.257}{\underset{P3}{Chewiness}} \quad \frac{0.280}{\underset{P4}{Chewiness}} \right\} \quad (16)$$

$$Overall\ acceptability = \left\{ \frac{0.239}{P1} \quad \frac{0.236}{P2} \quad \frac{0.274}{P3} \quad \frac{0.252}{P4} \right\} \quad (17)$$

4.2 Modified model

In the comprehensive fuzzy evaluation model the subset **Y** is obtained by maximizing-minimizing procedure. Based on the principle that the larger the membership the higher the quality, this model may be modified. In the modified model the best process condition is determined from the overall quality set, which is obtained by eqn (18). In this equation the matrix Rr is the normalized matrix R based on its rows, i.e. each quality attribute. The subset **Y** is obtained by maximizing procedure, which is used to identify the major contributing attribute for the best process condition and shown in eqn (19). The elements in matrix **Rc** are also normalized as in the original model.

$$Overall\ quality = X_m \times Rr_{m \times n} \quad (18)$$

$$y_j = \bigcup_{i=1}^{m} (rc_{ij}/x_i) = \max_m (rc_{ij}/x_i) \quad (19)$$

By applying this model the results for salmon are given as follows:

$$Rr_{4\times 4} = \begin{bmatrix} 0.261 & 0.249 & 0.253 & 0.237 \\ 0.234 & 0.248 & 0.244 & 0.275 \\ 0.266 & 0.253 & 0.245 & 0.237 \\ 0.246 & 0.293 & 0.211 & 0.250 \end{bmatrix} \begin{matrix} Colour \\ Flakiness \\ Juiciness \\ Flavour \end{matrix} \qquad (20)$$

$$P1 \quad\ P2 \quad\ P3 \quad\ P4$$

$$Overall\ quality = \left\{ \frac{0.252}{P1}\quad \frac{0.262}{P2}\quad \frac{0.237}{P3}\quad \frac{0.250}{P4} \right\} \qquad (21)$$

$$Ry_{4\times 4} = \begin{bmatrix} 0.302/0.210 & 0.279/0.210 & 0.308/0.210 & 0.276/0.210 \\ 0.243/0.250 & 0.249/0.250 & 0.267/0.250 & 0.288/0.250 \\ 0.223/0.270 & 0.205/0.270 & 0.216/0.270 & 0.199/0.270 \\ 0.232/0.270 & 0.267/0.270 & 0.209/0.270 & 0.236/0.270 \end{bmatrix} \begin{matrix} Colour \\ Flakiness \\ Juiciness \\ Flavour \end{matrix}$$

$$= \begin{bmatrix} 1.44 & 1.33 & 1.47 & 1.31 \\ 0.97 & 1.00 & 1.07 & 1.15 \\ 0.83 & 0.76 & 0.80 & 0.74 \\ 0.86 & 0.99 & 0.77 & 0.87 \end{bmatrix} \begin{matrix} Colour \\ Flakiness \\ Juiciness \\ Flavour \end{matrix} \qquad (22)$$

$$P1 \quad P2 \quad P3 \quad P4$$

By comparing the elements in each column in eqn (22), subset **Y** can be obtained.

$$Y_4 = \left\{ \frac{0.302}{\substack{Colour \\ P1}}\quad \frac{0.279}{\substack{Colour \\ P2}}\quad \frac{0.308}{\substack{Colour \\ P3}}\quad \frac{0.276}{\substack{Colour \\ P4}} \right\} \qquad (23)$$

From eqn (21) the best process condition P2(0.262) can be identified, followed by P1(0.252) and P4(0.250), and P3(0.237) sequentially. This agrees better with the result from the taste panel (eqn (13)). Eqn (23) shows that for P2 the major contributing attribute for the overall quality is colour. From eqn (22) it can also be seen that the major contributing attribute for the lowest overall quality produced by P3 is flavour. It is also noticed that subset **Y** is different to that derived from the original model.

The memberships in eqn (20) can also be used to compare the different process conditions based on an individual quality attribute. Although P2 is identified as the best process condition for overall quality P1 produces the best colour (0.261) and juiciness (0.266), and P4 the best flakiness (0.275).

For carrot, the best overall quality is from P2(0.253), followed by P3 and P4(0.251) and P1(0.244) sequentially (eqn (24)). Compared to the original model this result matches that from taste panel (eqn (14)) better. From eqn (25) the major contributing attribute for P2 is colour. Similarly, from matrix **Ry** (not shown in the text) the major contributing attribute for the lowest overall quality from P1 can be identified as flavour.

$$Overall\ quality = \left\{ \frac{0.244}{P1} \quad \frac{0.253}{P2} \quad \frac{0.251}{P3} \quad \frac{0.251}{P4} \right\} \quad (24)$$

$$Y_4 = \left\{ \frac{0.271}{\begin{array}{c}Colour\\P1\end{array}} \quad \frac{0.273}{\begin{array}{c}Colour\\P2\end{array}} \quad \frac{0.284}{\begin{array}{c}Juiciness\\P3\end{array}} \quad \frac{0.272}{\begin{array}{c}Colour\\P4\end{array}} \right\} \quad (25)$$

For chicken, it is found that for overall quality the best process conditions are P1(0.267) and P3(0.266), followed by P2(0.242) and P4(0.225) sequentially (eqn (26)). The major contributing attribute for P1 and P3 is flavour. Similarly, from matrix **Ry** (not shown in the text) the major contributing attribute for the lowest overall quality from P4 is juiciness. Compared to the result from the original model this result, again, does not match that from the taste panel well but it is improved.

$$Overall\ quality = \left\{ \frac{0.267}{P1} \quad \frac{0.242}{P2} \quad \frac{0.266}{P3} \quad \frac{0.225}{P4} \right\} \quad (26)$$

$$Y_4 = \left\{ \frac{0.279}{\begin{array}{c}Flavour\\P1\end{array}} \quad \frac{0.280}{\begin{array}{c}Flavour\\P2\end{array}} \quad \frac{0.291}{\begin{array}{c}Flavour\\P3\end{array}} \quad \frac{0.299}{\begin{array}{c}Flavour\\P4\end{array}} \right\} \quad (27)$$

5 Conclusion

Fuzzy mathematics is an easy and potentially useful tool for dealing with the uncertainty and multiple factors associated with the assessment of the organoleptic quality of food. The agreement between the results from the comprehensive fuzzy evaluation model and the taste panels for the best process condition varies. The reasons for this may be from both sensory analysis and the model. The original model was modified and the agreement has been improved. To apply this technique in food process development experimental validation and more research are required.

References

1. Zadeh, L.A. Fuzzy sets, *Information and Control*, 1965, **8**, 338-353.

2. Zhang, Q., Litchfield, J.B. & Bentsman, J. Fuzzy predictive control system for corn quality control during drying, *Proceedings of the 1990 Conference of Food Processing Automation, Am. Soc. Agric. Eng.*, St. Joseph, MI, 1990.

3. Brown, R.B., Davidson, V.J., Hayward, G.L., & Whitnell, G.P. Fuzzy process control for food processes, *Proceedings of the 1990 Conference of Food Processing Automation, Am. Soc. Agric. Eng.*, St, Joseph. MI, 1990.

4. Linko, P. Uncertainties, fuzzy reasoning, and expert system in bioengineering, *Annals- New York Academy of Science*, 1988, **542**, 83-101.

5. Unklesbay, K., Keller, J., Unklesbay N. & Subhangkasen, D. Determination of doneness of beef steaks using fuzzy pattern recognition, *Journal of Food Engineering*, 1989, **8**, 79-90.

6. Dohnal, M., Vystrcil, J., Dohnalova, J., Marecek, K., Kvapilik, M. & Bures, P. Fuzzy food engineering, *Journal of Food Engineering*, 1993, **19**, 171-201.

7. Zhang, Q. & Litchfield, J.B. Applying fuzzy mathematics to product development and comparison, *Food Technology*, 1991, **45**, 108-115.

8. Chen, Y., Liu, Y. & Wang, P. Models of multifactorial evaluation, *Fuzzy Mathematics*, 1983, **1**, 61-68.

9. Sheard, M. & Church, I. *Sous Vide Cook-Chill*, 19-21, Leeds, Leeds Polytechnic, 1992.

10. Young, H. *Factors affecting the quality and shelf life of cooked chilled foods with special reference to full meal vending*, Ph.D Thesis, Dorset Institute of Higher Education, 1986.

11. Reffensperger, E.L., Peryam, D.R., & Wood, K.R. Development of a scale for grading toughness-tenderness in beef, *Food Technology*, **10**, 627-630. (1956)

Ensuring food authenticity in the food and catering industry

P. Creed

Department of Food and Hospitality Management, Bournemouth University, Fern Barrow, Poole, Dorset BH12 5BB, UK

Abstract

Food adulteration has been the subject of the earliest food laws in the United Kingdom in order to prevent poisoning as well as fraud. Today, although ill effects from food are more likely to be due to micro-organisms combined with poor manufacturing and handling procedures, consumers are much more aware of their legal rights to obtain goods of the correct description. This awareness is also much more common now with caterers and food manufacturers who are introducing quality management systems which specify agreed contents and compositions for raw food materials. This paper discusses the authenticity problems which food manufacturers, retailers and caterers face, the steps which are currently used to quantify the scope and importance of these problems, the methods used to detect them for a range of foods and the practical steps which can be used in the industry to give confidence to themselves and the consumer.

1 Introduction

In recent years, food safety has often been the focus of attention for the food and catering industries because of the news media's interest in the subject. Due to the resulting pressure from consumers and consumer groups, action by government in the form of new rules and regulations have been introduced to complement existing general legislation on public health. Perhaps as a result of this, less attention has been placed on overcoming problems of food authenticity which before the causes of food poisoning were understood was the main focus of food legislation: in 1860, The Food and Drink Act was introduced to protect the consumer against the dishonesty and harm caused by adulteration of foods with bulking agents or poisonous ingredients. At this time, pepper was being bulked out with 'mustard husks, pea flour and juniper berries': tea leaves with 'dried and curled ash leaves': table wines flavoured with 'bitter almonds containing prussic acid': boiled sweets 'coloured with poisonous salts of copper and lead' and in 1851 'most bread contained alum and coffee was diluted with chicory, acorns or mangelwurzel (a kind of beet)' (Tannahill [3]).

Now that most parts of the food industry are introducing quality management systems, methods for making certain that the raw food materials entering the production systems are authentic, is assuming a much higher level of importance. This paper aims to discuss the problems which food producers, retailers and caterers face, the steps which are currently used to quantify the scope and importance of these problems, the methods used to detect them and what practical steps can be used in the food industry.

2 Problems of food authenticity

Many problems caused by foods not being what they appear to be, in other words, not authentic, can lead to prosecution under legislation, rejection of products being supplied to retailers or caterers or loss of custom through consumers feeling cheated. As manufacturers, caterers, retailers and consumers are prepared to pay a premium for foods perceived as having a higher level of quality, opportunities will arise for unscrupulous parties to supply inauthentic food. They can, for example, adulterate raw food materials with lower cost components which will not be detected easily or even manufacture counterfeit products, both activities which will lead to a loss of confidence by the consumer in the whole chain of food production. This can affect all companies in that sector whether operating responsibly or fraudulently. It should not be forgotten that the act of eating is a human activity which is so basic to survival that it implies a trust by the consumer that the manufacturer of that food has done everything possible to ensure that it is wholesome and authentic.

2.1 Counterfeiting

This has usually been associated with expensive non-food items such as watches and perfumes but recently in the United States, the FDA (Food and Drug Authority) has investigated the production of counterfeit infant food products and other non-food retail products like shampoo. Consumption of the low grade infant food led to illness in some cases. The counterfeiting involved the highly organised purchase of damaged and out-of-date brand-name goods intended for destruction and illegal acquisition of export goods or the production of the counterfeit goods which were all sold as genuine at suspiciously low prices to retailers who are now being sued for the consequences of their lack of care (Unger [1]).

2.2 Selling stale foods

Selling stale foods as fresh is another problem. Recent allegations in the United Kingdom have concerned milk past its sell-by-date which has been returned by retailers, being recycled by adding it to fresh milk (Nelson & Calvert [2]). In many areas of high unemployment, out-of-date goods are often available at low prices. Vacuum-packed meat and meat products can also be repackaged after cleaning up to remove signs of spoilage.

2.3 Bulking out

Bulking out a high value food commodity with a cheaper component in order to raise profits is another problem: premium quality vegetable oils such as maize, sunflower, groundnut and palm oils can be diluted with cheap rapeseed oil

(MAFF [4]): extra virgin oil can contain lower quality second pressings: instant coffee with bean husks and skins (MAFF [5]): dried durum wheat pasta can be extended with common wheat (MAFF [6]): fruit juices can be diluted with ingredients not derived from the flesh of the fruit but from the skins or pulp: genuine vanilla extracted from pods can be laced with the cheaper synthetic version, vanillin (Lamprecht et al [17], Prince & Gunson [24]): jams, fruit purées and fruit pie fillings can be bulked with cheaper fruit and not declaring it: and sheep or goat dairy products can contain undeclared cow's milk: and meat products may be extended by incorporating mechanically recovered meat - a paste obtained by squeezing protein from bones using high pressure. Water is a very cheap ingredient and can be pumped into cured meat products or incorporated into frozen foods to increase the weight and mislead the consumer by not declaring the true extent of the addition.

2.4 Substitution and misdescription

The substitution of high quality goods with cheaper versions can deceive and cheat the consumer through misdescribing the type of food or its country of origin: wines can easily be re-labelled and cheap spirits can be put into the bottles of expensive brands: mineral water can be artificially carbonated but declared as 'naturally carbonated': premium price rice (e.g.. Basmati, American long-grain) can have cheaper rice added: breaded fish products can contain cheap fish instead of the type described on the pack (MAFF [7]): and low quality frozen meats can be repackaged in genuine cartons from reputable meat packers which once contained higher quality meats. Abusing the consumer's perception of freshness by labelling and selling previously frozen and thawed meat as fresh chilled is yet another way of misdescribing food in order to make more profits. Consumer resistance to the irradiation of food also requires methods to detect undeclared treatment of a wide range of food products.

3 Overcoming the problems

The problems mentioned above are possible abuses and should not be taken as a view that the food industry is operating fraudulently but as all foods can be subject to fraud, the industry should be aware of these possibilities. In the United Kingdom, trading standards officers working with the Public Analyst's laboratories form the means of enforcing relevant legislation in the food authenticity area and with Environmental Health Officers are the deterrent to unscrupulous operators in the food industry. However, they can only function properly if they have enough staff and access to scientifically accurate techniques to detect adulteration in the cases under suspicion. Many sectors of the food industry have taken the responsibility of policing themselves by writing codes of practice for quality assurance systems which include food authenticity as one aspect of the system. Some examples are for fruit juice (BSDA [15]), rice (FAIM [10]) and coffee (Prodolliet [8]). These systems are often linked to international regulations from bodies such as the Codex Alimentarius Commission which, for example include checks on the authenticity of edible fats and oils (Bell & Gillatt [16]). European regulations and directives will also have a great effect.

As part of their brief to oversee the quality of the production processes for agricultural and food products, the Ministry of Agriculture, Fisheries and Food

(MAFF) set up a working party in 1992 to develop a surveillance system to clarify whether a particular food authenticity problem warranted further investigation. In order to provide a balanced view, MAFF food scientists are working with representatives from the food, retail and catering industries and from trading standards and consumer organisations.

This system is based on a three stage process: assessment of the priority of the authenticity problem: evaluation and development of reliable and accurate methods to detect the particular adulteration and developing a surveillance exercise in order to get a 'snapshot' of the current extent of the problem. The protocol developed to assess the priority takes account of many factors such as market factors (market value of commodity, its volume, number of purchases and market growth rate), consumer factors (perception of the particular food and susceptibility to the authenticity problem) and any evidence of adulteration. These factors are weighted according to their importance and to how many sectors of the food industry (manufacturing, catering and retail) could be affected. The resulting score provides an index of importance for progress to the next stage of evaluating methods for the detection of the particular adulteration. This can provide an impetus for funding research in developing and combining the appropriate analytical techniques to detect the particular adulteration. Once a suitably accurate and reliable method or combination of methods has been decided, laboratories are asked to tender for analysing the samples. After assessing the laboratories' reliability and accuracy in this work, the surveillance exercise can begin. Trading standards officers from regions representative of the whole country collect samples of all brands or materials supplied to manufacturers or sold to caterers and retailers and send them off for analysis to the chosen laboratories. The levels at which adulteration is considered to be deliberate takes into account that in food processing, it is often impossible to ensure that storage vessels do not carry over products from one batch to another, for example when changing from one oil type to another. The variability inherent to any analysis is also considered.

The results of the surveillance exercise thus provide a 'snapshot' of the extent of the authenticity problem over the United Kingdom. So far, reports have been published by MAFF on the authenticity of fish and fish products [7], instant coffee [6], pasta [5] and vegetable oils [4] and suggest that most authenticity problems come from a small number of producers who provide only a small part of the total volume of the particular food sold. Other areas of interest relate to the authenticity of thawed meat sold as fresh chilled, the authenticity of spirits and sugaring of wine.

At the European level, an EC research programme FAIM (Food Authenticity: Issues and Methodologies) intends to provide a forum for discussion of food authenticity issues, to collate information and to encourage collaboration between food scientists on methods of analysis (Flair-Flow Europe [9]).

4 Methods of detection

The scientific literature on analytical methods to detect adulteration in foods is vast so a few representative examples of the methods for detection of adulterants in a range of foods are given below. These methods include techniques based on DNA, immunochemistry, electrophoresis (including microgel and capillary methods), chromatography (including HPLC - High Performance Liquid Chromatography), NMR (Nuclear Magnetic Resonance), mass spectrometry, isotopic methods, NIR (Near Infra Red spectroscopy), FTIR (Fourier Transform Infra Red spectroscopy) and dielectric spectroscopy. These techniques are applicable to foods to varying degrees and are at varying stages of development for practical use.

4.1 Methods for orange juice

Orange juice is produced mainly by crushing the fruit in the country of production, evaporating a large proportion of the water to reduce the volume to be transported, freezing the concentrated juice and then on arrival in the country of consumption, reconstituting by adding water and packing into cartons or bottles. The scope for adulteration includes addition of: unauthorised sugars: extra water: pulpwash (water washed through the orange pulp after the initial extraction of the juice) and/or peel derived products: cheaper juices as extenders: mixtures of special amino acids and organic acids to mask poor quality starting materials: and colouring additives (BSDA [15], FAIM [18]).

Methods for detecting these problems include HPLC, NIR spectroscopy, isotopic and DNA methods. In HPLC, the relevant components are separated out by passing them through columns to be identified through peaks on the detection device by their residence time and quantified by the peak height. For NIR spectroscopy, interaction between near infra red radiation and the chemical bonds found in the organic compounds always present in foods produce absorbances to give spectra which with mathematical techniques (e.g. Fourier Transform) will allow quantification of the relevant compound (Scotter [39]). Isotopic methods make use of the ratio between isotopes of oxygen, carbon or hydrogen to detect additions of artificially-produced sugars, alcohol etc. and detect these ratios through mass spectrometry - separation due to differing atomic weights. DNA methods rely on the interaction between a specific DNA sequence and the target complementary sequence if present in the sample. Various techniques can then be used to amplify any bound DNA for species identification.

Ooghe et al. [25] have used HPLC to detect the pattern of polymethoxylated flavones in orange juice so that the addition of tangerine juice, for example, could easily be detected. Using a similar technique, small amounts of grapefruit, bitter orange, and/or bergamot added to pure orange juice were detected by the presence of flavanoid compounds not normally present in pure orange juice (Ooghe et al. [26]). In contrast, Marini & Balestrieri [33] using HPLC to obtain data on acidity, fructose, glucose and sucrose content, amino acids, flavanone glycosides, ascorbic acid, furfural, hydroxymethylfurfural, and beta-carotene concentration of test samples for comparison with authentic samples, found that deviations from the range of variation caused by climatic, growing and processing factors did not necessarily indicate that orange juices were not authentic. Brause [32] with a matrix approach using a combination of HPLC profiling, SNIF-NMR (Specific Natural Isotope Fractionation - Nuclear Magnetic Resonance), pulsed amperometric detection and other methods provided a reliable assessment of the presence of pulpwash and added sugars and hence the authenticity of orange juice.

NIR spectroscopy can be used for rapid screening of orange juice for authenticity (Scotter & Legrand [27], Twomey et al. [28]). One study using NIR data discriminated between three sources of juice using the statistical procedures of principal components analysis of 92 calibration samples followed by canonical variates analysis using up to 25 principal components (Evans et al. [29]). In another study using NIR on seven different types of fruit juice (orange, apple, red grape, white grape, black cherry, blackcurrant, grapefruit), countries of origin for orange juice samples were 98% correctly predicted: discrimination of orange juice samples derived using different processes gave an 81% correct prediction: NIR spectra discriminated between the seven fruit juices with 100% accuracy; and 85% of authentic and adulterated samples were correctly identified (Flair-Flow Europe [31]).

DNA fingerprinting also has potential uses for investigating fruit juice authenticity problems (Patel [30]).

4.2 Methods for honey

Honeys from particular geographical regions or collected from particular flowers can attract a premium price so there is scope to misdescribe the source of honey and also to add high-fructose corn syrups to bulk out the honey (FAIM [18]).

Methods used include chromatography, isotopic methods and NIR spectroscopy.

Using anion-exchange liquid chromatography with pulsed amperometric detection (LC-PAD), Swallow & Low [34] studied the addition of invert syrups (beet or cane) and high-fructose corn syrup to honey. Fingerprint oligosaccharides were shown to be present in these sweeteners, which were either not detectable or present only in low concentrations in pure honey. The technique will have more potential for authenticity testing when data on pure honey samples has been built up. The method was successfully used to quantitate oligosaccharides in 4 Canadian uniforal honeys from Canola, Alfalfa, Alsike and Trefoil (Swallow & Low [36]). Zalewski [35] used principal component analysis (PCA) of large data sets derived from analytical methods for the authentication of honey samples by determining sample position and relationship to the groups of rape, heather or honeydew honey.

Doner [37] used stable carbon isotope ratio analysis to detect adulteration of honey with high-fructose corn syrup.

Robertson et al. [38] reported on many uses of NIR including determining the purity and authenticity of origin of honey.

4.3 Methods for fish and fish products

As fish stocks decline, there is scope for manufacturers to substitute more expensive fish species in their products with cheaper fish species. Many products are breaded or reformed so the physical features of different species cannot be used for identification.

Three main methods available for fish species identification are immunological techniques, DNA technology and electrophoretic and isoelectric focusing techniques (MAFF [7]) but the first is considered the most realistic for laboratory use (Mackie [23]).

Immunological techniques make use of the antibody-antigen interaction where foreign proteins stimulate the production of antibodies to remove the intruding protein. This is exploited in the sandwich ELISA (Enzyme linked immunoabsorbent assay) technique where a coloured reaction is produced if specific proteins are present. This test can be used for many meat and fish proteins: to verify the identity of sardine amongst mackerel, herring and sild (Taylor & Jones [19]): to detect adulteration of reformed crustacean products such as breaded scampi by white fish (Taylor & Jones [20]): and to detect substitution of canned sardine with other species, adulteration of canned tuna with bonito, and adulteration of reformed crustacean meat with white fish (Taylor et al. [22]).

Electrophoretic and isoelectric focusing techniques work by separating out the fish flesh proteins in gels to provide a pattern or profile which is unique to a particular species. A development of this technique - SDS-PAGE (Sodium dodecyl sulphate polyacrylamide gel electrophoresis) which works with raw, cooked or heat treated fish proteins, has been used for the detection of the adulteration of raw, reformed breaded scampi with Pacific shrimp or tropical shrimp (Craig et al. [21]).

5 Practical methods for the food and catering industry

Buyers and quality assurance staff in the food and catering industries can often feel that the complications of ensuring the authenticity of foods are beyond them, especially those working in the smaller food operations. The multinational food companies with their own research laboratories will have their own comprehensive range of resources and methods for monitoring the authenticity of food materials but these procedures are likely to be confidential for commercial reasons or to keep them secret from those involved in food adulteration. Unfortunately this keeps much knowledge out of the public domain and away from those in the smaller food operations who may be most susceptible to buying inauthentic raw food materials.

Any quality management system has to incorporate some form of procedure to show 'due diligence' in checking that the raw food materials are within the tolerances set by their own specifications agreed with their suppliers and that the goods they manufacture will be accepted by their own customers further along the chain. Specified standards such as food temperature, the bacterial load of meats, integrity of wrapping etc. can be checked reasonably easily but adulterated vegetable oil or supposedly fresh chilled meat which has been previously frozen then thawed are more or less impossible to check without the expense and delay of sending samples to a specialised analytical laboratory.

Fortunately due to the research into new analytical techniques mentioned above and the speed of developments in electronics and computer technology which give high speed data processing for rapid statistical or mathematical analysis, many new devices are being developed which will enable the authenticity of raw material to be checked on-line or with minimal inconvenience. Examples of these are test kits which can enable the species of meat animal to be determined (Smith [11]): an electronic nose which can compare an aroma profile with reference data using artificial intelligence (Shiers & Farnell [12], Collins [42]): on-line video scanning of foods such as cereals, peanuts, peas, beans and other granular products, which can reject inauthentic items i.e. not conforming to predetermined size and shape criteria (Peddicord [43]): holograms incorporated into food packaging which ensure the origin of genuine goods (Anon. [13]): and foods fingerprinted with inert chemical markers which are detectable by immunoassay methods to provide traceability (Kevin [14]).

NIR is being developed to work on-line for food processing applications such as moisture, protein and fat/oil analysis of a wide range of food and agricultural products (Scotter [39]). The fructose and glucose content of fermenting liquids can also be measured on-line using FTIR (Bellon et al. [40]). A range of methods for on-line sensing of food quality and hence authenticity parameters has been reviewed by Giese [41] and Skjöldebrand [44].

Many of these methods should eventually be available at an economic price in portable forms which are practicable for everyday use by the food processor, caterer or retailer.

6 Conclusions

Rapid developments in analytical techniques coupled with sophisticated electronic devices will produce equipment for use with practical quality assurance systems that will enable the caterer, retailer and food manufacturer to check their incoming raw food materials. They should then have confidence that the food materials they receive are actually what they assumed them to be and hence worth the price charged for the final products when sold to the consumer.

References

1. Unger, H. The Global Conspiracy, *Food Processing*, 1996, **65**, (1), 23, 26.
2. Nelson, D. & Calvert, J. New dairy churning out recycled milk, *The Observer*, 19th November 1995, 3.
3. Tannahill, R. *Food in History*, Penguin, London, 1988.
4. MAFF (Ministry of Agriculture, Fisheries & Food - Food Safety Directorate), MAFF single seed vegetable oil surveillance exercise, *Food Surveillance Information Sheet No. 77*, November 1995, MAFF, London.
5. MAFF (Ministry of Agriculture, Fisheries & Food - Steering Group on Chemical Aspects of Food Surveillance), *Authenticity of Soluble Coffee, Food Surveillance Paper No. 46*, 1995, HMSO, London.
6. MAFF (Ministry of Agriculture, Fisheries & Food - Steering Group on Chemical Aspects of Food Surveillance), *Dried Durum Wheat Pasta, Food Surveillance Paper No. 47*, 1995, HMSO, London.
7. MAFF (Ministry of Agriculture, Fisheries & Food - Steering Group on Chemical Aspects of Food Surveillance), *Authenticity of Fish and Fish Products, Food Surveillance Paper No. 44*, 1994, HMSO, London.
8. Prodolliet, J. *AFCASOLE (The European Soluble Coffee Manufacturers' Association) Collaborative study on the HPLC determination of free and total carbohydrates in instant coffee*, AFCASOLE, June 19th 1992.
9. Flair-Flow Europe, Food Authenticity, *Flair-Flow-Reports*, 1995, F-FE 159/95.
10. FAIM, GAFTA (Grain and Feed Trade Association) Rice Research Steering Committee, *Food Authenticity News*, (2), November 1995.
11. Smith, D.M. Immunoassays in process control and speciation of meats, *Food Technology*, 1995, **49**, 116-119.
12. Shiers, V.P. & Farnell, P.J. The electronic nose: aroma profiling in the food industry, *Food Technology International Europe*, 1995, 168-171.

13 Anon. Holograms create impact, *Packaging Review South Africa*, 1994, **20** (12) 19, 21.
14 Kevin, K. BIOCOP!, *Food Processing - USA*, 1994, **55** (10) 89, 91-92, 94.
15 BSDA (British Soft Drinks Association), *The Fruit Juice Association Quality Scheme*, 1995.
16 Bell, J.R. & Gillatt, P.N. Standards to ensure the authenticity of edible oils and fats, *Food, Nutrition and Agriculture*, 1994, (11), 29-35.
17 Lamprecht, G., Pichlmayer, F. & Schmid, E.R., Determination of the authenticity of vanilla extracts by stable isotope ratio analysis and component analysis by HPLC, *Journal of Agricultural and Food Chemistry*, 1994, **42**, (8), 1722-1727
18 FAIM, FAIM Commodity Groups, *Food Authenticity News*, (1), March 1995.
19 Taylor, W.J. & Jones, J.L. An immunoassay for verifying the identity of canned sardines, *Food and Agricultural Immunology*, 1992, **4**, 169-175.
20 Taylor, W.J. & Jones, J.L. An immunoassay for distinguishing between crustacean tailmeat and white fish, *Food and Agricultural Immunology*, 1992, **4**, 177-180.
21 Craig, A., Ritchie, A.H. & Mackie, I.M. Determining the authenticity of raw reformed breaded scampi (*Nephrops norvegicus*) by electrophoretic techniques, *Food Chemistry*, 1995, **52**, (4), 451-454.
22 Taylor, W.J., Patel, N.P. & Jones, J.L. Antibody-based methods for assessing seafood authenticity, *Food and Agricultural Immunology*, 1994, **6**, (3), 305-314.
23 Mackie, I.M., Fish speciation. *Food Technology International Europe*, 1994, 177-180.
24 Prince, R.C. & Gunson, D.E. Just Plain Vanilla?, *Trends in Biochemical Sciences*, 1994, **19**, 521.
25 Ooghe, W.C., Ooghe, S.J., Detavernier, C.M. & Huyghebaert, A. Characterization of orange juice (*Citrus sinensis*) by polymethoxylated flavones, *Journal of Agricultural and Food Chemistry*, 1994, **42**, (10), 2191-2195.
26 Ooghe, W.C., Ooghe, S.J., Detavernier, C.M. & Huyghebaert, A. Characterization of orange juice (*Citrus sinensis*) by flavanone glycosides, *Journal of Agricultural and Food Chemistry*, 1994, **42**, (10), 2183-2190.
27 Scotter, C.N.G. & Legrand, A. NIR qualitative analysis - a new philosophy with special reference to rapid NIR screening for fruit juice authenticity, *Food Science & Technology Today*, 1994, **8**, (3), 167-171.
28 Twomey, M., Downey, G. & McNulty, P.B. The potential of NIR spectroscopy for the detection of the adulteration of orange juice, *Journal of the Science of Food and Agriculture*, 1995, **67**, 77-84.
29 Evans, D.G., Scotter, C.N.G., Day, L.Z. & Hall, M.N. Determination of the authenticity of orange juice by discriminant analysis of near infrared spectra. A study of pretreatment and transformation of spectral data, *Journal of Near Infrared Spectroscopy*, 1993, **1**, (1), 33-44
30 Patel, N.P. The use of DNA fingerprinting in food analysis, *Food Technology International Europe*, 1994, 171-174.
31 Flair-Flow Europe. Rapid tests for fruit juice authenticity, *Flair-Flow-Reports*, 1994, F-FE 126/94.
32 Brause, A.R. Detection of juice adulteration, *Journal of the Association of Food and Drug Officials*, 1993, **57**, (4), 6-25.

33 Marini, D. & Balestrieri, F. Variability of some analytical characteristics of orange juices, *Italian Journal of Food Science*, 1994, **6**, (2), 225-235.
34 Swallow, K.W. & Low, N.H. Determination of honey authenticity by anion-exchange liquid chromatography, *Journal of the Association of Official Analytical chemists (AOAC) International*, 1994, **77**, (3), 695-702.
35 Zalewski, R.I. Authentication of honey sample via test sample and principal component analysis, *Food Quality and Preference*, 1992, **3**, (4), 223-227.
36 Swallow, K.W. & Low, N.H. Analysis and quantitation of the carbohydrates in honey using high-performance liquid chromatography, *Journal of Agricultural and Food Chemistry*, 1990, **38**, (9), 1828-1832.
37 Doner, L.W. Verifying the authenticity of plant-derived materials by stable isotope ratio and chromatographic methodologies, *Journal of the Association of Official Analytical Chemists*, 1991, **74**, (1), 14-19.
38 Robertson, A, Hall, M.N. & Scotter, C.N.G. Near infrared spectroscopy - its use in rapid QC analysis and its potential in process control, *Food Science & Technology Today*, 1989, **3**, (2), 102-110.
39 Scotter, C.N.G. Use of near infrared spectroscopy in the food industry with particular reference to its applications to on/in-line food processes, *Food Control*, 1990, **1**, 142-149.
40 Bellon, V., Vigneau, J.L. & Sévila, F. Infrared and near-infrared technology for the food industry and agricultural uses: on-line applications, *Food Control*, 1994, **5**, (1), 21-27.
41 Giese, J. On-line sensors in food processing, *Food Technology*, 1993, **47**, (5), 88, 90-95.
42 Collins, M.A. Intelligent electronic noses - not to be sniffed at?, *Food Tech Europe*, 1994/1995, **1**, (5), 150, 152, 154.
43 Peddicord, T. Optimising production efficiencies through inspection, *Food Tech Europe*, 1994/1995, **1**, (5), 160, 162.
44 Skjöldebrand, C. The future for on-line quality control in the food industry, *Food Technology International Europe*, 1995, 115-118.

Two-stage beef roasting

M.A. Sheard, G. Xie

Food Research Group, Faculty of Cultural and Education Studies, Leeds Metropolitan University, Leeds LS1 3HE, UK

Abstract

Based on the fact that steam condensation always leads to high heat transfer rate, a two-stage roasting, involving steam first followed by forced hot air, was proposed. Its effect on heating time, lethality, cooking loss and texture of cooked beef joints was investigated experimentally in a computerised combination oven. Compared to forced hot air and combi-steam roasting the two-stage roasting effectively reduced the heating time with no significant difference in the connective tissue strength, cooking loss and pasteurising value but produced tougher muscle fibre. The roasting processes were simulated mathematically. Heat transfer coefficients at the three roasting conditions were determined experimentally. The values were 14.8 ± 1.7 and 15.9 ± 2.5 W m^{-2} K^{-1} for forced hot air and combi-steam roasting and infinite and 14.5 ± 1.3 W m^{-2} K^{-1} for the first and second stages of the two-stage roasting respectively. The errors for predicted heating times ranged from -5.5 to 3.9% and for pasteurising values (P_{70}) from -4.8 to -2.7%.

Introduction

In practice the choice of oven temperature in roasting represents a compromise between heating time and quality. The oven temperature during roasting might be regulated in a manner to optimize certain quality parameters without increasing heating time. Bengtsson et al.[1] suggested appropriate programming of oven temperature as a possible way to minimize cooking loss. Reichert[2] and Muller and Katsaras[3] cooked hams by the so-called ΔT-heating treatment, in which the temperature difference between the surrounding medium and the core of the product is kept constant. Significantly decreased cooking loss and increased organoleptic quality with an appropriately chosen ΔT-value was reported by Reichert[2]. Muller and Katsaras[3] demonstrated small insignificant improvements in cooking loss and colour intensity, and were able to lower the energy consumption of the heat treatment by 42%. Mielche[4] used four different roasting methods: (i) high temperature roasting with a constant oven temperature of 170°C; (ii) low temperature roasting with a constant oven temperature of 120°C; (iii) maximum temperature roasting with maximally 70°C 1 cm below the joint surface, and (iv) temperature difference roasting with constantly 50°C between the oven and 1 cm below the joint surface. He found there was no significant difference between the quality of the joints produced by three of the methods but the quality of the joints produced by the high temperature roasting was significantly poorer. The heating time varied with the method, the high temperature roasting needing the shortest time and the temperature difference roasting needing the longest time.

Townsend et al.[5] made model calculations of the temperature profile in an oven-heated cylindrical joint. An optimal temperature profile along the axis of the joint was defined and heating times and oven temperatures for simple and complex cooking procedures were optimized to minimize deviations from this profile. The simple cooking procedure consisted of cooking at a constant oven temperature with optional aftercooking at room temperature. The complex cooking procedure consisted of thawing at room temperature, two cooking stages at constant oven temperatures and aftercooking at room temperature. Only minor improvements of the temperature profile were found by using complex rather than simple cooking.

Efforts have also been made to predict the heating time of meat under different process conditions. One of the key parameters used in this prediction is the heat transfer coefficient (h) between heating medium and joint surface. Burfoot and Self[6] used PTFE cylinders heated in a convection tunnel to measure h values for different combinations of air velocity and temperature. The h values at 175°C varied from 14.5 to 19.2 W m^{-2} K^{-1} for air velocities between 0.15 and 0.60 m s^{-1}. Bengtsson et al.[1] found that the h value for slabs of meat varied from 3 to 15 W m^{-2} K^{-1} during natural convection heating. Godsalve[7] obtained the best agreement between predicted and measured heating times by using an h value of 8.5 W m^{-2} K^{-1} in a mathematical model of beef cylinders during heating. Values ranging from 5 to 10 W m^{-2} K^{-1} can be calculated using the correlations in Perry and Chilton[8] relating to h value in a natural convection oven. By comparing the experimental temperature-time histories with the analytical solution of the heat conduction equation for an infinite cylinder Mielche[4] reported h values of 14.3±0.4 W m^{-2} K^{-1} for beef cylinders roasted in a natural convection oven. Considering the effect of the reduction of cylinder radius caused by beef shrinkage, a mean radius before and after cooking was used in his h value estimation. Heating times were predicted with a maximum deviation of 10%.

The purpose of this study was: (1) to investigate the effects of different roasting conditions on heating time and the quality of beef joints, and (2) to predict heating times and lethality values by using the experimentally determined heat transfer coefficients.

Materials and method

Chilled rolled beef brisket joints, purchased from a local butcher's shop, were chosen as the sample for this study. Beef brisket is a relatively tough meat which has rarely been used in meat research. The reason for the choice of this type of beef was its cylindrical shape. The visible fat and connective tissue were removed and the joints were re-rolled to make them a cylindrical shape with diameter of 110.0±5.0 mm and length 90.0±5.0 mm. The average weight of the joints was 1006±10.2 g. The joints were stabilised at approximately 20°C before use. The roasting was carried out in a Rational CC6 computerised combination oven (Rational UK Ltd., Luton, UK). Three roasting conditions were investigated: (i) forced hot air, (ii) combi-steaming, which is claimed by the oven manufacturer to produce excellent texture of meat and poultry products and (iii) two-stage roasting (atmospheric steaming until the joint surface temperature reached approximately 97°C followed by forced hot air roasting). For roasting conditions (i) and (ii) the oven temperature was set at 175° C. For condition (iii) the oven temperature was set at 99°C for the first stage and then raised to 175°C for the second stage. All joints were roasted until the core temperature reached 74°C[9]. After roasting excess surface moisture was removed with absorbant paper and the joints were wrapped with aluminium foil, rapidly cooled and stored in a refrigerated cabinet at 3°C prior to texture measurement. All joints were weighed before and immediately after roasting and the cooking losses were calculated. For each roasting condition triplicate runs were conducted.

The texture of the cooked joints was assessed objectively by measurement of shear, compression and tensile force (along fibre direction). Peak force was used as the texture indicator for all three measurements which were carried out in a J J extension machine (J.J. Lloyd Instruments Ltd., Southampton, UK). Cooked beef samples of 10×10×10 mm (cube), 5×15×50 mm (strip) and 15×20 mm (D×L, cylinder); grips of TG17, TG26 and TG80 and load cells of 1000, 100 and 100 N were used for compression, tensile and shear force measurements respectively. In the compression measurement the beef cube was compressed from 10 mm to 5 mm perpendicular to the fibre direction. The numbers of samples used in compression, tensile and shear force measurements were 63, 40 and 55 respectively.

Temperatures of the heating medium, joint surface and core were measured by T-type (Cu/Con) fine wire thermocouples of 0.315 mm diameter (Labfacility Ltd., Teddington, UK) and recorded by a Squirrel 1205 data logger (Grant Instruments Ltd., Cambridge, UK) at time intervals of 1 min. For core temperature measurement a needle and thread, punctured through the centre of the joint from its curved surface, was used to guide a thermocouple junction to the geometric centre of the joint. For temperature measurement on the joint surface thermocouple junctions were wrapped with aluminium foil, formed into small rectangles (5.0×20.0 mm) and sewn onto the joint surface at the desired locations. Because of the large thermal conductivity of the aluminium foil and the close attachment between foil and thermocouple junction and between foil and joint surface, it is believed that these thermocouples were measuring joint surface temperature. The layout of the oven and the locations of thermocouples are shown in Fig. 1.

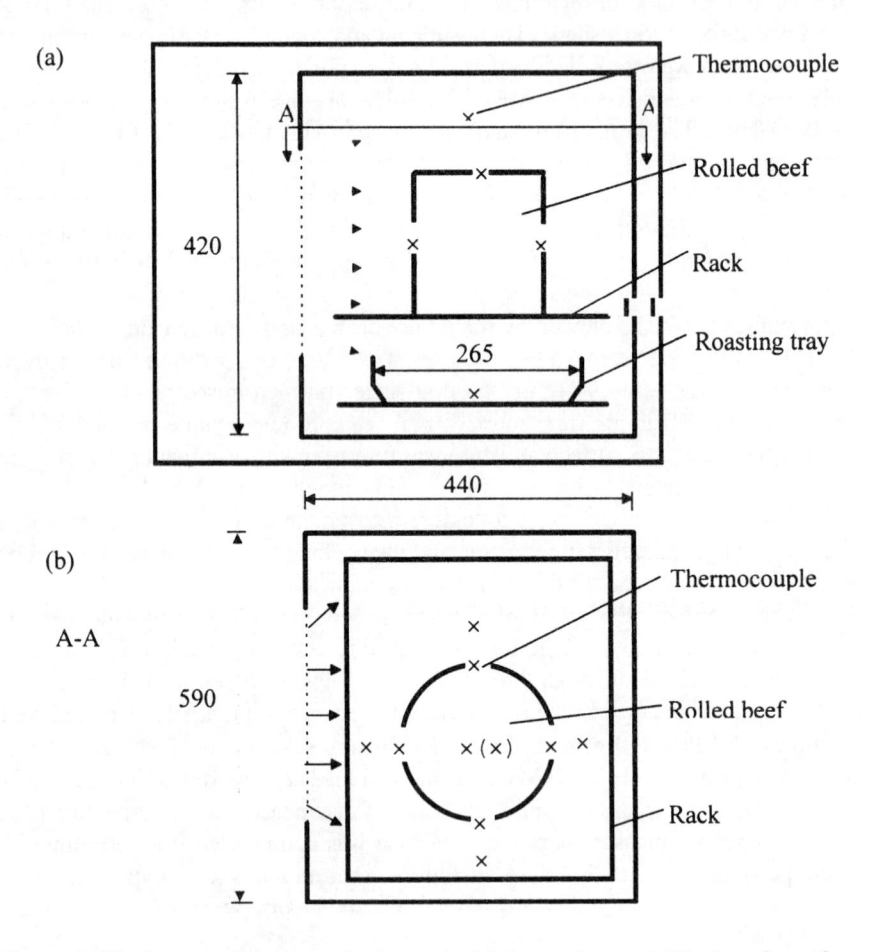

Figure 1. The layout of the oven (unit: mm): (a) front view and (b) view from the top (A-A view). × represents the thermocouples.

Theoretical calculation

Transient heat conduction into an isotropic and homogeneous finite cylinder is governed by a partial differential equation (eqn (1)) in cylindrical co-ordinates:

$$\frac{\partial T}{\partial t} = \frac{k}{\rho C_p}\left(\frac{\partial^2 T}{\partial r^2} + \frac{1}{r}\frac{\partial T}{\partial r} + \frac{\partial^2 T}{\partial y^2}\right) \quad (1)$$

The initial and boundary conditions were as follows (eqns (2) and (3)):

$$T(r,y) = T_o \quad at \quad t = 0 \quad (2)$$

$$\frac{\partial T}{\partial n} = \frac{h}{k}(T_h - T_b) \quad (3)$$

The differential equation was converted to finite difference equations and numerically solved using the Crank-Nicolson method, which has unconditional stability and convergence to the exact solution[10]. As a compromise between calculation accuracy and computing time, a time increment of 2.5 seconds and distance increments of 2.5 mm, were used in the numerical calculation. h values were determined by the optimisation method[11], i.e. changing h value until the correlation coefficient of linear regression between experimental and calculated temperatures reaches one, with thermal conductivity of 0.45 W m^{-1} K^{-1}, heat capacity of 3.35 kJ kg^{-1} K^{-1} and density of 1000 kg m^{-3} [4]. Validation experiments were done in order to compare the heating times between roasting experiments and predictions using the h values determined in the previous experiments, and to check the applicability of these h values for beef joints with different weight and dimensions and with different initial temperature. In the validation experiments three rolled beef briskets, one for each roasting condition, were prepared and stabilised in a refrigerated cabinet at 3°C overnight. For forced hot air, combi-steam and two-stage roastings the weights of the beef joints used were 1142, 1200 and 1151 g, and sizes were 150×60, 160×60 and 140×70 mm respectively.

The pasteurising value was estimated from:

$$P_{Tr} = \int_0^t 10^{\frac{T-T_r}{z}} dt \quad (4)$$

Simpson's integration method[12] was used to calculate the pasteurising value with a reference temperature of 70°C and z value of 10°C, (P_{70}), from the core temperature histories.

Results and discussion

The experimental results are summarised in Table 1. The analysis of variance showed that at the 95% significance level the shear force, tensile force and heating time were different and all other measurements were similar for the three roasting conditions. t-tests indicated that at the 95% significance level the tensile and shear forces of the beef joints roasted by the two-stage method were larger, and the heating time was shorter than the others.

Table 1 Effects of different roasting conditions on cooking loss, texture, lethality and heating time of beef joints roasted in a computerised combination oven

Parameters	Hot air mean±s.d	Combi-steam mean±s.d	Two-stage mean±s.d
Cooking loss (%)(n^a=3)	35.2±0.3	36.2±0.4	35.5±1.1
Texture (Newton):			
Shear force (n=55)	60.7±14.5	56.6±18.9	64.6±15.7
Compression force (n=63)	303.1±103.9	278.4±70.0	275.9±78.7
Tensile force (n=40)	14.2±1.7	15.9±2.5	20.4±3.0
P_{70} value (min) (n=3)	23.0±1.2	23.0±2.3	20.4±3.0
h (W m^{-2}K^{-1}) (n=3)	14.8±1.7	15.9±2.5	1st: ∞ 2nd: 14.5±1.3
Heating time (min) (n=3)	83.7±4.7	81.7±4.0	1st: 21.7±1.2 2nd: 46.3±2.5 Total: 68.0±1.7

^{a}: Number of samples used.

These results indicate that the two-stage roasting reduced heating time effectively with no significant differences in process lethality, cooking loss and connective tissue strength, compression force being commonly accepted as an indicator of connective tissue strength. However, the larger tensile and shear forces suggest that it produced tougher muscle fibre. Tensile force is a measure of muscle fibre strength and the shear force measurement involves a combination of compression, tensile and shear forces, with tensile force predominating. The peak shear force values are, therefore, related more closely to the myofibrillar component of toughness than to the connective tissue component[13,14,15,16,17].

Cooking loss is mainly caused by shrinkage of the muscle and water evaporation and drip from the joint surface. These are dependent on cooking temperature, pH and fibre length[18,19,20]. Cooking loss increases with increase in fibre length but the effect of fibre length decreases with increased heating time. This suggests that differences in temperature history could account for a great deal of the observed effect[20]. A high temperature at the joint surface layer will primarily increase the drip loss, while the evaporative cooking loss will increase almost linearly with heating time when roasting at constant oven temperature[1]. In the two-stage roasting the shorter heating time and the similar joint surface temperatures during the second stage compensate for the larger drip loss caused by the higher surface temperature during the first stage, resulting in similar cooking losses to the other two roasting conditions.

The effects of heat treatment on meat texture are very complicated. In general, cooking makes connective tissue more tender by converting collagen to gelatin; it also coagulates and tends to toughen the myofibrillar proteins. Both these effects depend on time and temperature, the former one being more important for the softening of collagen and the latter more critical for myofibrillar toughening[21]. Mielche[4] found that higher surface end-point temperature, shorter heating time and higher heating rate caused beef samples to have higher values of Warner-Bratzler peak force and M-force (the measure of fibre protein strength), while lower temperature and heating rate and longer heating time caused beef samples to be more tender. Compared to forced hot air and combi-steam roasting, it is the higher surface temperature, shorter time and higher heating rate in the two-stage roasting that results in the larger tensile and shear forces, i.e. tougher muscle fibre.

The advantage of two-stage roasting over the other two roasting conditions is the significant reduction in heating time. Compared to forced hot air and combi-steam roasting the two-stage roasting reduced heating time by 18.7% and 16.8% respectively. Fig. 2 shows the mean oven temperatures and mean joint surface and core temperatures, calculated from triplicate runs, at the three roasting conditions. For clarity and ease of comparison only the temperature-time histories for the first 30 min heating are shown in the figure. It can be seen that the rate of joint surface temperature gain within the first few minutes was different for the three roasting conditions. The rate was fastest in the two-stage roasting, caused by the steam condensing on the joint surface, followed by combi-steaming and forced hot air roasting. It can also be observed that, for all three roasting conditions, the rate of rise in surface temperature was low as it approached 100°C. This may have been due to surface water boiling and evaporation. The shorter heating time of the two-stage roasting is mainly the result of the rapid surface temperature rise during the first stage.

The h values determined in this investigation were based on the initial shape and size of the joints. The h values determined for forced hot air roasting are very close to those reported by Mielche[4] for a natural convection oven.. The values determined for both forced hot air and combi-steam roasting fall within the range reported by Burfoot & Self[6] for forced air roasting. Using the h values determined in this study the percentage errors between the simulation and original experiments ranged from -5.4 to 3.4% for the heating times and from -4.7 to -2.4% for P_{70} values and between the simulation and validation experiments the errors ranged from -5.5 to 3.9% and from -4.8 to -2.7% respectively. These results indicate that for beef joints with a weight between 1.0 and 1.2 kg the agreement between experiment and simulation for both heating time and P_{70} values is good. Prediction of heating time for roast meat is a complicated and difficult task due to changes in shape, size, physical and thermal properties of the meat and the evaporation of moisture during roasting. Some efforts have been made to develop a model combining heat and mass transfer and taking into account the changes in thermal and physical properties and dimension of meat in process simulation[22]. The predicted results have been improved but the calculation is complicated. The results obtained here suggest an alternative approach. Heat transfer coefficients for meat roasted under different conditions are determined experimentally based on the original shape and size of the joints.

The results can then be used to develop dimensionless number equations or empirical equations which include the shape and size of the meat. The heating time can be predicted by using the relatively simple model and these h values.

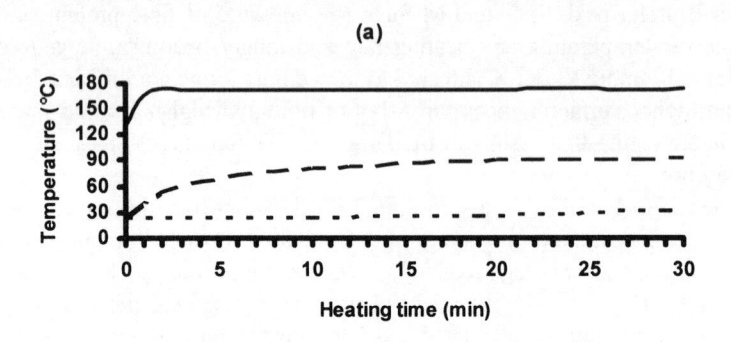

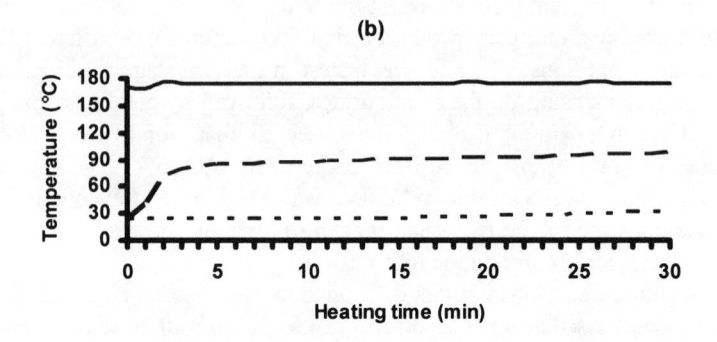

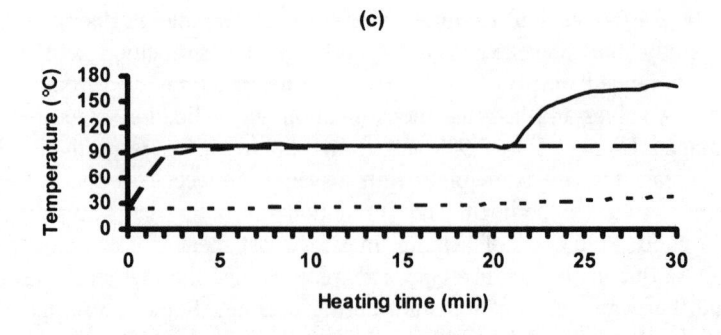

(a)

(b)

(c)

Figure 2. Temperature- time histories for heating medium (———), joint surface (— —) and joint core (- - - -) under three roasting conditions: (a) forced hot air, (b) comi-steam and (c) two-stage.

Conclusion

Compared to forced hot air and combi-steam roasting the two-stage roasting effectively reduced heating time with no significant effects on cooking loss, connective tissue strength and process lethality but produced tougher muscle fibre. h values for the three roasting conditions were not significantly different except for the initial steaming in the two-stage roasting. The predicted process times agreed well with the experimental results for all three roasting conditions.

Nomenclature

C_p: heat capacity (kJ kg^{-1} K^{-1})
k: thermal conductivity (W m^{-1} K^{-1})
h: heat transfer coefficient (W m^{-2} K^{-1})
L: cylinder height (m)
P_{Tr}: pasteurising value at a reference temperature (min)
P_{70}: pasteurising value at reference temperature of 70°C (min)
r: cylinder radius axis (m)
T: temperature (°C)
T_b: temperature at boundary node point (°C)
T_h: process medium temperature (°C)
T_o: initial food temperature (°C)
T_r: reference temperature (°C)
t: time (second)
y: cylinder vertical axis (m)
z: change in temperature required for a ten fold reduction or increase in the D value for destruction of microorganisms and spores (°C)
$\partial T/\partial n$: outward normal gradient of temperature (°C m^{-1})
ρ: density (kg m^{-3})

References

1. Bengtsson, E., Jakobsson, B. & Dakerskog, M. Cooking of beef by oven roasting: a study of heat and mass transfer, *Journal of Food Science*, 1976, **41**, 1047-1053.
2. Reichert, E. Die Delta-T-Kochung - ein neuer Begriff ? *Die Fleischerei*, 1980, **31**, 478-486.
3. Muller, D. & Katsaras, K. DeltaT-heat treatment of cooked hams: Technological and energy aspects, *Fleischwirtschaft*, 1985, **65**, 332-336.
4. Mielche, M. Effect of different temperature control methods on heating time and certain quality parameters of oven-roasted beef, *Journal of Food Process Engineering*, 1992, **15**, 131-142.
5. Townsend, A. & Gupta, S. Optimal roasting, *Journal of Food Process Engineering*, 1989, **11**, 117-145.

6. Burfoot, D. & Self, K.P. Predicting the heating times of beef joints, *Journal of Food Engineering,* 1989, **9,** 251-274.
7. Godsalve, E. *Heat and Mass Transfer in Cooking Meat,* Ph.D thesis, University of Minnesota, 1976.
8. Perry, H & Chilton, H. *Chemical Engineers Handbook,* 5th edn, McGraw-Hill, Kogakusha, 1973.
9. Burfoot, D. & Griffin, W.J. Effects of dimensions on the heating times and weight losses of cylindrical beef joints, *International Journal of Food Science and Technology,* 1988, **23,** 487-494.
10. Smith, D. *Numerical solution of partial differential equations: finite difference methods,* 2nd edition, Oxford University Press, Oxford, 1978.
11. Lebowitz, F. & Bhowmik, R. Determination of retortable pouch heat transfer coefficients by optimisation method, *Journal of Food Science,* 1989, **54,** 1407-1412.
12. Bajpai, C., Mustoe, R. & Walker, D. *Engineering Mathematics,* 2nd edition, John Wiley & Sons Ltd, Chichester, 1989.
13. Bouton, E. & Harris, V. A comparison of some objective methods used to assess meat tenderness, *Journal of Food Science,* 1972, **37,** 140-144.
14. Paul, C., McCrae, E. & Hofferber, M. Heat induced changes in beef muscle collagen, *Journal of Food Science,* 1973, **38,** 66-68.
15. Cross, R., Carpenter, L. & Smith, C. Effect of intramuscle collagen and elastin on bovine muscle tenderness, *Journal of Food Science,* 1973, **38,** 998-1003.
16. Voisey, W. & Larmond, E. Examination of factors affecting performance of the Warner-Bratzler meat shear test, *Canadian Institute of Food Science and Technology Journal,* 1974, **7,** 243-249.
17. Biswas, S., Radhakrishnan, T., Arumugam, P. & Ramamurthi, R. Studies on the relationships between 'tenderness' measures on myofibrillar and connective tissue components of heated and unheated mutton, *Cheiron,* 1989, **18,** 154-157.
18. Hamm, R.& Deatherage, E. Changes in hydration, solubility and charges of muscle proteins during heating of meat, *Food Research,* 1960, **25,** 587-592.
19. Bouton, E., Harris, V. & Shorthose, R. Effect of ultimate pH upon the water-holding capacity and tenderness of mutton, *Journal of Food Science,* 1971, **36,** 435-442.
20. Bouton, E., Harris, V. & Shorthose, R. Factors influencing cooking losses from meat, *Journal of Food Science,* 1976, **41,** 1092-1095.
21. Weir, E. *The Science of Meat and Meat Products.* (Ed. Amer. Meat. Inst. Found), Reinhold Publishing Co., New York, 1960.
22. Singh, N., Akins, R.G. & Erickson, L.E. Modeling heat and mass transfer during the oven roasting of meat, *Journal of Food Process Engineering,* 1984, **7,** 205-220.

Improvement of the quality of typical Hungarian wines by membrane filtration

K. Manninger, E. Békássy-Molnár, Gy. Vatai, M. Kállay
*University of Horticulture and Food, H-1118. Ménesi út 44.
Budapest, Hungary*

Abstract

Two Hungarian light wines, a red and a white of the same vintage, were investigated in a laboratory scale cross-flow membrane equipment using ceramic micro- and ultrafiltration membranes. The influence of pretreatment methods and operation parameters (recycle liquid velocity and membrane pore size) on the permeate flux quality of wines were determined. The sugar, alcohol, colloid, protein, polyphenol, extract tartaric acid and total acid content of the wine and its colour intensity were measured. The filtered products were checked by sensory analysis as well.

The ceramic membranes were found suitable for cross-flow filtration on Hungarian light wines, the cross-flow micro- and ultrafiltration can replace the pretreatment with chemicals.

Introduction

Sterilizing ultrafiltration and microfiltration has become an accepted practice within the wine making industry. Several wineries are using crossflow membrane filtration to clarify, stabilize and enhance their wines. The ultra- and microfiltration is being used to remove thermally unstable proteins, tannins, colour components, yeast, carbohydrate gums and oxidized pheolics[1,2].
It is a well know fact, that the quality and price of wines is influenced by the pretreatment as well [3].
The aim of our study was to investigate and compare the influence of different pretreatment methods and of membrane separation methods on the filtration rate and or the quality of Hungarian light wines.

Two Hungarian light wines, a red and a white of the same vintage were investigated. The applied pretreatment methods were as follows: a.)once racked, b.) racked + clarified, c.) racked + clarified + filtered by diatomaceous earth.

Ceramic micro- and ultrafiltration membranes were used in a laboratory scale cross-flow membrane equipment. The influence of recycle flow rate on the permeate flux and on the quality of wines was measured under constant pressure and temperature. The applicalibity of ceramic membranes in the filtration of Hungarian wines was studied.

Experimental

Experiments were carried out with single tube ceramic modules type 1T1 - 70 (SCT, France) 250 mm long with diameter 7/10 mm and filtration area of 55 cm^2. Two different modules were used: the ultrafiltration membrane was made of zirconia oxide with mean pore size of 50 nm, the microfiltration membrane was made of alumina oxide with mean pore size of 500 nm (0,5 μm). The two membrane modules were used alternately.

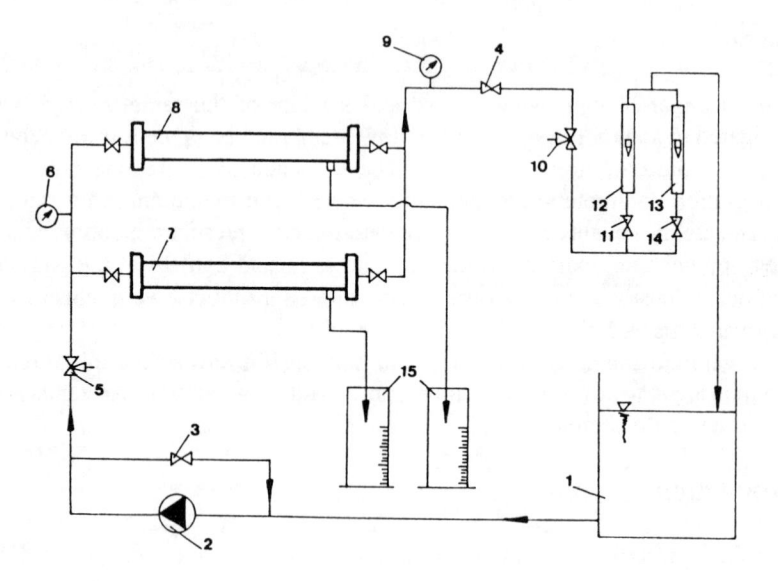

Figure 1: The experimental apparatus

1 – feed tank; 2 – rotary - vane pump; 3 – by - pass valve; 4 – pressure regulation valve; 5, 10 – discharge valves; 6, 9 – manomater; 7, 8 – membrane modules; 11, 14 – ball valves; 12, 13 – flow meters; 15 – permeate measuring cilinders

The experimental apparatus is shown in Fig.1. The feed solution (white or red wine) was set up into the 3 l feed tank. The recycle flow rate was circulated with a rotary vane pump, using a by-pass valve, the adjusted values were 100 l/h and 400 l/h. The transmembrane pressure i.e. the feed pressure was kept constant (300 kPa) with a regulation valve. The permeate flowrate was measured by measuring cylinders and stop-watch. The experiments were carried out at a room temperature of 20°C, which was adjusted by cooling the wine in the feed tank. The experimental runs were finished when the collected permeate volume reached 500 ml. The membrane module was cleaned after each run and the cleaning was repeated until the standard water flux was reached.

The permeate samples and the wines before filtration were analysed according to the Hungarian Standard. The next parameters were determined: concentrations of alcohol, sugar, sugar-free extract, total acid, tartaric acid, polyphenols, proteins, colloids, pH and colour intensity of the wine.

The characteristic parameters were measured by SPEKOL 11 type spectrophotometer. In the case of tartaric acid, ammonium-meta-vanadate was used as a reagent. The polyphenols were expressed in gallus acid using folin-ciocaltens phenol. The soluble proteins were determined by the Breadford method. The colour intensity was measured using 420 nm wavelenght in the case of white wine and 420 nm and 520 nm in the case of red wine.

The colloid content was precipitated after the well known alcoholic reaction.

Results and Discussion

The experimental results of the investigated pretreatment influence on permeate flux are presented in Figs. 2-4. The figures show the effect of different pretreatment methods: once racked (r); racked + clarified (rc): racked + clarified + filtered by diatomaceus earth (rcf) at a constant feed pressure of 300 kPa, constant feed temperature of 20°C and different recycle rates with both wines, white and red. The presented results are typical in all experimental runs.

The results shown in Fig. 2, using low recycle rate (100 l/h; 0,72 m/s) are in agreement with our earlier investigations [3]. The highest permeate flux was reached in the case of complete pretreatment (rcf) of wine and the lowest in case of once racked wine (f), the differences are significant. Using high recycle rate (400 l/h, 2,9 m/s) in almost all experimental runs only negligable differences could be observed: i.e. the pretreatment had no significant effect on permeate flux (Fig.3). In some cases (Fig.4.) the permeate flux after simple treatment (r) was not significantly lower than that of combined ones (rc and rcf). These conclusion are very interesting, because that means the crossflow micro- and ultrafiltration in some cases can replace the chemical pretreatment (clarifiers, diatomaceus arth, etc.) decreasing the waste disposal and saving energy in the wineries.

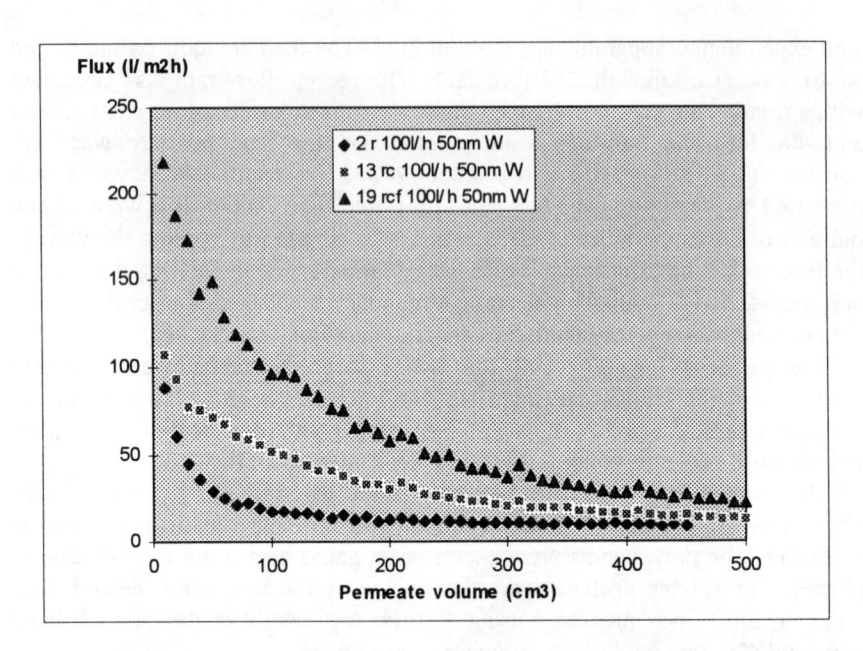

Figure 2.: Influence of pretreatment on permeate flux (white wine low recycle rate, ultrafiltration).

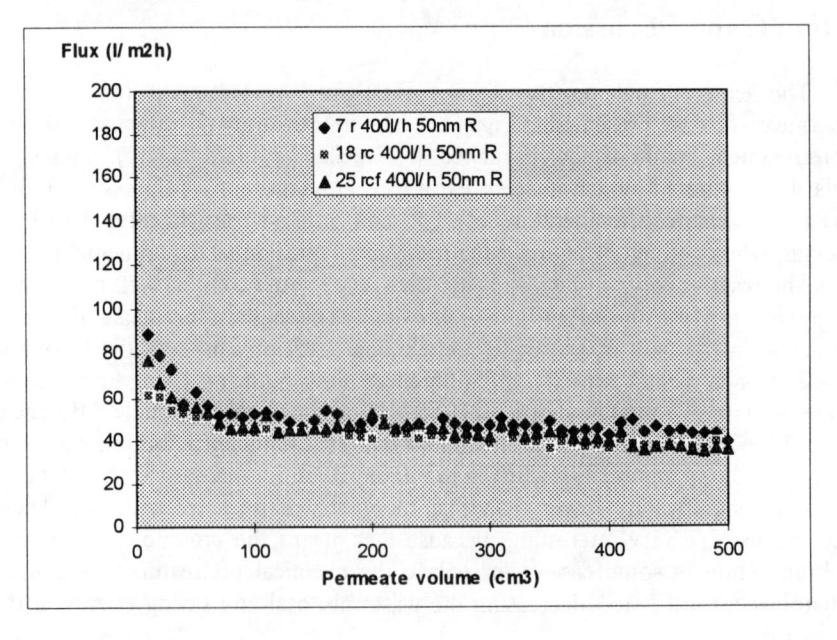

Figure 3.: Influence of pretreatment on permeate flux (red wine, high recycle rate, ultrafiltration)

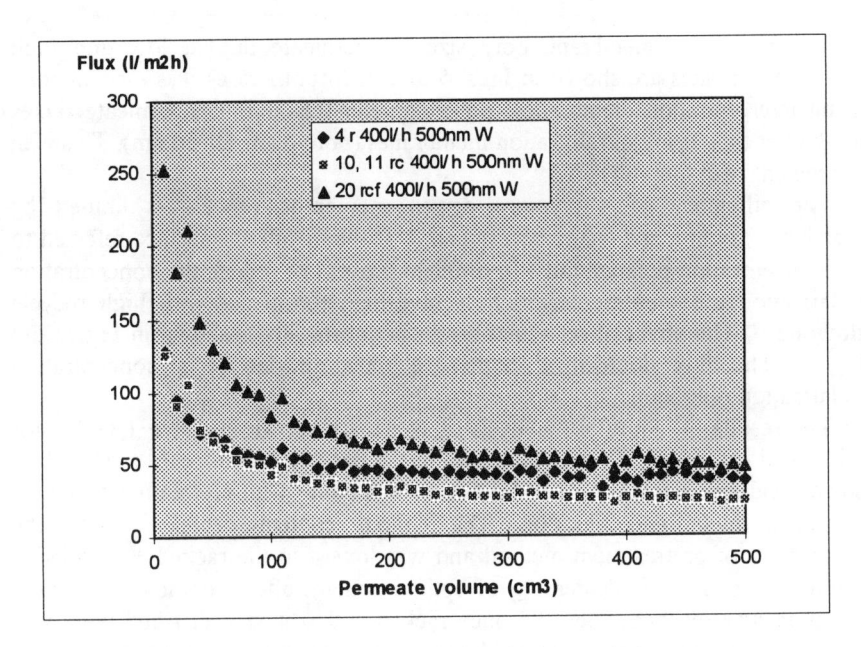

Figure 4.: Influence of pretreatment on permeate flux (white wine, high recycle rate, microfiltration)

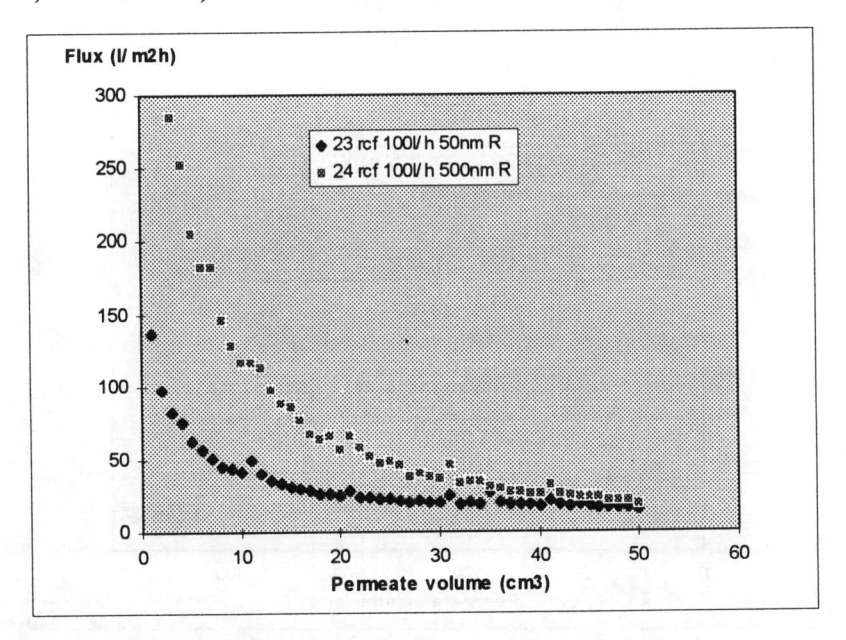

Figure 5.: Influence of membrane pore size on permeate flux (red wine, low recycle rate, racked + clarified + filtered)

The effects of membrane pore size on permeate flux at low and high recirculation rates are shown in Figs. 5 and 6. In both cases it is evident that, using microfiltration membranes (pore diameter 0.5 μm) the permeate fluxes are higher than using ultrafiltration membranes (pore diameter 50 nm). This is in agreement with literature[4-6].

The efficiency of a pressure driven membrane process is limited by membrane fouling and solute accumulation at the membrane surface referred to as "concentration polarization". In order to prevent or reduce the concentration polarization the simplest way is to achive high turbulence with high recycle flowrate[4-7]. The effect of recirculation on permeate flux is shown in Fig.7. and Fig 8. The flux doubled at higher recirculation due to a concentration polarization reduction.

The results of laboratory analysis of wines are shown in Table 1 and 2 for white and red wines. As was expected on the basis of literature[3,5,6] the alcohol, sugar, acid and extract content did not change either with pretreatment or with crossflow membrane filtration. The colloid content decreased with the complexity of pretreatment method and was lowest in the racked + clarified + filtered wine. Further decreasing of colloid content after crossflow membrane filtration was reached when the once racked and racked + clarified wine was analysed. The protein content also decreased with the increase of the complexity of the pretreatment, as well as with crossflow membrane filtration, while the decrease was most effective in case of ultrafiltration.

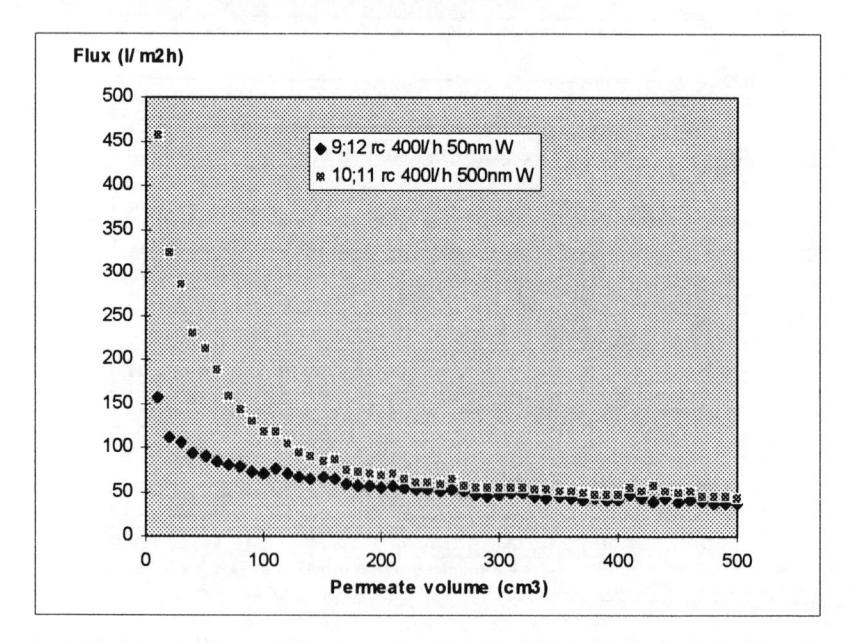

Figure 6.: Influence of membrane pore size on permeate flux (white wine, high recycle rate, racked + clarified).

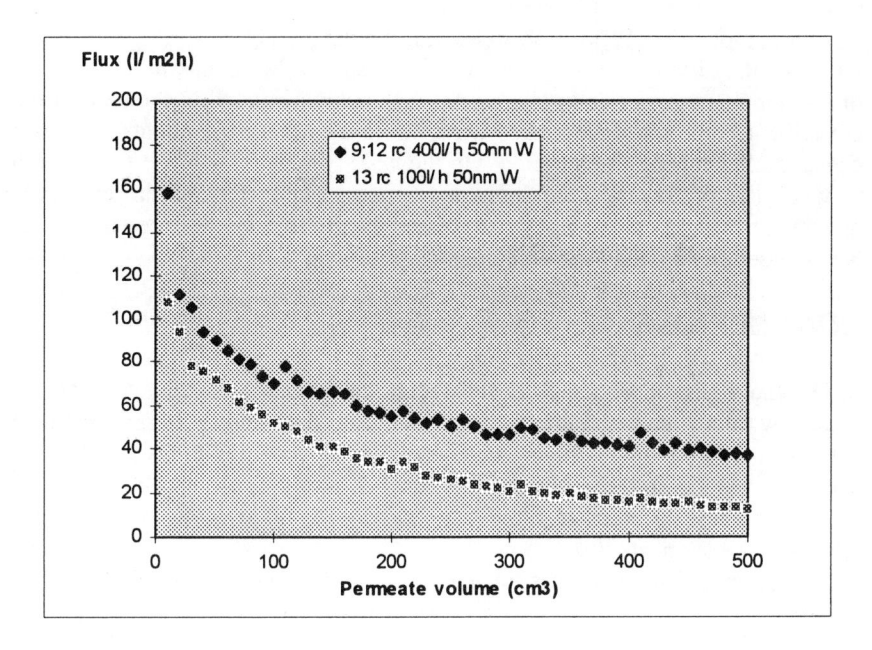

Figure 7.: Influence of recycle rate on permeate flux (white wine, ultrafiltration, racked + clarified).

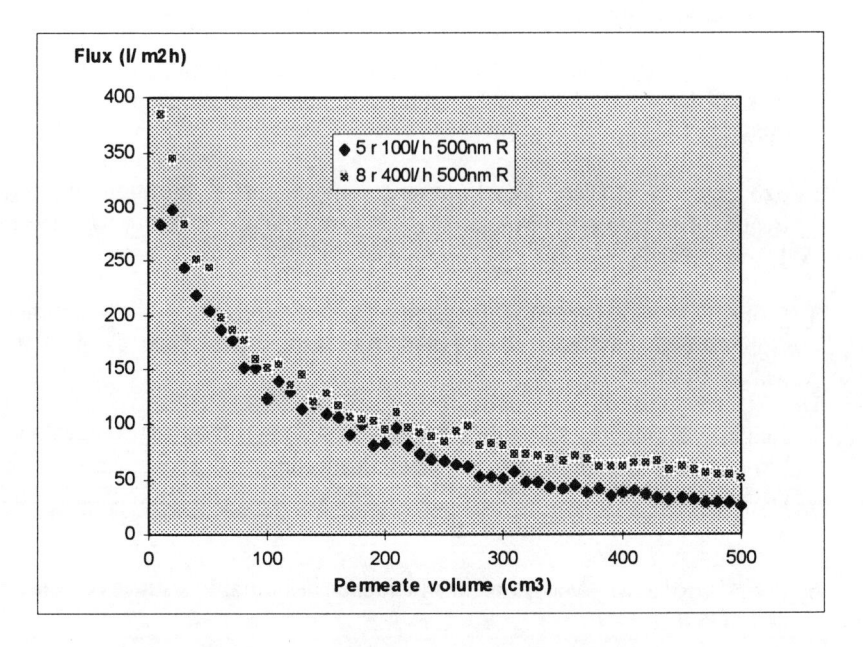

Figure 8.: Influence of recycle rate on permeate flux (red wine, microfiltration, once racked)

The colour intensity in both cases was similar before and after membrane filtration, the difference taking only a few percent, of deviation within the range of measurement error. The polyphenol was slightly increased in the case of the crossflow filtration of once racked wine, but in the other experiments it had not changed after membrane filtration. On the basis of sensory analysis the relation between the pretreatment methods and process parameters of the crossflow membrane filtration could not be expressed clearly, but all the filtered samples had a better quality, than the unfiltered wines.

CONCLUSIONS

On the basis of obtained experimental data can be concluded that the crossflow micro- and ultrafiltration in some cases can replace the pretreatment with chemicals (clarifiers, diatomaceus earth, etc.) decreasing the waste disposal and saving energy in the wineries.

The increase of recycle flow rate decreased the fouling and strongly increased the permeate flux.

The ceramic membranes are suitable for cross-flow filtration of wine.

REFERENCES

1. Wucherphennig, K., Dietrich, H.: Die Bedeutung der Kolloide für die Klarung von Most und Wein, Wein-Wissenschaft, 1989, 44 (1) p. 1-12.

2. Schmitt, A., Koehler, H., Miltenberger, R., Curschmann, K.: Vergleich verschiedener Cross-Flow-Filtrationssysteme. Weinwirtschaft-Technik, 1987, 123 (12) p. 11-12, 14-18.

3. Szövényi, E., Kállay, M.: Expiriences of tangential filtration in wine treatment in Hungary (Hungarian), *Magyar szôlô és borgazdaság*, 1991, **1**, p. 2-10.

4. Mohr, C.M., Leeper, S.A., Engelgan, D.E., Charboneu B.L.: Membrane Applications and Research in Food Processing, Noyes Data Corp., Park Ridge, 1989.

5. Junge, Ch.: Etude sur l'ultrafiltration et la microfiltration tangentielle (cross-flow microfiltration) de vins, Prodeedings of 68^{e}ASSEMBLÉE GENERALE "OFFICA INTERNATIONAL DE LA VIGNE ET DU VIN", p. 7-1 to 7-12, Paris, France, 5-9 Septembre, 1988.

6. Peri, C., Riva, M., Margarini, G.: Incidence des filtrations sur les coloides, ibid, p. 5-1 to 5-10.

7. Porter, M.C.: Handbook of Industrial Membrane Technology, Noyes Pub., Park Ridge, 1990.

Culinary Arts and Sciences 249

Table 1. Analysis data - White wine

Run No.	WINE	Recirc. flowrate (l/h)	Membran pore (nm)	Alcohol conc. (v/v%)	Extract (g/l)	Sugar (g/l)	Total acid (g/l)	pH	Tartaric acid (g/l)	Poly-phenols (mg/l)	Color (420 nm)	Protein (mg/l)	Kolloid g/l	Average flux (l/m2h)
-	WHITE racked feed (control)	-	-	11,54	20,3	0,9	5,2	3,64	2,04	348	0,179	18,8	0,920	-
3	WHITE rac.	400	50	10,04	16,5	0,9	4,2	3,64	1,08	235	0,176	12,4	0,220	63,66
4	WHITE rac.	400	500	11,56	19,0	0,9	4,7	3,65	1,28	291	0,182	10,4	0,675	49,32
1	WHITE rac.	100	500	10,89	18,0	0,9	4,8	3,62	1,32	200	0,206	15,5	0,300	21,01
2	WHITE rac.	100	50	10,52	17,6	0,9	4,5	3,64	1,21	189	0,186	10,9	0,250	17,42
-	WHITE racked+clarified feed (control)	-	-	11,58	19,8	0,9	5,2	3,67	1,79	275	0,267	17,6	0,875	-
10	WHITE rac.+clar.	400	500	11,22	18,8	0,9	5,0	3,67	1,48	272	0,181	8,4	0,575	98,77
11	WHITE rac.+clar.	400	500	10,73	18,0	0,9	4,8	3,67	1,35	266	0,182	7,5	0,550	89,80
9	WHITE rac.+clar.	400	50	10,68	18,0	0,9	4,8	3,65	1,38	263	0,169	6,6	0,375	56,00
12	WHITE rac.+clar.	400	50	11,08	18,8	0,9	5,1	3,67	1,40	268	0,186	6,9	0,275	61,32
13	WHITE rac.+clar.	100	50	10,38	18,0	0,9	4,7	3,66	1,38	248	0,196	7,5	0,450	34,16
-	WHITE racked+clarified+filtered feed (control)	-	-	11,48	19,6	0,9	5,1	3,66	1,43	261	0,241	14,7	0,675	-
20	WHITE rac.+clar.+filt.	400	500	11,14	18,5	0,9	5,2	3,63	1,42	263	0,247	11,6	0,675	75,36
22	WHITE rac.+clar.+filt.	400	50	11,19	19,0	0,9	5,1	3,64	1,41	261	0,205	4,6	0,900	62,13
19	WHITE rac.+clar.+filt.	100	50	10,64	17,8	0,9	4,9	3,60	1,20	267	0,261	15,1	0,550	65,32
21	WHITE rac.+clar.+filt.	100	500	10,68	17,8	0,9	5,0	3,62	1,36	256	0,257	9,9	0,675	77,13

Table 2. Analysis data - Red wine

Run No.	WINE	Recirc. flowrate (l/h)	Membran pore (nm)	Alcohol conc. (v/v%)	Extract (g/l)	Sugar (g/l)	Total acid (g/l)	pH	Tartaric acid (g/l)	Poly-phenols (mg/l)	Color (420 nm)	Color (520 nm)	Protein (mg/l)	Average flux (l/m2h)
-	RED racked feed (control)	-	-	11,44	28,1	9,7	5,6	3,32	3,39	1003	0,334	0,319	75,8	-
8	RED rac.	400	500	11,23	25,8	9,7	5,2	3,24	2,58	914	0,377	0,393	59,0	116,04
7	RED rac.	400	50	10,43	23,7	9,7	4,8	3,24	2,34	822	0,346	0,364	61,9	49,92
5	RED rac.	100	500	10,86	25,3	9,7	5,1	3,27	2,58	889	0,356	0,366	62,1	91,29
6	RED rac.	100	50	11,05	25,8	9,7	5,2	3,26	2,67	916	0,353	0,366	63,0	27,25
-	RED racked+clarified feed (control)	-	-	11,24	25,8	9,7	5,2	3,25	2,72	819	0,410	0,415	63,6	-
16	RED rac.+clar.	400	500	11,05	26,1	9,7	5,4	3,24	2,97	850	0,396	0,398	58,1	56,28
18	RED rac.+clar.	400	50	10,88	25,5	9,7	5,3	3,23	2,89	840	0,405	0,407	55,1	42,53
17	RED rac.+clar.	100	500	10,29	24,2	9,7	5,0	3,24	2,73	840	0,417	0,411	56,9	40,70
14	RED rac.+clar.	100	50	10,55	25,0	9,7	5,3	3,27	2,82	834	0,397	0,395	57,6	30,92
-	RED racked+clarified+filtered feed (control)	-	-	11,32	26,1	9,7	5,2	3,27	2,72	815	0,389	0,385	67,5	-
15	RED rac.+clar.+filt.	400	500	10,92	25,0	9,7	5,2	3,24	2,75	934	0,401	0,419	52,9	57,90
25	RED rac.+clar.+filt.	400	50	10,64	24,5	9,7	5,2	3,23	2,54	854	0,380	0,399	51,9	45,37
24	RED rac.+clar.+filt.	100	500	10,95	25,8	9,7	5,3	3,23	2,71	931	0,397	0,417	53,2	81,84
23	RED rac.+clar.+filt.	100	50	10,18	23,2	9,7	5,0	3,25	2,43	809	0,375	0,389	54,1	32,18

Demographic change and food product development: the challenge of an ageing UK society

A. Colquhoun, A. Reid, P. Lyon
School of Food and Accommodation Management, University of Dundee, 13 Perth Road, Dundee DD1 4HT, Scotland

Abstract

Scotland, in common with many other industrialised societies, has an ageing population. More people are surviving to deep old age but there are trends within trends. Women continue to outlive men and there is a noticeable growth in single-pensioner households. The vast majority of old people live independently in the community. People tend to retire earlier and, while the financing of old age is in transition, more are in receipt of occupational pensions to augment State sources. People entering retirement now will have experienced the major changes of lifestyle in the post-war decades - including increased travel opportunities and food diversity. It is important to recognise the scope and scale of the social and cultural changes of recent decades and their impact on the perceptions and tastes of older people.

In this paper we draw attention to the heterogeneity of this section of society and the potential for looking at them as a market for new products and services. From this perspective, it is possible that the ageing population could stimulate economic activity and with appropriate products/services their quality of life in general and their independence in particular could be enhanced. To this end, we examine some examples of food products and services developed for this important section of society.

An Ageing Population

Medical science has been outstandingly successful in eliminating the impact of some diseases which used to be major causes of early death[1] and now also offers increasingly effective intervention with serious injuries. Furthermore, improvements in nutritional status generally have helped extend the prospect of long life to more people. A number of changes have worked congruently to extend life expectancy. The elderly (60 years +) have increased significantly as a proportion of the population from 7.5% at the turn of the century to just over 20% in 1993 and are set to maintain that figure for some years to come[2] . The UK trend closely conforms to changes across the European Union[3]. It should be noted that there will be a greater proportion of not just elderly, but very old people in the population.

Lives of the old

Commentary on the old can give rise to an assumed homogenity but as with any other section of society, they may have little in common apart from their chronology, and even this may involve a span of three decades. The diversity of life experiences - from work and income to health circumstances and emotional support networks - will make their mark on old age as much if not more than any other period of life. However, certain cohort features are fairly predictable notwithstanding individual variation.

Women in the UK continue to outlive their male counterparts for a number of reasons ranging from the carnage of warfare earlier in the cohort's history to gender-related lifestyle characteristics. This imbalance was very clear in 1991 with a male:female ratio of 1:2 for 80-84 year olds and 1:3 for those people who were 85 and over [4]. There are similar but less pronounced imbalances for other 'senior citizen' age bands. Associated, in part, with this differential longevity is the growing incidence of households containing one person above pensionable age [5]. For Great Britain, this has risen from 7 per cent of all households in 1961 to 16 per cent in 1991 [6]. When we turn to 1991 UK Census data, it is clear that the sex differential in survival to old age is only part of the story. Women of pensionable age in one person households outnumber men by nearly four to one. In Britain, there were 2,603,504 women aged 60 and over living in households with no dependent children as compared with 698,785 men aged 65 and over living alone [7].

Central to the experience of old age are health circumstances, particularly as these affect prospects for independent living. With increasing age there is the likelihood that the long term consequences of lifestyle (eg factors associated with occupation, housing, diet, tobacco/ alcohol consumption) will be manifest in terms of physical disability or specific illnesses. Any limit to mobility is potentially a serious threat to continued independence especially where the old person lives alone.

The financing of old age in the UK is in transition with the promotion of Personal Pension Plans to supplement any occupational arrangements and discourage reliance on future state provision[8]. However, those currently retired and especially the very old are less likely to be beneficiaries of these newer pension strategies. Insurance-based Personal Pension plans were a product of the

1980s and, historically, occupational pension schemes were only available to limited sections of the workforce. Differential involvement, by worker age and social class, is still apparent [9]. For the very old, income will be primarily derived from private wealth and/or state pension/income support[10].

In summary, we can be sure that there will be more 'very old' people in the British population. As now, women will outnumber men and a significant number will live alone. Although many could not be judged well-off by the standards of the working population, not all will be solely reliant on state pension and/or income support. This latter point is important. There are those who look to the future with trepidation because of the perceived welfare burden. However, others caution against the pessimistic misreading of specific cohorts of old people [11] and look to the changed demography for new markets[12] and to new technological developments[13] for innovation of products and delivery.

Nutrition and Old Age

With old age there are a number of fairly predictable physiological/ biological changes which are important in any consideration of appetite, choice and the amount of food consumed. For example, the secretion of digestive juices is diminished; the mobility of the gastro-intestinal tract is decreased; sensory perceptions are diminished, and there is decreased absorption and use of nutrients. In conjunction with other socio-psychological changes - disengagement, bereavement, retirement and so forth - in this part of the life course, food and intake patterns can be noticeably affected. Social and biological aspects of ageing together need consideration by those involved in food preparation, food service delivery and food product development.

Some years ago, recognition of such factors led to the establishment of the *Nutrition Programme for Older Americans* [14]. The aim was to meet the nutritional and social needs of persons of 60 years and over who could not afford adequate diets; were unable to prepare adequate meals at home; were of limited mobility and were isolated and lacked the incentive to prepare and eat food alone.The original programme provided for a range of services but these have been amended over time and changes were made to include nutrition services and education as well as the provision of meals both within the community (congregate) and home delivered. Congregate meals were to be provided for older persons - particularly those on low incomes - with the aim of meeting social contact/support needs in addition to providing a balanced diet. Home delivered meals - for up to five days a week - were designed for those housebound through illness and disability and, as in the UK, provided the main social contact and the main source of food for many clients. Both home-served and congregate meals are designed to provide one-third RDA.

Old Age and Independent Living

Providing meals for older people in the community
The meals-on-wheels service is an interesting example of a longstanding and practical contribution to 'community care' before the term became both fashionable and the subject of intense political debate. Its origins[15] were in the Women's Voluntary Service (WVS)[16] in the 1940s and, despite arguments for direct local authority implementation to ensure an even provision, it has remained strongly linked to volunteer efforts. It has long been recognized as an important facilitator of independent living among the elderly [17]. Some years ago, it was noted that...' *recipients of meals-on-wheels tend to be very old (eg over 80) and living alone. National survey data shows that about one third of such very old people have no surviving children...*'[18]. Further recent growth in the numbers of the very old simultaneously makes good provision more vital and more precarious.

Meal provision was only part of the logic behind this service and there has been reference over the years to the 'latent' functions of social contact for the isolated and checking the need for health/social services intervention. These dimensions are still being signalled as important [19]. However desirable, this may be overstated as an element of everyday practice for as Dunn comments... '*the tendency to justify one activity (providing a mobile meal) in terms of another (providing contact) is dubious. The usually brisk delivery of a pre-cooked meal can rarely be a viable substitute for worthwhile social relations, and it is hazardous to regard it as such*'[20].

Equally, the service cannot be viewed as either the only strategy or the perfect solution to the problems of the independent elderly. First, the service itself may not be the best arrangement for some. There have been warnings of a need for better service co-ordination, the use of alternatives such as lunch clubs, for...'*greater choice and more flexible packages*'[21]. Moreover, there was the well-timed caution that......'*the provision of meals-on-wheels does not become counter productive in that it heightens the sense of social isolation and reduces the desire for, and availability of, other forms of social contact*'[22]. Second, the traditional format of the service - the delivered hot meal - is increasingly only one of several alternatives[23] [24] which might offer greater flexibility for individual needs and, in effect, better value for money.

However, it is in the changed context of social policy, demographics and household technology that major opportunities emerge for new initiatives in products and service delivery. An increasing number of local authorities are now using a frozen meal delivery service. As an example of the increasingly blurred distinction between public and private sector, *apetito* [25] offers a range of services.

* A complete contract service for production and delivery (taken up by Greenwich and Kingston Councils) where meals are delivered once a week, stored in the clients own freezer (available on loan from the company) and regenerated either in microwave or steamer (both of which are available on loan). Drivers take the following week's order - selected from a brochure of more than 60 meals - at the time of delivery.

* A service operated in conjunction with the Womens' Royal Voluntary Service (WRVS) whereby the latter do deliveries of frozen meals to clients on a weekly basis (eg Berkshire Council).

* Supplying frozen meals to a council which then regenerates them centrally and delivers hot meals to clients (Islington Council).

Other companies currently operating in the 'mixed economy' of meal provision for elderly persons in the community include *Wiltshire Farm Foods*. In addition to delivering meals to residential homes[26], they also offer a 'paid for' service to elderly clients who, whether in their own homes or in sheltered housing, are perhaps not able to qualify for their local authority meals-on-wheels service. Throughout the Tayside, Fife and Central Regions of Scotland, orders are collected with the prior delivery or posted/telephoned in to provide a fortnightly delivery of individually-portioned frozen meals. Some clients use this service to supplement the 'standard' meals-on-wheels service or whilst they are on waiting list. The Company's clients are predominantly elderly but the service is not restricted to that section of the population. For example, a warden in a sheltered housing complex buys the products and there is some interest from households where shift-work patterns preclude regular mealtimes.

There is a growing number of private participants in this field. *Compass* have a four year contract (from August 1994) to provide welfare catering for Avon County Council and *CCG (Commercial Catering Group)* have 'welfare contracts' in the London boroughs of Wandsworth, Redbridge, Richmond and Kensington/ Chelsea. The advantages of such schemes, especially for the full service -production and delivery of frozen meals - mode, are an increased menu choice, more flexibility about the time of the meal, avoidance of the problems of using school meals kitchens during vacations and easier quality control with greater consistency and precise portion control.

However, given the variability of personal circumstances in old age, such arrangements are not the best solution for everyone. Frozen/home regenerated meals require that the client has a fair degree of mobility, mental alertness and, even, interest in food. Client decisions have to be made, so the system is really only appropriate to the more independent elderly. Also, it is debatable whether such developments finally lay to rest the enduring concern for the 'latent' functions of home delivered meals - as a point of social contact - or actually ameliorate a growing gulf between rhetoric and practice. With a hot delivered meal, speed is of the essence for reasons of food quality and safety. Little time may be spent with the client in a 'drop and run' situation. With the delivery of frozen meals, do drivers make sure that the food is safely stored in the freezer as well as taking next week's order ? In which case 'strategic' contact could be at higher levels even if there is limited scope for anything else.

Perhaps it is not the state/method by which food is delivered that needs to be addressed here. Given the formal commitment to foster independent living in the community for as long as is feasible, one needs to question the role of social work monitoring if a twice-weekly hot meal is the major source of contact that the client has with the community. There has to be a continuing concern for those old people with chronic arthritis or the earlier stages of Alzheimer's disease. What arrangements exist for ensuring that the client has actually eaten the hot meal provided ? One example of the kind of support systems that may need to be increasingly considered by public authorities is the *Safe and Well* scheme run by the WRVS for Derbyshire County Council[27]. This involves a weekly visit or telephone call to check on the client's wellbeing to augment their role in the bulk delivery of frozen meals.

Products, portions and packaging
Not all old people, not even among the very old, will require meal provision to the extent described above. There is an increasing need for products which take account of smaller appetites and smaller households among the elderly population. Useful developments in this respect include the wider availability of 'portion packs' - previously restricted to the caterer - for butter, sauces and spreads, preserves, pick'n'mix cheese and the like. One step beyond these products is the snack market. Eden Vale Hot Choices, for example, are microwaveable snacks being launched with a range of four products - Chicken Tikka Masala, Spaghetti Americaine, Macaroni Cheese, Ham and Pasta with Tuna and Sweetcorn - and have a recommended selling price around £1.29. All small-portion packs can be convenient purchases for older consumers giving less waste, and more variety, albeit at a higher price.

Furthermore, developments in *functional foods* - food ingredients and products designed and promoted for provision of identified health benefits - provide significant opportunities for products specifically targeted at older people. It is estimated that the sales of such foods could reach 5% of total food expenditure - £30 billion for western Europe. For currently targeted social/age groups, products with identifiable health claims include high fibre breakfast cereals, high energy sport's drinks, probiotic yoghurts[28]. Ribena fruit and fibre, Tesco yoghurt with fibre and Pact spread containing n-3 polyunsaturated fatty acid would be specific examples. Perhaps more applicable to older people is the Nestlé relaunch of a reformulated version of their Build-Up fortified drinks and soups. These include five varieties of fortified milk shake powders and a hot chocolate drink which provides one third RDA of vitamins. and minerals. Hot chocolate jars (approximately 10 servings) retail at around £1.99 and soups at £1.15 for a box of two sachets.

This developing interest is testament to the statement that there are... *'major opportunities for innovation of added-value food products which have the potential to improve health and reduce risk of some important diseases'*[29]. Over the last two decades, the perception of 'healthy food' has focused on products which have reduced levels of 'what is bad for you'. So we have seen a plethora of low fat, reduced sugar, low salt products. With functional foods, the marketing message is the promotion of positive healthy eating and, in relation to the older market, another subtlety needs to be addressed.. *'products positioned here should be positive and upbeat, without falling into the trap of making sad attempts to recapture lost youth'*[30].

Positive and practical information for elderly persons in the community is needed. The United States Department of Agriculture extension services offer a model for intervention in terms of nutritional support/counselling for older adults and community workers - who are able to further disseminate information. Unfortunately, we have very few nutritionists working in Scottish communities although there are some Home Economists working for local authority and voluntary services. There is an important contribution to be made here by food retailers. Nutritionists at major food retailers, for example Tesco's Consumer Advice Centres produce information leaflets giving guidance for older people. Opportunities for direct interaction between consumers and advice providers must be seen as highly desirable if messages are to have a lasting impact on behaviour.

On a lighter, but commercially very important note, the growing market for *nostalgia foods* is related[31] to the growing number of older people in UK society. The mature market is significant not only because of its size but because there is some economic differentiation and a fair proportion of older people have sufficient resources to spend on non-essential - even frivolous - goods. Moreover, the idea that old age should be uniformly a time of decline, disengagement and disinterest is increasingly challenged at a practical level[32]. Borthwick's 'Golden Years' study[33] indicated that mature consumers have continued to incease their expenditure on food and drink throughout the recession, in contrast to their younger counterparts. The revival of old favourites - instant custard, malt loaf and spam for example - show that manufacturers are beginning to take seriously the potential of this market segment. Often, nostalgia food products can be produced without the lengthy preparation periods of yesteryear. Traditional foods in traditional packaging may be even more acceptable if they are microwaveable. An example of this might be a 'steamed pudding' which can be conveniently re-heated in a microwave oven.

Conclusions

There are important challenges for food producers, retailers and Meals-on-Wheels providers in response to the changing demographics of the UK and similar advanced industrial societies. These have yet to be fully exploited. Most food product development seems to be currently targeted at the young and families with dependent children. Imagination needs to be lifted beyond the provision of 1 or 2 portion packs to a more detailed consideration of the needs of an increasingly elderly society. Population ageing, and the socio-economic characteristics of the old, provide many opportunities for innovation and market development, but we have to be very careful to appreciate the heterogeneity of the old and ensure that the poor are not forgotten in the pursuit of more affluent sections.

Notes and References

1. See Victor, C., *Old Age in Modern Society: a Textbook of Social Gerontology*, Chapman and Hall: London, 1994, 91-96

2. See Table 2.3, *Annual Abstract of Statistics 1994*, HMSO: London, 1995

3. Walker, A., Alber, J. and Guillemard, A-M., *Older people in Europe: Social and Economic Policies*, Commission of the European Communities: Brussels, 1993

4. See Table 2.3, *Annual Abstract of Statistics 1993*, HMSO: London, 1994

5. Refers to the minimum ages the UK state retirement pension can be paid if there is eligibility, that is, 60 years for females and 65 years for males.

6. *Social Trends*, No 23, HMSO: London, 1993

7. 1991 Census, *Household Composition: GB*, HMSO: London, 1993

8. See, for example, Brindle, D., State pension for all may have to go, *The Guardian*, 8 April 1993; Thomas, R, Pension burden puts state welfare at risk, *The Guardian*, 26 April 1995.

9. *General Household Survey 1992*, HMSO: London, 1994

10. The very old may also experience the 'double-bind' of low/static income with rising costs linked to decreased mobility. See Age Concern, *Short Change: the Effects of Low Income on Older People*, Age Concern England: London, 1995.

11. See Thane, P., *Economic Burden or Benefit ? A positive View of Old Age*, Discussion paper 197, Centre for Economic Policy Research: London, 1987.

12. See, for example, *The Grocer*, New services geared to the elderly, 5 November 1988; Caro, M., The mature market: destroying the myths, *Lodging*, 14, 10, June 1989, 27-30,; Willman, J., With a greyer picture of the future in mind, *Financial Times*, 8 March 1994, 14.

13. Many technological developments - for storage, preparation and regeneration of food/foodstuffs - have come to the aid of the elderly. Microwaves, for example, are present in 45 % of all households with an economically inactive head; 30% of single-person households with one adult aged 60 or over, and 52% of all households of 2 adults with one or both aged 60 years plus (Tables 2.37 & 2.38, *General Household Survey 1993*, HMSO: London, 1995).

14. See Williams, S., *Essentials of Nutrition and Diet Therapy*, 6th edition, Mosby-Yearbook: St Louis (Miss), 1994

15. Means, R., *Community Care and Meals-on-Wheels: a Study in the Politics of Service Development at the National and Local Level*, School of Advanced Urban Studies Working Paper 21: University of Bristol, 1981.

16. WRVS after 1966 when the 'Royal' was added.

17. Dunn, D., *Food, Glorious Food: a Review of Meals Services for Older People*, Centre for Policy on Ageing: London, 1987.

18. Barker, J. and Noble, A., *Sustaining the Elderly: a Report on Meals-on-Wheels in an Inner London Borough*, Age Concern England: Mitcham, 1983, 5.

19. Clavey, J., Meals-on-wheels quality standard set, *Caterer and Hotelkeeper*, 15 October 1992, 14

20. Dunn, *op cit*, 13

21. Barker and Noble, *op cit*, 2

22. *ibid*, 4

23. Dunn, *op cit*, 30

24. For an outline of a student food product development project involving cook-chill meals-on-wheels, see Colquhoun, A. & Lyon, P., Chill and go: a 90s alternative for the independent elderly, paper at *XIV International Home Economics and Consumer Studies Research Conference*, Sheffield, 1994.

25. The UK branch of the company is based in Sevenoaks, Kent with the German parent company *apetito AG* already having extensive operations in Germany, Holland and France with a 1994 turnover of 239 million DM (£94.1 million). See Shrimpton, D., Who are appetito ? *Cost Sector Catering*, April 1994, 29.

26. The company supply multi-portion dishes to residential homes for cover with staff holidays and Halal meals/ Asian vegetarian diet meals for residents where facilities/expertise are not available for their preparation.

27. WRVS Food Services Ltd - a trading subsidiary set up to bid for contracts under CCT arrangements - was awarded the Derbyshire contract as they were felt to be the only independent provider capable of delivering the complete service - including the *Safe and Well* dimension. See Shrimpton, D., Frozen meals-on-wheels: friend or foe ? *Cost Sector Catering*, June 1995, 20-21.

28. Young, J., Future opportunities for functional foods, *Food Manufacture*, October 1995, 63/72 reviews Rebuk, K., Righelato, R. and Young, J., *Future opportunities for functional foods*, Leatherhead RA/Research Information: Leatherhead, 1995.

29. *ibid*, 72

30. Darrington, H., Focus on functional foods, *Food Manufacture*, April 1995, 25/27/29, 29

31. Nostalgia has also been linked to rapid social change and even the depressing 'flywheel of routine'.

32. This is clearly in evidence with a recent newspaper report (Francis, M., Still kicking, *Sunday Times*, 21 January 1996) on Age Concern's *Ageing Well* initiative designed to stimulate activity and promote good health via the activity itself and its associated social contacts.

33. *Food Manufacture*, Golden oldies - a growing market, August 1995, 14.

Posters

The effects of natural antioxidants on the quality and storage stability of ready-to-serve foods

M. Karpinska, J. Borowski, M. Danowska

Institute of Human Nutrition, University of Agriculture and Technology, 10-718 Olsztyn, Poland

A number of studies have indicated that certain antioxidants have a carcinogenic effect on living organisms. This investigation studied the effects of natural antioxidants utilising meat balls produced from turkey meat. The natural antioxidants in this case were derived from a combination of sage and a mixture of other spices which included red and black pepper, garlic and marjoram. The products were coated in breadcrumbs and fried in soybean oil then stored in a refrigerator at a temperature of 4°C for 2 to 4 days. Chemical analysis was carried out on the raw material before frying and after storage utilising the Folch lipid fraction technique. Experimental results demonstrated the effects of product composition on the intensity of the oxidation process in lipid fraction. The addition of both the sage and spice mixture retarded the oxidation process, Nevertheless, sage proved to be a more effective antioxidant than the mixture of spices.

Application of yellow mustard mucilage in foods

W. Cui, N.A.M. Eskin, H. Liu, R. Pereira

Departments of Foods & Nutrition and Food Science, University of Manitoba, Winnipeg, Manitoba, Canada R3T 2N2

Mucilage extracted from yellow mustard using a modified procedure was found to exhibit rheological properties similar to that of xanthan gum. Close examination of the mucilage resulted in the isolation of three fractions exhibiting shear thinning properties at low concentrations. One of these fractions was purified and identified by one and two dimensional NMR spectroscopy as a water-soluble 1,4-linked ß-D-glucan. Subsequent research identified significant synergistic interactions between yellow mustard mucilage and the galactomannan, locust bean gum. Significant interactions were observed in a mixed blend of mustard mucilage and locust bean gum in the ratio of 9:1 with respect to dynamic viscosity and storage modulus. Using this ratio, it was possible to produce a mayonnaise product with comparable properties to commercial products. Further studies examined the use of mustard mucilage as a stabilizer for the production of ice cream. Mustard mucilage performed as an effective stabilizer comparable with commercial stabilizers.

Sedimentation phenomenon in canola oil

H. Lui, R. Przybylski, N.A.M. Eskin

Departments of Foods & Nutrition and Food Science, University of Manitoba, Winnipeg, Manitoba, Canada R3T 2N2

The occasional development of a haze in canola oil represents a potential problem to the quality and acceptability of the oil. Extensive studies identified the major component of the haze-causing material as wax esters containing long chain saturated fatty acids and alcohols ranging from C16:0 to C32:0. Chemical composition and thermal properties of a canola oil sediment were compared to a sediment obtained from a commercial winterization of this oil. Both sediments exhibited similar melting temperatures of 74.9C although the commercial sediment contained higher amounts of fatty acids and alcohols containing greater than 24 carbon atoms in the chain. The effect of temperature and sediment concentration on the clouding time of canola oil showed that storage at 5C resulted in the shortest clouding time. Further work examined crystalline conditions on sediment formation in canola oil. The final crystal size depending on the cooling rate with the crystal habit being rod-like changing to a round and leaf-like shape at low cooling rates (<0.5C/min). Using acetone, a preliminary method was developed for predicting the development of this phenomenon in canola oil.

Your roasted chicken quality problem solution - might be brine solution

S. Zalewski

Research and Development Centre for Catering Equipment "Gastromasz", 85-021 Bydgoszcz ul. Gdanska, 134 Poland

A decrease in the quality of meat, fish and poultry dishes prepared from frozen products is a well known. We investigated thawing methods (air, water, microwave, brine and direct) for roast chicken to determine advantages and disadvantages of sensory quality, product yield and protein loss. Results show brine to be the best for sensory quality due to juiciness; salt distribution and yield. The only disadvantage was a slightly higher (0.5%) protein loss. We then optimised brine thawing parameters (time, temperature, salt concentration) for the same measures. The recommended parameters for 1.2-1.4 kg frozen chicken are: 3.5, 4 and 5% NaCl for 8, 7, and 5 hours respectively at 4-6°C with a brine/chicken ratio of 3:1. A brine thawing tank has been developed and patented for commercial use. Thermal insulation ensures correct thawing temperatures and a mixing pump even distribution of the solution. Effects of brine treatment on chicken before thermal processing have also been investigated on fresh chickens prior to roasting. Increased yields of 4% and quality by 2 points on a 10-point scale, as compared to traditional dry salting, have been shown. The only disadvantage was higher (2.5% vs 1.5%) protein loss. Fresh chicken brining processes have been optimised and the following parameters are recommended for commercial use: 5, 6 and 7% NaCl in brine for 3, 2.5 and 1.3 h respectively. The process can be performed in a patented thawing tank providing 10-20% crushed ice or a cooling system is used to maintain a temp. <6°C.

A study of the functional properties of soy proteins for emulsion type foodstuffs

I. Alexieva

Department of Public Catering, Higher Institute of Food and Flavour Industries, 4000 Plovdiv, Bulgaria.

Some functional properties of a soy protein isolate (SPI, 90.5% protein content) obtained by a limited enzyme hydrolysis with thrypsin hydrolysates (SEH, SEHP, and SH with DH = 3 - 9%) were studied. The processes of coalescence were investigated. Protein concentration enhances stability of oil droplets to coalescence. The native and modified proteins differ in their absorption at the oil-water interface and in the formation of interface films. SEH shows better stabilising properties displayed in higher "life-time" characteristics of the oil droplet at the flat interface than the native isolate. The samples with higher DH have considerably lower stabilising properties. It was shown that emulsions with SPI exhibit slightly lower dispersal characteristics than SEH. The limited hydrolysis favours changes in the lipophilic hydrophilic balance that ensure emulsions with SEH to have smallest droplet size (2.0-2.9μm). SEHP and SH develop similar sized oil droplets. Emulsions, stabilised with soy protein isolate and its hydrolysates show non Newtonian behaviour, the first being with higher viscosity. Emulsions with soy hydrolysates have rather similar viscosity characteristics, effected by the oil phase volume and protein product content.

A study of model emulsions stabilised with pectins
I. Alexieva
Department of Public Catering, Higher Institute of Food and Flavour Industries, 4000 Plovdiv, Bulgaria.

The emulsifying and stabilising properties of pectins were studied through the dynamics of coalescence of oil droplets. The effects of some technological factors (type and concentration of pectins, oil phase volume) on basic emulsifying characteristics were investigated. Apple pectins with different esterification degrees were tested: native with high (HED = 73.2%); middle (MED = 54.0%) and low esterification degree (LED = 45.0%). It was found that the increase of pectin concentration led to a decrease of coalescence constants (k) depending on the esterification degree. It was also shown that an equation, $\lg N/No = -k.(t - t_o)^{3/2}$, adequately describes the time of coalescence of oil droplets at the flat oil-pectin solution interface. The effect of pectin concentration on coalescence constants can be shown by the mathematical expression $k = 3.731 - 3.105.x_1$; R = 0.74. It was proved that pectin concentration (x_1) and oil phase volume (x_2) effect stability and rheological characteristics. A mathematical model, describing the effect of x_1, x_2 on emulsion stability index was developed.
(IES = 2.374 + 27.06 x_1 + 1.37 x_2; R = 0.95). The role of x_1, x_2 and x_3 (shear rate) on emulsion viscosity (η) can be described by the following equation:
$\eta = \pm a_o + a\, x_1 + b\, x_2 - c\, x_3$.

The influence of a plasticizing system and physical properties of formed masses on the quality of "convenience food" production

A. Półtorak, A. Neryng
Warsaw Agricultural University, Department of Engineering and Technology, Poland

The constant increase in the demand for "convenience" type foods can be seen, and for this reason, the demand for machines capable of producing such food has also increased. When adapting machines of Western construction, one needs to take into account the habits of the Polish consumer. As a result, time-consuming and costly studies are required to determine the ideal working parameters of the machines and correct operating procedures. So as to decrease the costs of research, developmental trials for "convenience foods" are undertaken using mathematical models. Factors influencing the process of automatic production may be divided into two: the type of plasticizing system used, and the physical properties of the products used. Evaluation of a plasticizing system was undertaken in a production machine, Rheon Cornucopia KN 100, typically used to produce pattie, ravioli, noodles and stuffed rolls. Masses used for production were non-Newton liquids. Measurement of physical properties, e.g. density, were performed using traditional, as well as instrumental methods, with the application of endurance machine Instron 4301 equipped with specially designed capillary viscometer for the measurement of viscosity.

Influence of some physical parameters of dough on the quality of machine-formed convenience foods

A. Wierzbicka, A. Neryng, E. Biller
Warsaw Agricultural University,
Faculty of Human Nutrition and Home Economics, Department of Technic and Technology, Poland

The rising demand for convenience foods in Poland means that methods of rapid production are necessary for popular foods such as pierogi. This poster investigates the effect of various formulations of dough (varying ingredient proportions) on its physical parameters which affect its ability to be used in automatic forming equipment (Rheon Cornucopia KN100) for manufacturing one type of pierogi - a cottage cheese mixture filling surrounded by dough. Measurements were made on the dough and filling using Extensograph and Farinograph equipment (for elasticity, resistance to stretching and consistency), on an Instron machine (for load and stress at maximum and yield points, resistance to squeezing, energy) and colour measurement using a Minolta Chromameter. Four groups of dough characterised by varying amounts of liquid, fat and eggs were studied. The expected performance of these formulations was then tested with the automatic forming machine. It was found that products conforming to national standards for proportion of filling without defects of cracking, leakage, uneven thickness of dough over filling or shape came from doughs using a higher proportion of fat and eggs.

The eating quality of the meat of four different slaughter masses of Bonsmara cattle

I.C. Kleynhaus

Technikon, Pretoria, Private Bag X680, Pretoria 0001, South Africa

To describe and compare the eating quality of four different slaughter masses of Bonsmara cattle, ninety-two Bonsmara bull calves were allocated to the following slaughter groups: commencement of experiment (reference group), 345 kg, 410 kg and 475 kg slaughter masses. The wing rib cuts were cooked according to a standardised dry heat cooking method and the M.longissimus thoracis of the wing ribs was evaluated by a ten member trained sensory analysis panel, using an eight-point category scale for the different attributes of meat quality. Cooking losses and shear force resistance measurements were also determined on the wing rib. Significant differences were found in aroma, sustained juiciness, first bite, muscle fibre and overall tenderness, amount of connective tissue residue, as well as shear force resistance. The meat from the reference group was most tender. When comparing the other three slaughter masses, the meat from 475 kg slaughter masses was the most tender (first bite and overall tenderness) and showed the least resistance to shear. Slaughter mass also significantly influenced all the cooking related and average percentage thawing and cooking related attributes except for drip density.

This study showed that the mass of the carcass affects the eating quality, the cooking-related attributes, as well as the average percentage losses in thawing and cooking.

Cooking variables and sensory quality characteristics of South African venison with reference to springbok (*Antidorcas marsupialis marsupialis*)

D.M.J. van Rensburg

Technikon, Pretoria, Private Bag X680, Pretoria 0001, South Africa

Ninety-one springbok carcasses were divided into three age groups : young, mature and very old, to be aged as follows : whole sides with the skin on, whole sides without skin, primal cuts with bone-in and deboned primal cuts in vacuum bags. The ageing periods where 3,7,10,14 and 21 days post mortem, respectively, at 0-4°C. The cooking methods applied were oven roasting for the legs and the loins, casseroling for the neck, breast and flank-cuts and grilling for the shoulder patties. Samples were evaluated for aroma, juiciness, tenderness, texture and flavour. Thawing loss & Total volume drip loss of the deboned cuts were significantly more than for the other ageing methods, therefore less acceptable. Sensory evaluation have shown significantly higher scores for carcasses aged for 3-7 days with the skin on and for cuts obtained from mature and older animals. Future research needs to address: packaging methods, optimum freezing periods and nutritional values.

Section 2.3: Information Technology

Towards a model of computer use in the hospitality industry

M. Peacock

Business School, University of North London, Holloway Road, London N7 8DB, England

Abstract

The hospitality industry adopts a wide variety of "poses"[3] to advanced technological change. Managers at five hundred hospitality establishments in London and the South East of England were questioned during the course of the research. A variety of suggested influences on the innovatory process are broken down and analysed. Structural and cultural influences are stressed and computer confidence, attitudes to computerisation and computer use itself are separated for the purposes of the project. The role of computer confidence in the adoption of advanced technological "solutions" is questioned, together with the link between advanced technological change and business effectiveness.

Introduction

As computers become a commonplace feature of the hospitality industry, differences become more apparent in the variety of ways advanced technology is introduced and utilised. Smaller hotels and restaurants are still largely refusing to countenance computers. Some hotels are committed to a rolling programme of increasing expenditure and upgrades of their existing systems. In other establishments management are grudgingly accepting head office sponsored innovations, while in other places they are actively resisting them.

For the last four years, the variety of approaches that hospitality enterprises can use with computerisation has been examined with postal surveys, telephone interviews and face-to face interviews of just under five hundred enterprises (both hotels and restaurants). It is often suggested that the industry is slowly coming to see computerised automation as the way forward, as hospitality managers develop confidence and faith in new technology. "Computer controlled robotics will be used to prepare, cook and in some cases, serve a variety of food products" (Payne[8]). Research suggests that the picture is more complicated than this.

Technological innovation is not simply the product of structural factors. Although it is difficult to find an uncomputerised large hotel, it is inadequate to suggest that once a hotel has reached a certain size then it will be computerised. The key players in any innovation are the owners and the managers, and any change is mediated through the ideas and actions of these key players. To develop the model we looked at three separate areas: computer confidence, attitude to computers and use of computers. Although often assumed to be closely related the project emphasised the differences between these three.

Computer confidence

Computer confidence is growing. How long managers feel it will take to learn the new computer system has decreased significantly over the last four years, as managers and operatives become more familiar with the hardware and the software. There is no causative link between computer confidence and gender, but there is a link between computer confidence and education (see Table 1). Computer confidence was measured by the length of time a manager felt it would take to learn a new system. Managers educated to HND or degree level were half as likely to feel that a new system would take more than a year to learn. There is no evidence to suggest that this confidence varies with the grade of qualification and managers with postgraduate qualifications are not significantly more computer confident. The exposure to computer systems that a student receives at college gives the future manager the confidence to deal with a variety of systems.

Table 1
Computer confidence against access to higher education

		Computer Confident	
		Yes	No
Experienced	Yes	88.1%	11.9%
Higher			
Education	No	73.7%	26.3%

Chi-Square Significance 0.01938
Percentages shown are for education.

Education has an interesting effect on the computer literacy of London's hospitality managers. Women managers have higher levels of literacy because they also have a higher proportion who are qualified to degree level. Non-white British have lower levels of computer literacy and this group is the most educationally disadvantaged.

However, statistical analysis suggested that the influence of ethnicity on computer confidence was separate from the education effect. Response rates from Chinese and Indian restaurants were significantly lower than from other survey segments of the market, and while the data suggested that these could be highly profitable establishments there was a statistically significant lack of computer confidence (see Table 2).

Table 2
Ethnicity against computer confidence

	Computer Confident	
	Yes	No
White British	88.5%	11.5%
Non-white British	50.0%	50.0%
Non-British European	73.1%	26.9%
Outside Europe	73.8%	26.2%

Chi-Square Significance 0.01240
Percentages shown are for each ethnic group.

Attitude towards computers

Attitude to computers was defined by the expressed willingness to invest in new hardware or software. While, occasionally, the research did find a manager unwilling to invest because they claimed their existing system was perfect, users fully utilising their systems could always suggest additional purchases which would enhance performance.

One significant finding was that there was no link between computer confidence and a favourable attitude towards computers. We had assumed that the more confident the manager, the greater their willingness to invest in new systems. This is not the case. There are a significant number of managers who, while computer literate, do not feel positive about computers. Of the group of managers currently using computer systems, those unwilling to invest in future systems were marginally more computer confident (88.9%) than those looking to increase expenditure on information technology (82.8%). If computer confidence varies with exposure to information technology, then it might be possible to suggest that in some cases managers start off as unconfident non-users, become enthusiastic users and end up as disillusioned confident users.

Computers and innovation carry positive connotations. To question these terms suggests fear of change, technophobia, perhaps even a manager "unwilling to adapt." Yet there is a significant number of confident, using managers who do not see the systems providing what they want and they question the benefits of further investment.

Another significant factor affecting attitude towards computers was whether the manager was also an owner. For both users and non-users when owners are compared with managers, they are significantly less favourable to new technology. This effect was not produced by differences in education, differences in how flexible owners and managers are, or differences in scale of the establishment. If establishments already using computers are analysed, owners are 8% less likely to be investing in more hardware or software. If the establishment is not using computers then the owner is 6% less likely than the manager to be planning computerisation. This resistance is important as the owner-manager of the small business is "critical in determining the rate and manner of technological adoption" (Dickson et al[5]).

If an owner is going to introduce a computer system, then it is highly likely that they will feel that they have lost direct personal control, either because their enterprise has grown, or because they are no longer physically present. The most highly-computerised small restaurant owner in London was as an ex-airline pilot who had lived in Japan: the exception that proves the rule (Peacock[9]).

When compared with managers, owners have a significantly more autocratic management style: they are twice as likely to classify their management style as autocratic. They also have very low formal monitoring of accounts. This could be influenced by time constraints, or by government revenue collecting activity. Both autocratic style and lack of financial monitoring are linked with a resistance to computerisation. Another factor is the hospitality entrepreneur's desire for some degree of independence and freedom. The McDonalds-style restaurant management systems run counter to their culture and they are prepared to risk some accounting "fuzziness" in exchange for a sense of autonomy in their business.

A favourable attitude to computers was also linked to how flexible the manager was. Managers were asked how easy they would find it to work in a different type of hospitality enterprise and their responses compared to their attitudes towards computers. the more flexible they saw themselves the more likely they were to have a favourable attitude towards computers.

A favourable attitude towards computers was also linked to attitudes towards investment. The manager who saw investment as simply following instructions (not a small group in the hospitality industry) were more likely to be pro-computers. However the two most pro-computer groups were those who saw investment in terms of cost reduction or operational efficiency. Large numbers of managers saw investment appraisal in terms of "money available" or "need", and these more unsophisticated approaches were linked with more negative attitudes towards computers.

Use of computers

Unlike computer confidence and attitude to computers there is a link between use of computers within an establishment and the attitude of the managers. Users are twice as likely to be favourable to further investment. The more highly-computerised the establishment the more favourable the individual manager is going to be towards technology, even with the significant numbers of anti-IT computer literate managers. This means that the influences outlined above — the differences between owners and managers, investment appraisal approaches and flexibility — all influence technological innovation. What does not significantly impact on this process is computer confidence and its associated factors: education and ethnicity. It is true that highly-computerised establishments have a disproportionate number of managers educated to higher education levels, but both computer use and the educational background of managers is due to the structural factors outlined below.

Size really does matter when it comes to computer use: large hospitality enterprises find that they simply cannot manage without their computer systems. Sector too is important. It is not just that hotels generally employ more people than restaurants, hotels are simply more likely to purchase and use an information system. The independence of an establishment was traditionally associated with low levels of technological innovation, but the study found a number of large, independent restaurants which were amongst the most innovative users of advanced technology in our research.

No link was found between computer use and location, but non-British ownership of companies has also been linked to higher levels of investment in information systems (Daniel[4]). The data on this was unclear: if a McDonalds franchise can identify itself as a British-owned business then ownership is, to some degree, separate from its information technology strategy. What was clear from the data was that the state of non-British owned establishments was linked to the centralising ethos and hardware policy of the corporate headquarters. The dominance of "master-slave" design configurations tell their own story.

Hotels and restaurants using advanced technology do not automatically come complete with a set of highly-educated, computer literate managers. The two may coincide but there is no causal link. Highly-computerised establishments have flexible managers who are able to relate investment to business targets such as cost reduction and operational efficiency. Ownership will be exercised very indirectly and non-personally. Overall in the survey, the most computerised establishment is a large hotel where ownership expresses itself as a set of quantifiable business targets for the general manager.

Irony

The irony for this research is that the computer confident but cynical hospitality manager has a case. Profit movements, sales and productivity figures were examined in depth but no relationship was found between any of these and computer use. Undoubtedly there are establishments where investment in information technology is ensuring improved profits. On the other hand there are installations where the benefits have failed to cover the cost of the investment. Highly-computerised establishments are more successful, but not because of their computer use. Small, uncomputerised restaurants are less successful because they are small, not because they are uncomputerised.

Trade journals and advertising literature trumpet the virtues of information systems. However, the growing number of hospitality managers who are not technophobic yet harbour significant reservations about the effectiveness of current systems, has to question the constant and costly updating of information systems.

Mismatch

One possible reason suggested by the research for the lack of impact of information technology in the hospitality industry was a mismatch between organisational culture and effective use of technological innovation. Kasavana[6] encapsulated current applications philosophy when he claimed that "the new elements of success appear to be controls, controls and controls." Throughout the research managers consistently endorsed this philosophy when justifying investment. The accounts-driven, head-office-sponsored network of an international chain, through to the restaurant management system of a local independent, all emphasised the control of subordinates. It was noticeable that the language used favoured "control" connotations. Often, hospitality executives would talk of information systems as "the new discipline."

This emphasis on control is the product of the organisational culture of the industry. It has already been noted that the relatively more autocratic style of owners, as opposed to managers, and that this is linked to less favourable responses to information technology. However, over one-fifth of the hospitality managers in the project classified themselves as "autocrats". The hospitality industry is a highly autocratic industry. Also noted above is the centralising tendency of multinational enterprises, with IT strategy concentrated in the centre.

This autocratic, inflexible culture leads to an organisational inertia, which was quite striking when it was combined with technological innovations. Managerial resistance to organisational change was a feature of most new systems installed in the hospitality industry. At the most basic level, it was exceptionally rare for job titles to be changed even when there was a fundamental shift in the way tasks were performed. There is a tendency for IT specialists to be appointed at managerial levels with responsibilities which cut across traditional levels of demarcation, but their power and influence are limited to the extent that their role is not based on a specific function: they have a servicing role.

Other factors also foster organisational inertia. Child and Loveridge[2] note that in the banking industry "grading systems and a traditional gender-based division of labour inhibit organisational innovations". The same is probably more true about the hospitality industry, with (for example) the manner in which hotel reservations clerks are characterised as "the girls" inhibiting any move towards empowering these operatives as salespeople.

The end product of this organisational inertia and limited process of technological innovation has been neatly summarised by Leitch[7]: "expensive technology + old organisation = expensive old organisation".

Information technology is inextricably bound up with innovatory organisational possibilities for several reasons. Firstly, computers and their operators can carry out a much wider range of functions: permitting much more flexibility when matching roles with humans. Secondly, the manner in which information technology can be linked circumvents the need for a centre: location, other than the point of service, becomes irrelevant. Finally, staff costs decrease in relation to capital costs, so it becomes easier to experiment with different organisational possibilities (Child[1]). This will be the agenda for the hospitality industry of the future.

Summary

Computer use is the product of interplay between size, sector and managerial attitudes towards information technology. These attitudes are the product of approaches to investment, management style, flexibility and whether the manager is also an owner.

Computer confidence, with its associated factors of education and ethnicity, is not inhibiting technological innovation amongst current hospitality industry users. There is a significant number of managers who are both computer confident and unwilling to invest further in new technology. This is symptomatic of the failure of technological innovation to significantly impact on business performance in the industry.

What is inhibiting the effective utilisation of information technology is an autocratic and inflexible culture which is incapable of responding imaginatively to the possibilities that technological innovation can bring. At the moment, the innovators within the hospitality industry are characterised by a "naive expectation of a mechanistic relationship" (Child & Loveridge[2]) between technological innovation and business performance. Little or no consideration is placed on the organisational context for the technology and the end result is the expensive hospitality organisation.

References

1. Child, J. Information technology and organisation, *Innovation and Management: International Comparisons*, eds J. Child & T. Kagono, De Gruyter, 1988
2. Child, J. & Loveridge, R. *Information technology in European services*, Blackwell, Oxford, 1990.
3. Clark, P. & Staunton, N. *Innovation in technology and organisation*, Routledge, London, 1989
4. Daniel, W.W. Workplace industrial relations and technical change, Frances Pinter, Shaftesbury, 1987.
5. Dickson, K., Smith, S., Reid, L. & Woods, A. Technological innovation and small service sector enterprises, paper to *14th National Small Firms Conference*, November, 1991.
6. Kasavana, M.L. Computers and multiunit food-service operations, *Cornell H.R.A. Quarterly*, 1994, June, 72-80.
7. Leitch, P. The virtual organisation, *Flexible Working*, 1995, Vol. 1, No. 1, 19-20.
8. Payne, B. Computer Robotics, *European Hotelier*, 1995, Third Quarter, 45.
9. Peacock, M.J. *Information technology in the hospitality industry*, Cassell, London, 1995.

An information system for determining an optimal nutritional intake

G. Karg, S. Kreutzmeier

Institute for Social Economics of the Household, Technical University, Munich, Germany

Abstract

Nutrition in private households poses a problem as to which dishes should be chosen for certain individuals and meals, during a planned period. To help solve this problem, an information system has been developed where a computer generated menu plan fulfils the following conditions: the preferences for dishes and the requirements for energy and nutrients of every single household member are met, at the least cost.

In searching for an optimum nutritional intake, there are two options available to determine types and quantities of dishes. In option one, the types of dishes for various meal times are predetermined by the user. The quantities of the dishes are then determined by the computer. In option two, the types and quantities of dishes are determined by the computer according to the above conditions.

The information system can be used both by nutrition counsellors and private households to determine an optimum nutritional intake.

1. Problem

The German Nutrition Report indicates that the German population eats too much in total and too much fat, sugar and salt (DGE[1]). This is because the majority of people select dishes, to a great extent, according to taste and/or cost, and to a lesser extent according to the energy and nutrient contents of the dishes. An inappropriate diet over a period of time can lead to nutrition-related diseases (DGE[1]). Since nutrition takes place primarily in private homes, an effective improvement in nutrition begins there. An improvement is only possible if the dishes are chosen in such a way that appropriate nutritional conditions are met which in general refers to appreciating the benefits and cost of good nutrition.

The benefits are both a subjective and objective. The subjective part is characterised by the extent of the individual household members' preferences and aversions to different dishes. The objective part is characterised by the energy and nutrient contents of dishes. The conditions for nutritional benefit are such that only combinations of dishes are selected that meet target values for the subjective and objective benefit (Karg[2]). The target values for the subjective benefits are set by the user and indicate what types of dishes, how frequently (lower and upper limits), the individuals would like them in the planned period. The target values for the objective benefit are set by the DGE (Deutsche Gesellschaft für Ernährung = German Society of Nutrition) and state what nutrients, in what quantities (lower or upper limits), the individual person needs each day of the planned period.

The nutritional cost comprises, for example, the food cost and the cost of preparing foods for the dishes. The cost condition states that the nutritional cost must meet a target value or be as low as possible.

As the set of dishes available, and the conditions to be met are large, households need help with this decision making process. Support for this is offered by the information system discussed here. It consists of a planning model for optimum nutritional intake, a database and programs to carry out the necessary tasks.

2. Model for planning an optimum nutritional intake

The decision model consists of variables and of conditions which have to be met by the variables.

2.1 Variables

The variables are the type and/or the quantity of dishes.

2.2 Conditions

Conditions refer to all variables and to single variables.

2.2.1 All variables
Conditions which have to be met by all variables are calculated as subsidiary conditions and the main condition.

The subsidiary conditions are constraints which are determined by the benefit which has a subjective and an objective component. The subjective benefit indicates individual likes and dislikes of household members, the objective

Culinary Arts and Sciences 287

benefit represents the energy and nutrient content of the dishes (Karg[2]). These constraints are formulated as conditions which have to be satisfied.

The main condition is also called the target function and represents the cost of the resulting menus. The cost contains mainly food costs and time costs for preparing dishes. In the operation, either the food cost or the time cost is optimised.

2.2.2 Single variables
Conditions referring to single variables define whether the types and/or the quantities of the dishes are fixed or variable.

Option 1
In this option, the types of dishes are fixed and the quantities of the dishes are variable. The types of dishes for the meal times of the planned period are determined by the user before optimisation. They are chosen arbitrarily according to the household preferences and then the portion sizes of the chosen dishes are determined (Karg[3]). The menu plan should meet the energy and nutrient requirements of the household members and should have a minimum cost.

Option 2
In this option both the types and quantities of dishes are variable and are determined in the optimisation process. We distinguish between a successive and a simultaneous determination.

In the successive determination, the optimum types of dishes are firstly chosen for one person in the household, for fixed quantities. Then for the other people the optimum quantities of the selected dishes are determined (Baur[4], Baur & Karg[5]). Simultaneous determination can then follow Armstrong & Sinha[6] Steinel[7]. Armstrong & Sinha[6] used binary and continuous variables. The first variable (binary) indicates whether the dish with the usual portion is chosen (1) or not (0). The second variable has a value greater than 0 if the binary variable is 1. The value of the second variable is continuous and indicates how much the quantity of the dishes are greater than the standard portion. Quantities of dishes smaller than one portion are excluded. Steinel[7] determines the kind of dishes independent of standard portion sizes and offers therefore more flexibility than Armstrong & Sinha[6] (Steinel[8]).

To simplify understanding we have named the models as follows:
If option 1 is used, we speak of the model with fixed types of dishes and variable quantities. If option 2 is used, we speak of the model with variable types and quantities of dishes.

In addition we distinguish between pure and mixed models. Pure models incorporate only one of the option, mixed models incorporate both.

3 Database

The database consists of two different types of data: the data independent of the user and the data dependent on the user.

3.1 Data independent of the user

These data can be grouped into primary and secondary data.

3.1.1 Primary data
The primary data are:
- nutrient requirements for different groups of individuals
- nutrient contents and food wastage when preparing dishes
- prices of food items
- recipes
- preparation times for dishes
- classification of dishes into groups of dishes.

<u>Nutrient requirements for different groups of individuals</u>
The nutrient requirements of various groups are summarised in the recommendations of the DGE. The DGE distinguishes between recommendations, guidelines and estimated values.

Recommendations are given if the breakdown of the daily nutrient requirements for different individuals is known. The recommended daily allowance is, in general, the mean value plus two standard deviations in the corresponding population.

Guidelines are given if it is more important to avoid excess consumption of a nutrients rather than a deficiency, for example, guidelines for energy, fat and sugar.

Estimated values are given if exact recommendations cannot be made due to insufficient research knowledge.

Only recommendations and guidelines are used for the database because the estimated values do not yet indicate the requirements with the desired precision and certainty. (DGE[9])

The requirements can be divided in short and long term nutrient requirements. A short term requirement is for energy which has to be met every day without major variations. The supply of all other nutrients does not have to be met every single day but over a longer period of time (Steinel[7]). Nutrient requirement data are in the database for energy and the following nutrients:

- protein
- fat
- carbohydrates
- fibre
- water

- vitamin A
- vitamin D
- vitamin E
- vitamin B1
- vitamin B2
- vitamin B6
- vitamin C
- niacin
- folic acid

- sodium
- chloride
- potassium
- calcium
- iron
- iodine.

Nutrient contents and food wastage when preparing dishes
These data are provided by the BLS (Bundes-Lebensmittel-Schlüssel = Federal-Food Item-Code). The BLS is a computerised nutrient database containing data on approximately 160 nutrients from 12,000 food items, including fresh foods (with and without waste), prepared dishes and convenience products.

The database of the information system does not contain all 160 nutrients. Only those are selected for which data on "Nutrient requirements of different person groups" are available.

Food Prices
If users cannot supply data on food prices, the database offers price data for the main food items collected by the Kuratorium für Technik und Bauwesen in der Landwirtschaft (KTBL = an institute gathering and maintaining data in the areas of agriculture and household management). The latest data refer to the year 1994.

Recipes for the dishes
Recipes for the dishes contain the ingredients and method of preparation. The database contains data that have been taken from the AID (Auswertungs- und Informationsdienst für Ernährung, Landwirtschaft und Forsten = Analysis and Information Service for Food, Agriculture and Forestry) data for nutritional analysis in private households (AID[10]).

Preparation times for the dishes
The preparation times for the dishes are available from the KTBL where the preparation times are divided into a fixed and a variable component. The fixed part is independent of the quantities of the food items that have to be prepared. The variable part increases proportional with the quantities of foods. (Steinel[7])

Classification of dishes into groups
The classification defines which dishes are assigned to which dish group. Examples for groups are: soups, salads, desserts. The necessary information for

the classification is taken from the AID data set for the analysis of nutrition (AID[10]).

3.1.2 Secondary data
Secondary data computed from the primary data consists of::
- the nutrient contents of dishes
- the food cost
- the time to prepare dishes.

These data are determined as follows:
The **nutrient contents of the dishes** are calculated from the nutrient contents of food items and the recipes of the dishes. The **food costs of the dishes** are calculated from food prices and the recipes of the dishes. The **time (total amount of time) to prepare dishes** is calculated from the preparation times of the dishes and the recipes for them.

3.2 Data dependent of the users

The data dependent on, and supplied by the users are:
- age, gender, size and weight of household members,
- likes and dislikes of the household members for dishes and various meals and
- the meal times for the planned period when individual household members eat at home.

4. Computer Programs

A large amount of data has to be processed to determining an optimum nutrition in a private household. To facilitate this process, programs are used which allow the following tasks to be undertaken easily:
- management of the database and
- solving mixed-integer linear programming problems.

The program for the management of the database is 'MS-ACCESS' and the program for solving mixed-integer linear programming problems is 'What's Best!'. Using interfaces both programs are combined into an information system.

5 Results

An optimum nutritional intake can be determined for given households and planned periods. Households differ in the number and type of individuals, for example, single households and two parent families with one or more children. The planning period will generally be one week. The optimum menu plan shows the types and quantities of dishes to be eaten by different members of household at given meal times in the planned period.

References

1. Deutsche Gesellschaft für Ernährung (DGE) e.V. (ed.) *Ernährungsbericht 1992*, DGE, Frankfurt/Main, 1992.
2. Karg, G. Modelle zur Bestimmung einer optimalen menschlichen Ernährung, *Hauswirtschaft und Wissenschaft*, 1982, **30**, 34-47.
3. Karg, G. Bedarfsgerechte und kostengünstige Ernährung im amilienhaushalt, *Entscheidungsbereich Haushalt*, Darmstadt, 1980, **257**, 49-63.
4. Baur, E. Optimale Menüs für ausgewählte Systeme der Schulverpflegung und Schülergruppen, München, Technische Universität, Fakultät für Landwirtschaft und Gartenbau, Diss. 1981.
5. Baur, E. & Karg, G. Optimale Menüs für ausgewählte Systeme der Schulverpflegung und Schülergruppen, *Hauswirtschaft und Wissenschaft*, 1982, **30**, 118-127.
6. Armstrong, R.D. & Sinha, P. Application of quasi-integer programming to the solution of menu planning problems with variable portion size, *Management Sience*, 1974, **21**, 474-482.
7. Steinel, M. *Normativer Kosten-Nutzen-Vergleich verschiedener Ernährungsformen im privaten Haushalt*, Studien zur Haushaltsökonomik Bd.8, Lang, Frankfurt/Main, 1992.
8. Steinel, M. Modelle zur Bestimmung einer optimalen Ernährung in privaten Haushalten, *Zeitschrift für Ernährungswissenschaft*, 1993, **32**, 79-92.
9. Deutsche Gesellschaft für Ernährung (DGE) e.V. (ed.) *Empfehlungen für die Nährstoffzufuhr*, Umschau Verlag, Frankfurt/Main, 1992.
10. Auswertungs- und Informationsdienst für Ernährung, Landwirtschaft und Forsten e.V. (AID) (ed.) *Daten zur Analyse der Ernährung*, Bonn, 1984.

Poster

Computer and dietary assessment

C.E.A. Seaman
Centre for Food Research, Queen Margaret College, Edinburgh EH12 8TS, Scotland

As computers become faster, more powerful and more readily available, their usage in food related research is increasing. Common computer applications include nutritional analysis, nutrition education, the collection and analysis of data in the sensory evaluation laboratory and the statistical analysis of data from food research projects. In addition, information and shareware programs relating to food are common on the Internet and can be downloaded for personal use.

In the development of computer programs for the non-professional to analyse dietary information, the user-friendly interface; that part of the software package which is shown on-screen and which largely determines the ease with which the program can be used, size of database and cost are all important considerations.

This interactive poster outlines and demonstrates the development of nutrition education software Nutri-Test. Nutri-Test was developed and validated at Queen Margaret College for Master Foods, a division of Mars UK. It provides a useful example of a package designed to appeal to consumers and schoolchildren and a cost-effective way of delivering healthy eating information, incorporating a quiz and providing the opportunity to analyse a 'typical days diet'.

Section 2.4: Nutrition and Dietary Intake

Nutritional changes in Eastern Germany since 1990

G. Ulbricht
German Institute of Human Nutrition, 14558 Bergholz-Rehbrücke, Germany

Abstract

The fall of communism was accompanied by fundamental economic and social changes in Eastern Germany. In addition to this, the nutritional behaviour of many people also changed. The development of nutritional patterns in the new "Läender" was investigated using several techniques involving questions in individual households and canteens, particularly in the Land Brandenburg. Results are described in three areas: food consumption, meal arrangements and nutrient intakes. Trends from 1990 to 1994 are also outlined.

1. Introduction

The change in Eastern Germany and the reunification in 1990 changed individual food and nutrition habits. These changes were investigated and described in several studies [1- 8]. Results from three studies, which were undertaken or interpreted by the Central Institute of Nutrition/the German Institute of Human Nutrition Potsdam-Rehbrücke [3,5,8] have been selected and three are discussed here, particularly:

     1. changes in food consumption,
     2. changes in meal arrangements, especially eating away from of home,
     3. changes in energy and nutrient intake.

The results characterise the most important developments in nutrition in Eastern Germany since 1990, although their validity is only proven for the groups of people investigated. The statistics demonstrate that it may be incorrect to apply these results to the population as a whole.

However many results from different surveys agree with each other. This observation suggests that the groups of people investigated behave in a very similar way. Therefore, most of our results may be generally applied.

2. Methods

The results of all studies are based on the questioning of consumers, or from households records. Between 150 and 1000 people or households were included in each study. The selection of interviewees and the recording of information were carried out using different methods [3,5,8]. Therefore, the data from these studies were analysed separately, normally using SPSS. Results for nutrient intakes were calculated from food consumption data, utilising data from Bundeslebensmittelschlüssel, Version II.

3. Results

3.1 Changes in food consumption

Figure 1 and Table 1 show the most important changes in food consumption since 1990, compared with the consumption in the former German Dempcratic republic (GDR). The bar chart (Figure 1) shows, for selected products, the percentage of people who ate more (right side of the vertical line) and the percentage of people who ate less than before (left side of the vertical line). The columns show the frequency of increase and frequency of decrease for each product during the survey.

As shown in Table 1 we can divide the foods into three categories. The first category contains products which have greatly increased in demand. In the first period after reunification (1990/91), more products belonged to this category than in the later period (1992/93). This was a result of an unsatisfied demand from the people in the former GDR. Furthermore, the improved taste of products and the efforts towards a more healthy diet, were the deciding factors for the much higher consumption of, for example, fruit (including tropical fruit), vegetables, margarine, wholemeal bread and yoghurt. During the following survey, the intensity of demand diminished and several products fell into the second category, for example, canned fruit and vegetable, cocoa products and yoghurt.

There was an increase in the number of products within the second category in the subsequent surveys compared with the previous ones because of a decreased intensity of demand changes. A transition of products from the first to the third category, or in reverse, order did not happen.

Table 1. Categories of food consumption changes 1989 - 1993

Category of consumption trend	Survey[1]	Product No.	Identification
1. Category: increase of consumption frequency > 20 %	I (P 1989 - 90)	1, 2, 3, 4, 6, 13, 15, 19, 20, 24, 26, 27	1 tropical fruit, fresh 2 fresh fruit (without 1) 3 fresh vegetable 4 canned tropical fruit 5 canned fruit(without 4) 6 canned vegetable 7 potatoes 8 wholemeal bread 9 mixed-grain bread 10 white bread 11 cereal products 12 pastries 13 cocoa products 14 sweets 15 margarine 16 cooking and salad oil 17 butter 18 milk/cream 19 yoghurt 20 cheese 21 quark 22 beef 23 pork 24 poultry 25 sausages 26 fish products 27 non alcoholic soft drinks 28 alcoholic beverages
	II (P 1990 - 91)	1, 3, 15, 20, 21, 24, 27	
	III (B 1989 - 92)	1, 2, 3, 8, 13, 15, 19, 20, 21, 24, 27	
	IV (B 1992 - 93)	1, 2, 3, 8, 15, 20, 24	
2. Category: moderate change of consumption frequency or equality (< ± 20 %)	I (P 1989 - 90)	5, 7, 9, 11, 12, 14, 16, 18, 21, 25, 28	
	II (P 1990 - 91)	2, 4, 5, 6, 7, 9, 10, 11, 12, 13, 14, 16, 18, 19, 22, 25, 26, 28	
	III (B 1989 - 92)	7, 10, 11, 18, 22, 25, 26	
	IV (B 1992 - 93)	7, 9, 11, 13, 18, 19, 21, 23, 25, 26, 27	
3. Category: decrease of consumption frequency > 20 %	I (P 1989 - 90)	10, 17, 22, 23	
	II (P 1990 - 91)	17, 23	
	III (B 1989 - 92)	9, 12, 17, 23	
	IV (B 1992 - 93)	10, 12, 17, 22	

Notes: [1] P = Potsdam
B = Land Brandenburg

Only a few products, for example, butter, pork, beef and on the last survey, white bread, had to be placed into the third category. Butter appeared in this category in all four surveys. The significant and persistently diminishing demand for butter, and the decreased consumption of pork and beef are due particularly to substitutions which were based on an increased consumption of margarine and poultry. In this case, the lower prices of these commodities play an important part.

3.2 Meal arrangements

The typical daily meal arrangement in Germany consists of 3 main meals: (1st breakfast, lunch and supper). In several cases there are 2 additional meals (2nd breakfast [elevenses] and "vespers", [high tea]) as well as occasional snacks between the meals. Remarkable differences in nutritional behaviour should be noted between working days and weekends. In the former GDR, about 90 % of the adults had three to four daily meals [9]. Supper had the highest frequency. The majority (more than 90 % of the sample) took one hot meal per day, either at noon (88 %) or in the evening (12 %). Restaurant meals were about 5 percent. As nearly all the people were working, canteen meals were very frequent. In 1985/88 they amounted to more than 8.5 millions main meals on weekdays. Thus, before the change, nearly every second person participated in canteen food.

Since 1990, canteen food has changed fundamentally. Companies have, in many cases, developed from a social institution to an independent firm operated as a free enterprise. Subsidies for canteen food were stopped and the prices of meals went up. Therefore, and because of the reduction in the number and size of enterprises, the interest in canteen meals has also decreased. Table 2 shows the beginning of this process.

Table 2. Eating away from home (study in Potsdam 1990/91, 212 adults)

	Answers of interviewees (%)				
	Canteen meals	Restaurant meals	Fastfood in Restaurant	Fastfood at a kiosk	Brought in food
Before and after the change no participation	22.2	19.2	67.9	46.0	14.4
More than before	1.9	0.9	15.1	21.6	37.0
Exactly as before	27.8	27.2	8.0	20.2	42.3
Less than before	29.7	44.1	3.3	7.5	2.9
Participation behaviour and never after the change	18.4	8.5	5.7	4.7	3.4
Σ	100	100	100	100	100

Canteen and restaurant food have been reduced (lower lines): and fastfood as well as brought in food have been consumed more as before.

Further surveys established that in 1993 only 70% of adults in the new "Länder" preferred a daily hot meal [10]. The participation in canteen food has decreased more and more. According to preliminary estimates for 1995, a maximum of two million meals per day were produced in firms, enterprises and educational establishments. This represents a quarter when compared with the canteen food in the former GDR.

3.3 Trends in energy and nutrient intake

The observed trends in food consumption have had positive effects on the intake of several nutrients. This was shown in the Study in Potsdam 1990/91 [7]. Since the change in Eastern Germany, the intakes of ascorbic acid, calcium, vitamin A and carbohydrate have increased, whereas they were lower for fat, cholesterol, iron and energy. These trends were nearly the same for men and women.

In another research project, we evaluated food consumption data of a continuous family budget survey in the new "Länder" from 1987 to 1994. The selected data in Figures 2 and 3 on eating at home (without canteen and restaurant meals), which are valid for 51 - 64 years old men, in households (pensioners), demonstrate valid changes also for many other groups of people. Figures 2 and 3 consider only nutrients with the most significant changes from 1987 until 1994 (except for protein and fat). These nutrients were energy and carbohydrates (decrease), vitamins A (especially carotene), vitamins E and C (increase) and vitamin B_1 (decrease), calcium (remarkable increase) as well as polyunsaturated fatty acids (remarkable increase) and cholesterol (decrease). These results confirm the trends which have been described in the first study.

4. Conclusions

1. Nutritional changes were most conspicuous in 1990 and 1991. After that, the intensity of change was slower. Butter, beef, pork and wholemeal bread decreased, tropical fruit, cheese and margarine increased.
2. After 1990, fastfood meals and the takeaway food increased, where as complete canteen food meals decreased.
3. The trends for energy and nutrient intakes agree with the aims of a healthy diet, for example, less energy and cholesterol and more calcium, vitamin C and unsaturated fatty acids. However, these developments are insufficient for an improvement in the overall nutritional situation.

References

1. Donat, P., Knötzsch, P.: Ein Lebensmittelmarkt im Wandel. AID Verbraucherdienst 35 (1990), S. 223-229.
2. Ulbricht, G., Friebe, D., Bergmann, M., Schmidt, G., Seppelt, B.: Änderungen im Ernährungsverhalten in den neuen Bundesländern. AID Verbraucherdienst 36 (1991), S. 235-240 und 38 (1993), S. 248-254.
3. Grunert, K.G., Grunert, S.C.: A comparitive analysis of the influence of economic culture on East and West German consumers' subjective product meanings. MAPP working paper no. 14, Handelshojskolen i Arhus, 1993.
4. Hruby, J., Ulbricht, G.: Change in Former Communist Countries and its Impact on nutrition Behaviour: Examples from two Central European Countries. Current Research into Eating Practices. Contributions of Social Sciences. AGEV Publication Series, Vol. 10. Supplementum to Ernährungs-Umschau, Frankfurt 1995, S. 201-206.
5. Friebe, D., Schrödter, R., Kornelson, C.: Forschungsbericht zum Projekt "Verbraucherverhalten im Land Brandenburg 1993/1994" des Deutschen Instituts für Ernährungsforschung Potsdam-Rehbrücke im Auftrag des Ministeriums für Ernährung, Landwirtschaft und Forsten des Landes Brandenburg. Bergholz-Rehbrücke 1994.
6. Thiel, Ch., Do Minh, T., Heinemann, L.: Vitaminaufnahme vor und nach der Wende in Ostdeutschland Vitamine-Mineralstoffe-Spurenelemente 9 (1994), S. 32-34.
7. Ulbricht, G., Schmidt, G., Seppelt, B.: Veränderungen der Energie- und Nährstoffaufnahme der Bevölkerung im ersten Jahr der Deutschen Einheit - Ergebnisse einer Modellstudie in Potsdam. Zeitschrift für Ernährungswissenschaft 35 (1996) (in print).
8. Aufgliederung der Einnahmen und Ausgaben ausgewählter privater Haushalte. Ergebnis der laufenden Wirtschaftsrechnungen. Herausgegeben vom Statistischen Bundesamt, Wiesbaden 1994.
9. Dlouhy, W.: Wichtige Ernährungsgewohnheiten der erwachsenen Bevölkerung der DDR. Forschungsbericht des Instituts für Marktforschung, Leipzig 1982.
10. Kauf- und Ernährungsgewohnheiten in den neuen Bundesländern. Eine Untersuchung des Instituts für Marktforschung Leipzig bei 1000 Haushalten in den neuen Bundesländern. Marktforschungsbriefe, no. ½ 1994. Herausgegeben von der Centralen Marketinggesellschaft der Deutschen Agrarwirtschaft mbH (CMA), Bonn.

Culinary Arts and Sciences   303

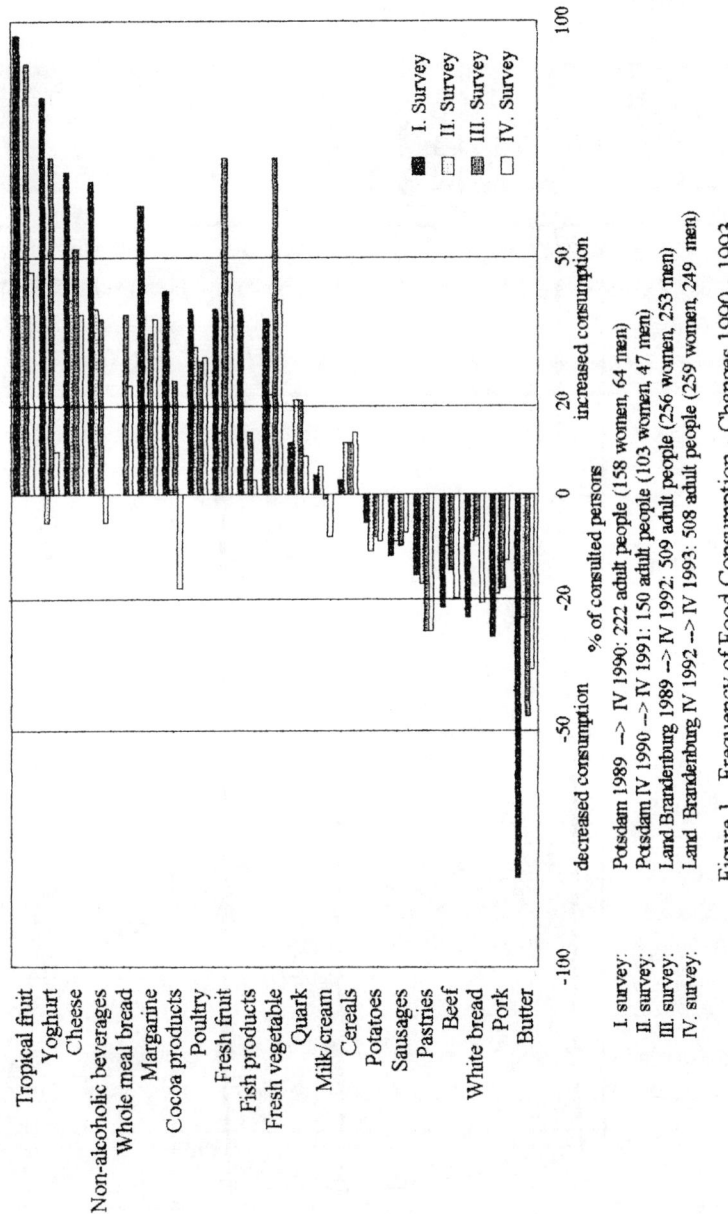

Figure 1. Frequency of Food Consumption - Changes 1990 - 1993

I. survey: Potsdam 1989 --> IV 1990: 222 adult people (158 women, 64 men)
II. survey: Potsdam IV 1990 --> IV 1991: 150 adult people (103 women, 47 men)
III. survey: Land Brandenburg 1989 --> IV 1992: 509 adult people (256 women, 253 men)
IV. survey: Land Brandenburg IV 1992 --> IV 1993: 508 adult people (259 women, 249 men)

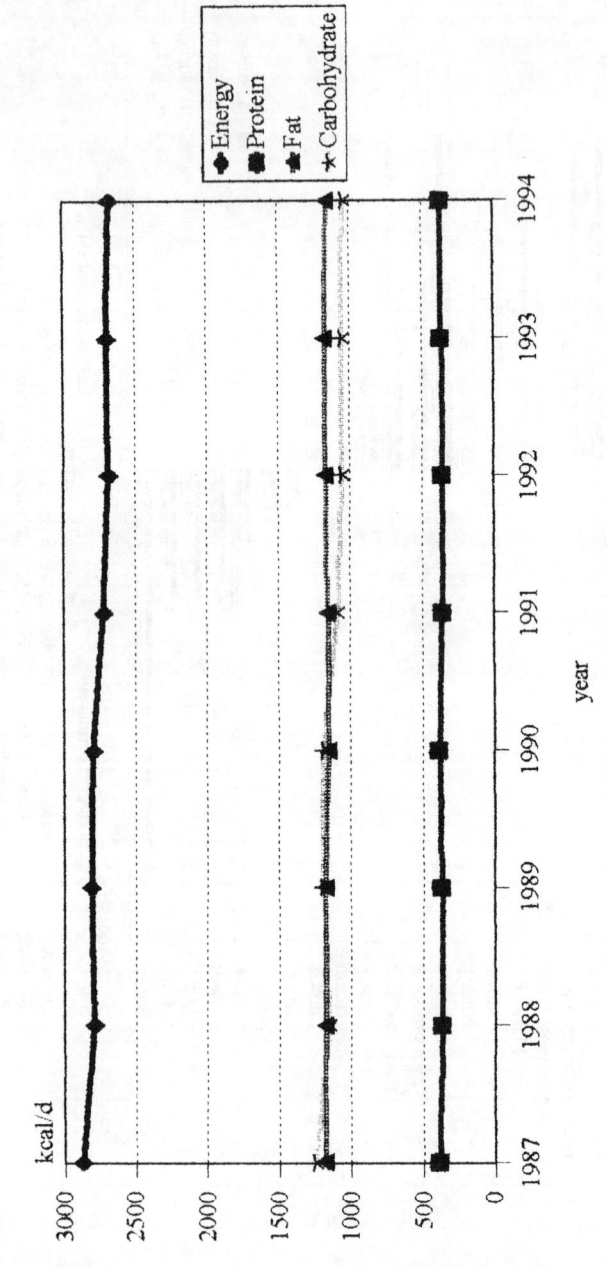

Figure 2: Energy intake per person, men 51-64 years

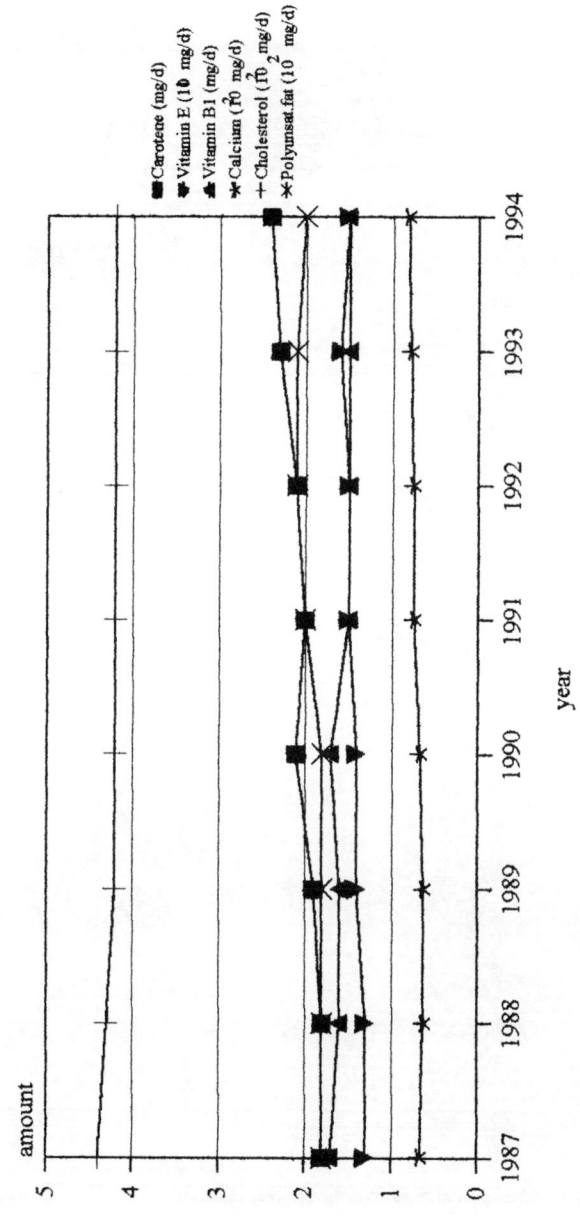

Figure 3: Intake of selected nutrients in Eastern Germany from 1987 until 1994 (men 51 - 64 years old)

Dietary intake in the Czech Republic

A. Aujezdská, D. Müllerová, L. Müller

Department of Hygiene, Faculty of Medicine in Pilsen, Charles University, 301 66 Pilsen, Czech Republic

Abstract

High energy intakes among the population of the Czech Republic, which consumes a high proportion of fatty meats and whole milk products, is reflected in the nutritional status of the inhabitants. This is due partly to insufficient nutritional education and partly to inadequate food labelling. The available database which contains the Czech Food composition data is limited and contains a poor variety of nutritional analyses. With this in mind, we examined 900 three-day dietary records of 4 groups within the population. We observed an excessive consumption of energy and fat, and a low intake of vitamin C in a group of managers and businessmen. A group of young women (students and mothers) displayed lower intakes of energy, limited intakes of iron, calcium, vitamins A, B_1, and B_2 compared with the Czech Recommended Dietary Allowances (RDA).

1. Introduction

1.1 Czech meal patterns in the period 1960 - 1990

During the last thirty years (1960 - 1990) consumption of meat in the Czech Republic has increased by up to 170% compared with the levels of the 1960's (Table 1). This is primarily due to an increase in the consumption of pork which has risen by 160%. At the same time, the intake of eggs has doubled and the intake of milk products has also increased. Compared to 1960s, the intake of potatoes and vegetables dropped by 20%. During the thirty years, fat intake increased by 1/3 which is not conducive to the good health of the population. In the traditional Czech kitchen, butter and lard have always prevailed. In 1960, animal fat represented 45% of fat intake. In 1990 total fat intake increased, but animal fat represents only 36% of total fat intake[1].

Table 1. The development of food consumption in kg per pers Czech Republic measured using a balance method [2].

The Increasing Trend (without fat)			The Decreasing Trend		
year	1960	1990	year	1960	1990
meat (all)	56.8	96.5	cereal	125.9	114.9
pork	31.5	50	potatoes	100.3	77.9
poultry	4.7	13.6	vegetables	84.7	66.6
fish	4.7	5.4	fruit	61.9	59.7
eggs	179	340			
fat and oil (all)	21.4	28.5			
butter	6.1	8.7			
lard	7.5	6.9			
plant fat	7.7	12.8			
milk, dairy (all)	173	256			
sugar	36.3	44			

The decreased consumption of vegetables, sustained low consumption of fruit, together with high energy intakes, particularly from fat, had a negative influence on the health of the Czech population.

High cardiovascular morbidity and mortality, the incidence of obesity, diabetes mellitus and cancer in the Czech population is among the highest in Europe (Table 2). Hyperlipidemia among middle aged men is over 50%[1]. In 1993 standardised mortality rates for men and women were the third lowest among post communist countries. Compared with neighbouring Austria, where the mortality situation was practically the same as in the Czech Republic until the early 60s, Czech standardised mortality in 1992 was 42.8 % higher in men and 33.5 % in women[2].

Table 2: Czech population: Standardised mortality rate (European standard) by selected causes of death [2,3]

Cause	Year	
	1960	1990
all causes - males	1500	1560
all causes - females	1010	940
Coronary heart disease SMR (both genders, 0 - 64 years) / 100000	47	86
Colorectal cancer SMR (both genders, 0 - 64 years) / 100000	7.5	13.7

1.2 Czech meal patterns in the period 1990 - 1995

The transition from a centrally planned to a market economy has brought substantial changes in the consumption of fat. The consumption of butter has decreased and has been substituted with margarine (20% polyunsaturated), blends of butter or edible oils. Margarine has become popular because it is less expensive than butter. The consumption of lard, which was the most traditional solid shortening, has decreased, and has been replaced with cooking fats (mainly, partially hydrogenated rape seed oil) or fats whipped with nitrogen[4]. Thus it can be seen that the consumption of edible oils has increased.

The patterns of food consumption have changed dramatically and, for example, less red meat but more white meat and rabbit is consumed. In addition, total milk consumption has reduced and there has been a switch from full fat 3%, to whole milk 2%, or skimmed milk 0.5%, as well the selection of low fat yoghurts and cheeses. This corresponds with the data from the Institute for Economy in Agriculture and the Czech Statistical Office (IEA&CSO)[1] and is shown in Table 3. Established Czech Recommended Dietary Allowances for the average population are presented for comparison in Table 4.

Table 3. Nutritional Evaluation of Food Consumption in the Czech Republic[1]

Nutrient	Unit	1989	1993
Energy	MJ	11.7	11.5
Protein total	g	92.2	87
animal origin	g	56.8	47.7
vegetable origin	g	35.4	39.3
Fat total	g	124.6	113.5
Fat % of total energy	%	39.4	36.5
Carbohydrate	g	338.6	358
Calcium	g	914	774
Iron	g	15.4	15.3
Vitamin A	mcg	858	761
Vitamin B1	mg	1.25	1.27
Vitamin B2	mg	1.54	1.45
Vitamin C	mg	40.7	44.6

Table 4. Recommended Dietary Allowances for average individuals

Nutrient	Unit	1992
Energy	MJ	9.62
Protein total	g	74.5
animal origin	g	38.5
vegetable origin	g	36
Fat total	g	69.6
Fat % of total energy	%	27.4
Carbohydrate	g	342.3
Calcium	g	874
Iron	g	15
Vitamin A	mcg	869
Vitamin B1	mg	1.1
Vitamin B2	mg	1.5
Vitamin C	mg	78.6

2. Methods

The new political and economic situation is stimulating a significant social change and a multilateral development in the previously uniform lifestyle of our society. This level of societal differentiation can be seen through one's nutritional behaviour. The 4 groups examined in 1994-1995 were undergraduate medical students (males and females), a newly rising group of businessmen and managers, mothers, 6 months after the birth of their child who were not lactating, and pensioned adults (males, females).

The examination included their health history, anthropometric measurements: body weight, body height, body mass index (BMI, body weight in kg divided by square body height in m), percentage body fat measured by callipers on 4 body sites. Analysis of plasma cholesterol level was carried out with the help of Reflotron.

Dietary data was obtained from three-day estimated food records of the varieties and amounts of all food consumed during two working days and one weekend. In addition to energy, the nutrients studied included: protein, fat, carbohydrate, calcium, iron, and vitamins A, C, B_1, and B_2. Nutrient values were calculated using software based on the Medical Faculty's nutrient composition database. Dietary daily intake examined was the mean intake and standard deviation in all samples. Iron and vitamin A estimates are by their nature very skewed. This is because food and nutrient intakes vary from day to day. For this reason, means and standard deviation should be used but interpreted with caution.

3. Results

3.1 Selected dietary intake findings.

Results, means and standard deviation (in brackets), are presented in Tables 5 and 6.

Mean energy intake was highest in the group of managers, 11.8 MJ (2800 kcal) compared with the other groups from data of Czech population calculated by IEA&CSO[1]. Mean protein intakes ranged from 0.9 to 1.2g per kg body weight. Protein accounted for about 12.8-14.7% of total energy intake.

The highest mean total fat intake, 95 g, was also found in the group of managers. Total fat contributed a lower percentage of total energy during our research (31%) compared with 36.5% during the survey of (IEA&CSO)[1] in 1993. A shift in the total fat intake could have occurred over time. Our results are extremely low and this could be caused by systematic bias in our food composition database or by under-reporting of fatty meals. From this point of view, some studies have documented that food consumption is under-reported by as much as 25% and occurs more often in women, overweight and weight conscious individuals[6,7].

Mean iron intakes were consistently higher in males than in females, coinciding with similar patterns observed in mean total energy intakes. Mean iron intakes in groups of females and elderly males ranged from 9.1 to 10.8 g per day.

Table 5. Basic anthropometric data, plasma cholesterol levels and daily intakes of energy and nutrients, mean (SD) for males.

Males		Undergraduate students N=208	Managers N=133	Elderly N=117
age	years	23 - 25	25 - 58	60 - 84
cholesterol	mM	4.4 (0.8)	6.11 (0.9)	5.9 (1.4)
BMI	kg/m²	23.3 (2.3)	26.1 (3.2)	27.2 (3)
Body fat	%	14.2 (4.9)	18.6 (5.2)	24.6 (6.3)
Energy	MJ	11.5 (3.4)	11.8 (4.1)	8.9 (2.7)
Proteins (all)	g/kg BW	1.2	1.13	0.9
of animal orig.	g	49.3 (20.5)	49.1 (17.9)	38.3 (15.7)
of vegetable orig.	g	43.4 (16.6)	46.3 (22.4)	35.9 (17.7)
Fats (all)	g	88.6 (34)	94.8 (37.1)	68.5 (25.4)
Carbohydrate	g	384.1 (119.3)	391.7 (164.1)	293.6 (128.7)
Calcium	g	853.2 (457)	690.8 (354)	724.8 (314)
Iron	g	14.8 (5.1)	14.4 (6.2)	10.4 (4.4)
Vitamin A	mcg	841 (867)	807 (1241)	1063 (2336)
Vitamin B₁	mg	1.4 (0.5)	1.3 (0.5)	1 (0.4)
Vitamin B₂	mg	1.7 (0.6)	1.7 (0.7)	1.3 (0.5)
Vitamin C	mg	51.3 (46.9)	35.1 (31.3)	34.2 (33.9)

Table 6: Basic anthropometric data, plasma cholesterol level and daily intake of energy and nutrients, mean (SD) for females.

Females		Undergraduate students N =273	Mothers N=124	Elderly N=117
age	years	23 - 25	18 - 38	60 - 84
cholesterol	mM	4.6 (0.9)	-	6.4 (1.4)
BMI	kg/m²	20.7 (1.9)	22.8 (3.7)	27.2 (18.8)
Body fat	%	23.5 (3.9)	-	32 (6.7)
Energy	MJ	8.3 (2.7)	8.2 (3.4)	7.4 (2.4)
Proteins (all)	g/kg BW	1.07	1.1	0.9
of animal orig.	g	33.5 (14.5)	38.8 (20.5)	33.8 (16.5)
of vegetable orig.	g	30 (12.4)	31.8 (18.3)	31 (14.6)
Fats (all)	g	66 (27.9)	67.9 (36.4)	53.3 (23.5)
Carbohydrate	g	281 (93.2)	298 (279)	252 (90.3)
Calcium	g	713.7 (356.6)	717.4 (531.1)	673 (379.7)
Iron	g	10.7 (4.2)	10.8 (4.5)	9.1 (3.6)
Vitamin A	mcg	726.8 (583.2)	842.1 (975.4)	791.3 (2278)
Vitamin B$_1$	mg	0.96 (0.36)	1.01 (0.50)	0.89 (0.4)
Vitamin B$_2$	mg	1.12 (0.46)	1.3 (0.81)	1.08 (0.59)
Vitamin C	mg	59.8 (42.4)	51.4 (41.7)	41.2 (35.7)

Some studies have shown that calcium intakes play a role in colon cancer, high blood pressure and kidney stones, and are instrumental in the prevention of the osteoporosis[8]. Recent literature has identified the importance of consuming calcium while bones are still forming to maintain peak bone mass[9]. The US RDA is 1200 mg for adolescents and adults up to 25 years and 800 mg for adults over the age of 25[10]. Osteoporosis is an important health problem in Czech elderly. From this point of view, the results presented for calcium intake, which range between 670 to 850 mg, are not very optimistic.

The mean intakes of vitamin C from food were (34.2-59.8 mg) indicating that a large portion of the Czech population did not meet the Czech RDAs. Mean vitamin C intakes were also lower in males, than females.

3.2 Social characteristics and nutritional evaluation of individual social economical groups

3.2.1 Managers and businessmen

The group of managers examined are characterised by their high social position, good finances, job stress, lack of free time, high prevalence of smokers and low nutrition education. The dietary patterns of this group are typified by high total energy intakes, higher intakes of dietary protein, low intakes of vitamin C and calcium from their food. Higher body fat, blood pressure and cholesterol levels are the important markers of their health state.

3.2.2 Undergraduate medical students

Their social characteristics include: high achievers, limited finances, good well being, and a feeling of job uncertainty. Students of medicine -males- showed a very good state of health as well as optimal dietary patterns. Students-females- have a lower BMI, which is in agreement with their slightly lower energy intakes. Mean intakes of calcium, iron, vitamin A, B_1, and B_2 from food are dangerously low.

3.2.3 Mothers

Groups of mothers often have limited finances, are concerned about their babies and often lack nutritional education. Their meal pattern is similar to the meal pattern of female undergraduate students.

3.2.4 Elderly

The group examined were self-sufficient elderly associated in Senior Citizens Clubs. Dietary intakes of calcium, iron and vitamins A and C were found to be insufficient. These correspond with low daily intakes of foods such as milk and dairy products, fruit and vegetables.

4. Conclusion

The long term, unfavourable, characteristics of the Czech diet: high consumptions of fat and insufficient amounts of vegetables, has slightly decreased in the last 5 years in favour of a more healthy diet with lower intakes and a better ratio of total fat. The intake of vitamin C for most of the population is not optimum. Also important are the borderline intakes of iron in women during their fertile period. Price increase have not only lowered intakes of fat, but also intakes of calcium. Economic differences in social groups after 1989 have created various dietary models. The traditional model is becoming highly dependent on the regulation of market prices and the finances to pay for it. From the point of view of health, future Czech diets must increase the consumption of fruit, vegetables, and low fat milk products, and continue the trend of decreasing the consumption of animal fat products, especially pork and eggs.

References

1. Andìl, M. Dietary carbohydrate intake in the Czech Republic, in *Seminar on Carbohydrates and Human Health*, International Life Sciences Institute, ILSI Europe, Prague, Czech Republic, 1995. Working Document.
2. Statistické informace, *Spotøeba potravin na 1 obyvatele ÈSFR, ÈR, SR*, Prague, roèník 1990.
3. Brázdová, Z. *Dietary guidelines for Czech population in the beginning of the 21st century*, Masaryk University, Medical School, Brno, Czech Republic. In Book of Abstract, Seventh European Nutrition Conference, Vienna, Austria, 1995.
4. Pokorný, J. *Trends in fats and oils consumption in the Czech Republic for the prevention of cardiovascular diseases*. Department of Food Science, Prague Institute of Chemical Technology, CZ - 16628 Prague, Czech Republic. In Book of Abstract, Seventh European Nutrition Conference, Vienna, Austria, 1995.
5. Pánek, J. & Pokorný, J. Recommended Dietary Allowances in Czech Republic, Prague Institute of Chemical Technology, Prague, Czech Republic 1992.
6. McDowell, M.A.et al. *Energy and Macronutrient Intakes of Persons Ages 2 Month and Over in the United States: Third National Health and Nutrition Examination Survey, Phase 1, 1988-91*. In Advance Data No. 255, 1994. U. S. Department of Health and Human Services, 6525 Belcrest Road, Hyattsville, Maryland 20782.
7. Schoeller, D.A. *How accurate is self -reported dietary energy intake?* Nutr. Rev. 48: 373 - 9. 1990
8. Sandler, RB, Slemenda, C., La Porte, R. et al. *Postmenopausal bone density and milk consumption in childhood and adolescence*. Am J Clin Nutr 42:270-4.1985
9. Arnaud, CD. & Sanchez, SD. *The Role of Calcium in Osteoporosis*. Annu Rev Nutr 10:39-414.1990
10. Recommended Dietary Allowances. 10th ed. Washington: National Academy Press. 1989.

The development and characteristics of Bulgarian national cuisine

I. Alexieva

Department of Public Catering, Higher Institute of Food and Flavour Industries, 4000 Plovdiv, Bulgaria

Abstract

The effects which factors such as history, social and economic conditions, and the influence of other nation's, religions, traditions, and customs have had on the characteristics and the style of cooking in the development of Bulgarian national cuisine are considered in this paper.

The historical development of the Bulgarian nation and their influence on traditional culinary methods are explored. The role and the effects of Thracians, Slavs and Proto-Bulgarians and the influence of other nations through various historical events on the style of cooking, variety of dishes and culinary terminology are also examined. The influence of Christianity, its customs and celebrations and their effects on traditions in the national diet and cuisine are shown as are the characteristics of the old Bulgarian cuisine.

Introduction

Bulgarian national cuisine reflects dietary traditions that have developed over 13 centuries of history. It has been influenced in many different ways including, geography, climate, history and economic factors as well as demographic changes, communications and other nationality's ways of life, for example, traditions, habits, religions and culture. Some of these factors which have influenced and effected the development of cookery and which have contributed towards the distinguishing characteristics of Bulgarian national cuisine are discussed.

Historical aspects

Before the foundation of the Bulgarian State

The formation of the Bulgarian State is a long and complicated ethnic-genetic process, the foundations of which were laid at the end of 9th and the beginning of the 10th century. The three dominant ethnic groups involved were: Slavs, Thracians and Proto-Bulgarians, each of whom contributed to the development of the basic characteristics of the national style of cooking.

Thracians are the oldest ethnic group to have played an important role in the Bulgarian ethnogenesis. The scant written documents and historical excavations give an idea of the food and methods of cookery they used. The Thracians made their living by hunting, fishing and farming, growing wheat, millet, flax and hemp. Fruit, vine growing and gardening were also developed and wine was widely produced. There are many traces showing that animal breeding was also known. Theophrat (372-287 BC) describes walnuts, chestnuts, pear and apple trees, hawthorn, vines, wild cucumbers, pomegranates, olives and broad beans among the plants, growing on the lands of Thracians. "...Thracians can make different types of wine - both exciting and calming. They prepare tasty bread from chestnuts. They like feasts and they drink not only wine, but "boza" (barley ale, a sweet barley drink with slightly sour taste)..." (Pliney, AD 23-79)[1]. According to Ksentophan, the Thracians ate a lot of vegetables, but meat and milk were their main dishes. They were nicknamed "milkmen" for the variety of dairy products in their diet. Historical documents, show that they were familiar with cheese making, butter, curd and yoghurt. Viron adds "... The Thracians eat meat, fats and pork. They breed pigs...". Another Roman author says: "...they eat honey, milk and cheese...". Amiam Mitsein (4th century) claims that Thracians live longer due to their diet. They restrain from 'harmful foods'. Their diet included milk, fruit, dried and fresh; vegetables and herbs. Furthermore, the Thracians used mineral water for healing purposes during the Roman conquest.[1]

Slavs are the second ethnic group, that participated in the formation of the Bulgarian Nation and are renown as farmers and animal breeders. From several written sources it is evident that Slavs dried and salted pork for inclusion in their winter diet. They bred sheep and goats, were very good apiarists, ate honey and dried fruits ("oshav") and drank "medovina" (mead). The Slavs knew of tubers and roots, nuts, and also herbs and plants, suitable for both eating and healing.

Many celebrations were connected to their religious beliefs but most of the Slavs' celebrations have been integrated with those of the Thracians and Byzantines along with various traditions and methods of cooking.

Proto-Bulgarians are the most dominant ethnic group to have participated in the formation of the Bulgarian Nation. They were steppe people, spending most of their time on horse back. Proto-Bulgarians supplied their tribes with game, fish, horse meat, veal, pork and mutton from the animals they bred, used milk and made "kumis" (fermented mare's milk). According to new data, Proto-

Bulgarians carried with them "trahana" (resembling couscous) and today a similar dish is still found in the national cuisine. They added to their diet dairy and meat products, edible fruits and plants which they found during raids. They prepared meat either by roasting on a spit using live coals or by boiling in a pot. Archaeological data show that the primary sources of food for Proto-Bulgarians were hunting, stock breeding and basic farming, which became better developed at a later period.

Living together, the ancient inhabitants of Bulgarian lands (Proto-Bulgarians, Slavs, Thracians, Byzantines, etc.) shared not only their cultures, traditions and customs, but also their methods of cooking and types of dishes produced.

Bulgarian State (681-1396)

The information about the food of the Bulgarians in earlier times is scant and one can only guess from remains and historical monuments[2]. Excavation of ancient mounds, hillocks and caves give an idea of what the inhabitants used to eat in the 8th century. Traces of feasts and treats such as the remains of pottery, amphora etc. can be found but it is believed that these relics are from pagan times, before the conversion of the Bulgarians to Christianity in 865. During excavation of the King's palace in Pliska, animal bones, remains of pitchers, cups, pots, jars and bowls were found, and these give an idea of the cookery methods and the type of foods used. There are some historical remains, indicating that ancient Bulgarians used honey. By the 6th century Proto-Bulgarians were well acquainted with roasting and historical evidence shows that in the 9th century Khan Omurtag made prisoners of war share his meals and eat roast meat. There were a lot of small dairy houses and water mills scattered throughout the country and many bakery products were known.

It is difficult to get any idea of the First and the Second Bulgarian Kingdom's cuisine from rare documents. There are some historical records showing that during the reign of King Boris III and Simeon II, Kliment Ohridski trained 3500 young men in the secrets of plant cultivation. The work of Theophilake Ohridski provides information on the cultivation and grafting of fruit trees during the 9th century demonstrating that fruit was often present on the table. The letter of King Boris III to the Pope Nickolai in Rome (866) adds to the knowledge of Bulgarian traditions in which the King asks several questions: "...What kind of animals and birds are allowed to be eaten?...": to which the Pope answers,
"...Everything that lives and moves (fish, birds, animals, vegetables), ... all the animals, that have meat, which are not harmful for the human can be used for eating..."
"...How many times and how long does man have to abstain from meat during the year?...".
"...Man has to keep away from meat on the days of fast and to eat as little meat as possible during the other days. Only sucklings are allowed to drink milk in

Lenten days. During the Lent hunting is forbidden and no fats or meat are allowed...".
The role of vegetables and herbs for a healthy diet are also discussed.

The Byzantines rhetor Grigorii Antioh (12th century) describes in his letters "...the everyday life of the ordinary people is for moaning, because the bread is made from bran or millet and it is half-done, because it has been baked in the fire; the wine is sour and disgusting. The salted fish smells bad. Meat is absent from the table, because no one kills the domestic animals used to help cultivate the land. Bread is full of whole wheat ears, flakes and chaff and what is worse, it is dry and as hard as a stone. It is of such a low quality that it scratches your palate and mouth". The author also mentions millet "kasha" (porridge), "kachamack" (hominy), bread and milk as characteristic dishes for the peasants at that time. In the book of King Borill (13th century) "Sinodikat", recipes for meat and poultry dishes can be found. The Arabic writer Ahmed Ibn Fadlana (10th century) says that in the King's Palace guests were treated to roast meat and varieties of other dishes. During the feast a special drink "sedge", made from honey was offered. The studies of Academic Stranski disclose that the ordinary Bulgarian fed mainly on fresh or dried ("sushevo") wild fruits (corneal cherry, strawberries, raspberries, blackberries, figs, wild pears) and the herbs, sorrel and thyme. The hermit Ivan Rilski (canonised after his death as a Saint) recommends in "Lecarstvenik", a book of medicines, using infusions of briars, hawthorns, blackberries and herbs. According to Stranski, wheat, barley and millet were the main source of carbohydrates in the diet. Butter, lard, and suet were the main sources of animal fat whilst walnuts were the usual source of vegetable oil. Olives were rarely eaten, and at this time sunflower, sesame and rape were unknown. The main sources of protein were provided by dried and salted meat with the rare inclusion of fresh meat.

Under the Turkish yoke until the Liberation (1396-1878)

The Turkish conquest, in addition to being lengthy and complicated, had a pronounced effect on food production, the peculiarities of national cookery and the distinctive character of nutrition in Bulgaria. A number of documents exist in which ways of life, farming and animal breeding are described and these contribute towards a better understanding of the customs, traditions and styles of cooking.

Turkish documents used for collecting taxes are valuable sources of information on the types of foods produced in Bulgaria. A tax known as "jushur" (tithe) was levied for growing foods such as wheat, beans, lettuce, cucumbers, melons, fruit trees, vines and for breeding pigs, sheep and cows. A higher tax was established on pigs bred specially for Christmas and as a result, Bulgarian customs and celebrations have unofficially recognised pork, forbidden by the Muslim Religion. Water mills, dairies, oil factories, "bozadjiinitsa" (a factory for making "boza"), "halvadjiinitsa" (a factory for making of "halva" - sweet desert made from ground baked sesame and sunflower seeds), "petmezadjiinitsa" (a factory for making treacle) were also taxed. Beans brought from America to Bulgaria are mentioned for the first time in the tax documents for the sale of vegetables.

The descriptions and the stories of travellers, explorers, diplomats and writers, who have visited Bulgaria add to the knowledge of the development and the distinguishing features of the Bulgarian cuisine. Hogier de Bouzbec writes in the 16th century: "... the Bulgarians do not take good care about their stomachs. They eat bread, salt, garlic, yoghurt. They add to their milk, cold water and use it as a drink or they add bread and use it as a main meal...this drink gives not only a pleasant feeling in the mouth, but it also quenches the thirst...when travelling, the Bulgarians eat boiled or dried meat, various fruits, yoghurt, cheese, dried fruits (apples, pears, plums, quince) and also peaches, dates and grapes...The bread is baked in hot ashes. They call it "pogacha". Its dough is made from flour and water, without yeast. When fresh it is very tasty...".[1]

The chronicles of Pierre Bellon du Mann in the 16th century mention the diet of monks, "...they start their meal with fresh onions and garlic. The dinner consists of salty olives and boiled broad beans; no matter whether they are ill or not there is no custom to add water in the wine...".[1]

Evlia Chelebey in the 17th century writes, "...in this part of the Empire there are a lot of vineyards, orchards and vegetable gardens, melon fields and water mills. In many inns and guest houses I was given "chorba" (soup) with groats and "kurban chorba" (boiled mutton). On Fridays I used to eat pilaff (braised rice) and "yahnia" (stewed vegetables and meat). I was treated on lamb kebabs, well-flavoured stewed lamb, fish and refreshing "oshav- medovina" (poached dried fruits with honey). When I visited the villages and the dairy houses I ate milk and yoghurt, cream, honey with cream, fresh cheese, whey, curds and "kashkaval" (yellow cheese). The fresh white bread is very tasty. In some regions there are many "halvadjiinitsi" and "bozadjiinitsi". Many herds of sheep, goats, cows and buffaloes are to be seen on the meadows...".[1] Historical data show that Bulgarians were efficient hunter-gatherers as well as proficient fishermen, molluscs being among the shellfish consumed.

The publications of K. Irechek, "Travels through Bulgaria", give some indication of the national cuisine, "...The dishes in the towns are not always well prepared. Foods of plant origin prevail over meat ones and the names of the dishes are mostly Turkish. Sour "chorba" with chicken, lamb or mutton is served. The meat is either roasted or grilled on a spit, or stewed with vegetables. It can be grilled in thin slices (steak) or dried on the sun ("pastarma"). Domestic birds and fish are also eaten... Rice is the dominant vegetable. It is offered with minced meat ("pilaff") or stuffed in vegetables (peppers and marrows), vine or cabbage leaves ("sarmi"). Vegetables are eaten in enormous quantities: okra, aubergine, tomatoes, celery, peppers. Maize is used for "kachamak"; sweet deserts are not prepared but in summer a lot of yoghurt is eaten...".

The studies of Sh. Shishkov[6] on the diet of the people living in the Rhodopes Mountains region, show that bread, was made from rye, maize and oat flour with yeast. The everyday dish was "kachamak". "Kasha" was prepared from boiled flour, walnuts, cheese, milk or butter. "Marudnick", "katmi" (crumpets and muffins, fried on a hot earthen plate or baked on a griddle over an open fire), "klin" (baked pie of layers of pastry with minced meat as a filing, like lasagne), couscous, pasta, "trahana", etc. Often the dish "sarmi" was made with rice or groats. Green or dried beans were the most widely used vegetables for "yahnia". Amongst the desserts were "oshav" (compote from boiled dried fruits), "sherbet" (sorbet), and "langur" (grape juice). The favourite meat dish was "Cheverme" (lamb grilled on a spit). Meat was roasted on live coals, in hot ashes, on a hot stone plate, or boiled with vegetables, rice and potatoes. The people from this region liked to drink butter-milk, water with either vinegar and sugar, or honey, coffee, but no alcoholic beverages.

At the end of this period the first articles and books on nutrition and cooking were written. These included: "How to keep our health" (S. Dobroplodni, 1865); "Calendar for Human Health"; "A Book about Food and Health" (Chr.G. Danov 1873); "Greediness and Its Harm"; "The Role of Food" (P.Pragov, 1871). Although these books give an imprecise and incomplete account of the situation with many naive explanations, they have, however, made a significant contribution to the foundation of Culinary Science and Technology in Bulgaria.

Communications. The influence of different nations on the peculiarities of the Bulgarian national cuisine.

In the early stages of development, Bulgarian cuisine was rather limited in the type of the products used, styles of cooking utilised and the varieties of dishes produced. Today it is very rich and varied, due to the influence of many nationalities (nearly 30), which have passed through the land that now makes up Bulgaria. These nations affected not only the culture but also the culinary traditions. For example, until 14th century vegetables such as tomatoes, pepper, potatoes, that are now widely used, were unknown. Tea was brought from China, black pepper from India, cloves from Zanzibar, and chocolate from Mexico.

Bulgarian cuisine has been greatly affected by the Turks and to a lesser extent the Greeks, due to their long and direct contact with the Bulgarian population. Amongst the dishes of Turkish origin are: heavy red "zaprajka", vegetables such as onion, peppers, and tomatoes, beef, mutton or goat's meat fried in tallow and stuffed in a sheep's stomach; "yahnia" - boiled vegetables with or without meat which is now one of the most popular dishes in the Bulgarian cuisine; kebabs ("ovarma kebab, orman, kavurma, adjem kebab, taskebabs") which are dishes resembling Hungarian goulash; "guvech", "turlu guvech"- vegetables and meat roasted in an oven in special earthenware; "musaka" - chopped vegetables and minced meat, covered with batter, baked in an oven; "damar" (tripe), "sharden" (boiled offal); "chorba"; "tarator", a cold soup made of yoghurt, diluted with water, chopped cucumbers, walnuts, garlic and sunflower oil; Turkish delight; "halva"; "baklava", "kadaiff" - baked vermicelli, poured with highly concentrated sugar syrup; "ashure" - a wheat pudding; "sutliash" - a rice pudding[2,3,5]. Most of the sweet and savoury dishes still exist in the modern Bulgarian diet and use the same names.

Comparatively little is taken from Greek cuisine. Southern fruits, for example, lemons, chestnuts, olives and pomegranates are used although some authors believe these to be of Turkish origin. The word "trapesa" (a laid table) is of Greek origin and the dish "plakia" (from the Greek πλακε, meaning small pan) is considered to have emanated from Greek cookery. It consists of boiled vegetables, without thickening, seasoned with garlic and lemon, served cold and was made by Bulgarian peasants at harvest and threshing-time.

French cuisine exerted its influence on Bulgarian cuisine after the liberation from Turkish rule. It introduced the idea of cooking meat, garnish and sauce separately. The French style of cooking affected specific tastes and for the first time attention was paid to the presentation of the dishes and the method of service. New culinary words were also added to the terminology, for example, "hors-d'oeuvres", "entree", "galantine", "soufflé", "au gratin" and "vol-au-vent". Some technical terms are also of French origin - deep frying (from the French word "friture"), "tranche" (slice), and "flambee" (flame).[5,6,7]

English cuisine has also played a considerable role in the terminology and development of new dishes. A number use English terms, for example "beefsteak", "steak", "sandwich", "toast", "ham and eggs", and "pudding". From German cuisine come: "Schnitzel", "Butterbrot" (butter and brot - bread), meat products - "Schinken", "Lebercase", and "Lebervurst". Italian cuisine has provided names of foods and dishes, for example, "salame" (salami); "salsa" (tomato puree); "cervalata" (a sausage with meat and brain); "insalata" (to salt) salad. The Russian influence on the Bulgarian cuisine is fairly strong. It began long before the liberation from Turkish rule through immigrants and continued thereafter. Dishes include smoked fish, salmon, caviar, and "borsch" (soup with all kinds of vegetables); "pelmeni" (with different fillings); and "ruska salata" (Russian salad).

The influence of other national cuisines on the traditional Bulgarian style of cooking is felt more recently with the development of better communications.

Religion, traditions and habits

Christianity together with its festivals and celebrations, have had a considerable effect on Bulgarian national cuisine. These effects include restrictions on the type of foods eaten: for example, meat and meat products are limited or abstained from during certain days or periods of Lent. The traditions of fasting every Wednesday and Friday and through the Long Lent as well as some celebrations have led to the development of a variety of dishes. The existing festivals and customs have also created particular traditions in the style of cooking and the national diet. Bulgarians have developed and established special traditions for the cooking and serving of dishes for the Christmas and Easter celebrations. Examples are given below for culinary traditions at celebrations, connected with Christianity:

Christmas Celebrations. The day before Christmas, 24th December, is called Fortune Day. At this time a special ritual bread is made from a rich dough and decorated with symbols such as a cross, suns, flowers, fruits and animals. A silver coin is placed in the dough and it is used to tell fortunes. A candle is lit and the oldest person in the family places incense on a ploughshare or tile which are then burnt. Afterwards the bread is then eaten with honey or garlic. Everything that the land has produced for the farmer is placed on the table. All the dishes, of which there must be nine, must be meatless . The dishes that best characterise the feast are boiled wheat and beans, "sarmi", peppers, stuffed with rice or groats, "oshav", "zelnic" ("banitsa" with cabbage or leek), "tikvenic" (layered pie with pumpkin and walnuts), baked pumpkin with honey and walnuts.

Christmas. At Christmas the entire family gathers around the table and after a 40 day period of lent, meat dishes are at last served. Pigs are slaughtered, specially for Christmas and roast meat dishes are eaten. The meat is spit-roasted over live coals or baked in an oven. Traditional dishes are "sarmi" (sour cabbage leaves stuffed with rice and minced meat); "pastarma" with sour cabbage; grilled pig's liver; "karvavitsa" (a sausage with rice, liver, spleen etc.); a roast duck or turkey, stuffed with rice apples, chestnuts and raisins; roast turkey with sour cabbage; and "banitza" with cheese or meat.

Winter feasts are rich in rituals and traditional dishes play a considerable role. ***Andrea's Day*** (30th November) marks the beginning of the Winter Feasts. It is believed that the day starts to grow as much as a wheat grain, from this day and that is why in the evening before the Andrea's day maize grains are soaked in water. On the following day they are boiled in a new pot. Wheat, millet, barley, dried beans, lens, and peas are also added. Some of the grains are thrown into the sky through the chimney for the wheat plants to grow very high. Then the whole family eats the boiled grains. This food is not given to guests, because it is believed that by doing this, the rich harvest will remain at home and will not be taken away by the guests. Popcorn, being the symbol of growth and germination of the seeds, is also made.

It is believed that St. Varvara protects children from disease hence on the day of ***Saint Varvara*** (4th December) a ritual bread is made by women who have small children. This is baked the day before as no grains can be cooked on the day. The bread is covered with honey, pumpkin jam or sugar and given to friends or neighbours.

The day of ***Saint Nickolas*** (6th December) is celebrated in the name of the protector of fishermen and sailors. It is the only day during the Christmas Lent, when fish or a soup made from fish is allowed to be eaten. The traditional dish is baked fish (usually carp), stuffed with walnuts, raisins, groats, or rice. Some of this dish is also given to the neighbours. According to tradition the fish must have scales and it is a sign of a bad luck if scales fall on the ground and somebody steps on them. Some meatless dishes which are offered include beans, cabbage and "sarmi". Ritual bread, decorated with dough symbols, is baked on this day. The meal is blessed before the feast starts in order to banish evil.

Ignagden (20th December) is the day, when the New Sun is born. Customs include fortune telling and each family prepares a ritual bread, boiled maize, wheat, beans and lentils, dried fruits, cabbage and pickles.

Easter Feasts. Boiled eggs are painted on Holy Thursday in all shades of colour, but the first one, as a symbol of health, good harvest and luck, has to be red. On Good Friday, a special ritual Easter-cake is made.

Characteristics of the Bulgarian national cuisine

Many factors, over a long period of time have influenced the characteristics and peculiarities of Bulgarian National cuisine.

The methods of cookery are confined to boiling, stewing and roasting. Products with either a low water content (grains, legumes) or tough in texture (meat from older animals) are usually boiled. Meat, potatoes, cabbage and legumes are generally stewed, either in water or oil, in earthen pots. The cookery process takes place at considerably lower temperatures for a longer period of time. Roasting is carried out at higher temperatures for a shorter period hence that is why only young animals and some vegetables (potatoes, pumpkin and fruit) are roasted or grilled.

The fat used comes from the animal being cooked. The only animal fat used in the earliest times was lard and tallow and the tradition of preparing meat dishes without adding extra fat has remained until now. It is also for this reason that fatty meat is used and the meat is cooked in its own juices. The process ensures that dishes prepared in this way taste delicious and have a strong meat flavour. Butter is however used for some dishes with vegetables (green beans, pears, nettles, spinach and mushrooms), and also low fat meat such as chicken and lamb "kachamak". Today sunflower oil is used in Bulgarian cuisine to replace animal fat. Traditionally, meat from older animals is prepared with dried beans, onion, and cabbage, either fresh or sour using fats such as tallow or lard. Meat from younger animals, chicken, lamb, veal, and leaf vegetables are cooked in suet or butter.

Appropriate herbs and spices are used to accompany the main dish. If the main dish has a strong flavour and/or colour, no spices are added. When foods are stewed, they change their sensory properties and for this reason spices, such as dried red or black pepper, allspice and bay leaves are added. Some fresh spices are used with legumes: for beans and lentils - mint and savoury; "tarator" - dill; vegetable marrow - dill; tomato salad - parsley.

Sharp or acidic dishes are not characteristic of Bulgarian cuisine. It is wrong to assume that the national dishes of Bulgaria are all strong and very spicy, although the Turkish and Arabic influences may be seen in some dishes.

Flour is not often used as a thickening agent allowing the natural flavour and taste of the main product in the dish to come through.

Milk and yoghurt form a part of many dishes: nettles are cooked in milk; spinach is served with yoghurt; and yoghurt is the main constituent of the cold soup "tarator".

One of the peculiar characteristics of Bulgarian cuisine is that all ingredients are cooked together not separately. The sauce, meat and vegetables are cooked together as an integral process, and provides the typical taste and flavour properties which characterise Bulgarian cuisine.

The scientific literature covering the development and characteristics of the Bulgarian national cuisine is sparse. A number of questions, concerning the national characteristics of the diet still remain unanswered and the style of cooking has not been fully analysed and studied. The present paper has tried to fill in some of the gaps in our current knowledge.

References

1. Archive Documents, Central National Library St. St. Kiril and Metodii, Sofia.
2. Vakarelski, Chr., *Bulgarian Ethnography,* NI Publ., Sofia, 1977.
3. Georgiev, G., et al., *Bulgarian Ethnography*, t. 1-3, BAN, Sofia, 1983.
4. Konstantinov, P., *The Eternal God - A Book About Folk Believes And Celebrations in Bulgaria*, Sofia, 1994.
5. Petrov, L., et al., *Bulgarian National Cuisine*, Zemizdat Publ., Sofia, 1983.
6. Radeva, P., A. Kirilova, *Bulgarian Feast Cookery - Traditions and Recipes*, Sofia, 1991.
7. Shishkov, G., *Technology of Dishes*, NI, Sofia, 1957.

The nutritional quality of vegetables processed by traditional cook-serve and meals assembly techniques

A. Walker, A. West, J. Lawson
The Hotel and Catering Research Centre, The Department of Food, Nutrition and Hospitality Management, University of Huddersfield, Queensgate, Huddersfield HD1 3DH, UK

Abstract

The results of the study are the findings of a comparison between a laboratory simulated meals assembly system with optimum cooking (15 minutes cook) and an abused (30 minutes cook) laboratory simulated traditional cook-serve system. More vitamin C was lost in a meals assembly system compared with the optimum and abused traditional cook-serve systems. These higher losses were attributed to the exposure time of the Brussels sprouts to higher temperatures. Some warmholding is unavoidable in hospital catering, the results of this study suggest the losses of vitamin C are not critical if the warmholding time is minimised.

1 Introduction

This study is part of ongoing research at the Hotel and Catering Research Centre, The University of Huddersfield, UK., in which the quality of traditional cook-serve, cook-freeze, conventional cook-chill, cook-chill sous vide and meals assembly systems within hospitals is investigated. Work involves laboratory based trials and evaluation of hospital catering systems taking into account cooking practices and controls. The results presented here are the findings of a quantitative comparison between laboratory simulated traditional cook-serve and meals assembly systems.

Food processing involves time and temperature cycles including cooking and cooling cycles as well as varying periods of warmholding. As the number of process stages increase so does the potential for nutrient losses. It is therefore important to assess at which stages losses occur with a view to developing improved processing techniques.

1.1   Aims

1. To compare the effects of different catering systems on vitamin C losses in Brussels sprouts.
2. To observe and then simulate optimum and abused cooking conditions for Brussels sprouts in a traditional cook-serve system and investigate the resulting losses of vitamin C.
3. To simulate the 'approved' conditions for processing Brussels sprouts in a meals assembly system and investigate the resulting losses of vitamin C.
4. To compare the results of vitamin C loss in Brussels sprouts in a meals assembly system with those experienced in both an optimally operated traditional cook-serve system and an abused one.

1.2   Development of catering systems in the National Health Service (NHS).

Food service in hospitals is traditionally difficult in providing food of an acceptable sensory and nutritional quality, especially when food usually has to travel to the patient. Hospital catering systems in the UK. prior to the publication of the Platt, Eddy and Pellet report[9] were based on traditional cook-serve techniques, which involved a period of warmholding prior to service. This report highlighted many problems with the catering provision in hospitals including high wastage, poor controls and low quality food.

The National Audit Office[7], suggested that catering departments have within their control a number of factors which could improve patient satisfaction. These factors include the length of time meals are ordered in advance, the appearance and presentation of the food, the amounts served and the temperature of the food at the time it is served. These problems cannot be removed or solved entirely because of the nature of the hospital environment. The majority of patients are captive customers as there are no alternatives to the hospital catering facilities. Feeding patients of different ages, illnesses and with special dietary requirements adds to the complexity of feeding, together with the short time some patients may spend in hospitals.

This need to transfer the technology of food manufacturing to the traditionally labour intensive catering industry was instigated by the pioneering work of the Catering Research Unit at Leeds University. This change was driven by the rising labour costs in the 1960's and early 1970's and problems highlighted in the Platt, Eddy and Pellet report[9]. These instigated the move towards more systematic methods of food production within NHS catering, utilising the technology already used in food manufacturing. Subsequently a number of systems have been developed to overcome these problems mainly in forms of cook-freeze and later cook-chill.

An enhancement of these systems was the development of meals assembly in which the caterer largely purchases pre-cooked frozen and chilled products from commercial manufacturers. A similar system was proposed by the Platt, Eddy and Pellet report[9] but it is only in recent years that such a system has become feasible because of product availability and relevant enabling technologies.

1.3   Cook-Freeze

The first technologically based catering system to be developed in this country was the cook-freeze system. This consists of a centralised unit producing food, portioning and blast freezing it to -5°C within 90 minutes and then storing it at -18°C until required. It is then transported to the site of regeneration and consumption. This system eliminates the need for food to be transported hot and therefore should solve the problem of losses in nutritional value caused by long warmholding.

1.4   Cook-Chill

The cook-chill system is a modification of the cook-freeze system and is said to involve less recipe modification, as well as being cheaper to install and operate. The system is the same as cook-freeze up to the point of chilling. Food items are blast chilled to an internal temperature of 0-3°C within 90 minutes. The chilled food is then stored for up to 5 days including the day of production and consumption. Microbiological safety is the main concern of cook-chill which can only be ensured with careful control of the system.

1.5   Meals Assembly

Since the developments of cook-freeze and cook-chill further advances in technology have led to an extension of these systems in the introduction of the meals assembly system. This is a system which concentrates on food service rather than production as the caterer utilises food from the food manufacturing industry.

Meals assembly utilises frozen, chilled and fresh foods coupled with controlled assembly, regeneration and service of meals to the patient. Frozen products, for example, vegetables, are tempered in controlled thaw cabinets before portioning. The controlled thaw cabinets operate at 8°C and raise the food to a temperature of between 0-3°C. All products must be consumed or disposed of within 24 hours on completion of thawing. The assembly of meals takes place in a temperature controlled environment. Foods are portioned and packaged with the minimum of delay into ovenable board containers for each ward and loaded into chilled insulated boxes. These boxes are then transported to ward level where ward hostesses unload them, check the temperatures (<5°C) and load into the regeneration trolley. The food is then regenerated (>75°C) and service commences within 15 minutes of completion of regeneration and is finished within a further 30 minutes. (DoH[2]).

2 Methodology

2.1 Materials

The Brussels sprouts were supplied by Brake Brothers Food Service Limited. The Brussels sprouts were UK in origin and Campden grade A. They were selected from one processing run to help eliminate compositional variability which can occur due to growing location and environmental factors such as soil type and climate. (Abrams[1], Harris[3]). The Brussels sprouts were mechanically harvested from the plants and transported to the processor in half tonne wooden boxes where they were graded, blanched, cooled, manually inspected and frozen in a flow freezer. Packed in one tonne pallatainers they were delivered to Brake Brothers Food Service Limited where they were repackaged into 2.5 kg bags. On arrival at the University of Huddersfield the Brussels sprouts were stored at -18 °C in a chest freezer.

2.2 Traditional Cook-Serve System

The cooking times selected demonstrate optimum (45 minutes maximum) and abused (120 minutes maximum) processing conditions observed in hospital catering practice.

Three kg of frozen Brussels sprouts were cooked in each of 3 gastronorm steamer trays in a combination oven on steam mode for 15 or 30 minutes with a 5 minute preheat before the Brussels sprouts were added. The core temperature of the Brussels sprouts after steaming was 85 ± 2°C. The Brussels sprouts were then held in a dry heat bain-marie for a total of 90 minutes at an average temperature of 75 ± 2°C. Random 50g samples of Brussels sprouts were taken for analysis at the following stages:- blanched/frozen, after cooking for 15 and 30 minutes and after warmholding intervals of 15, 30, 60, and 90 minutes.

2.3 Meals Assembly System

The cooking parameters selected simulate 'approved' current practice in meals assembly.

Frozen Brussels sprouts were thawed over an 18 hour period in a refrigerator at 0-3°C. On thawing the internal temperature was between -1- +2°C.

300g of thawed Brussels were put in to each of 12 ovenable board containers with 45 ml of water to generate more steam to aid the cooking process. The lids were then heat sealed and the containers placed in a bulk regeneration trolley at an ambient temperature of 22 ± 2°C. (Fig 1). Hot water at 80°C was also put into six lidded ovenable board containers to simulate a full load.

The regeneration trolley was turned on for 38 minutes plus 2 minutes standing time at an air temperature of 144°C. The cooked Brussels sprouts were then warmheld for 15 and 30 minutes in the same trolley, at an average air temperature of 102°C. Random 50g samples of Brussels sprouts were taken for analysis at the following stages:- blanched/frozen, thawed, after regeneration and after warmholding intervals of 15 and 30 minutes.

2.4 Vitamin C Determination

Fluorimetric analysis measures both active forms of vitamin C (ascorbic acid and dehydroascorbic acid) and was carried out according to the ISO 6557/1 analysis procedure.

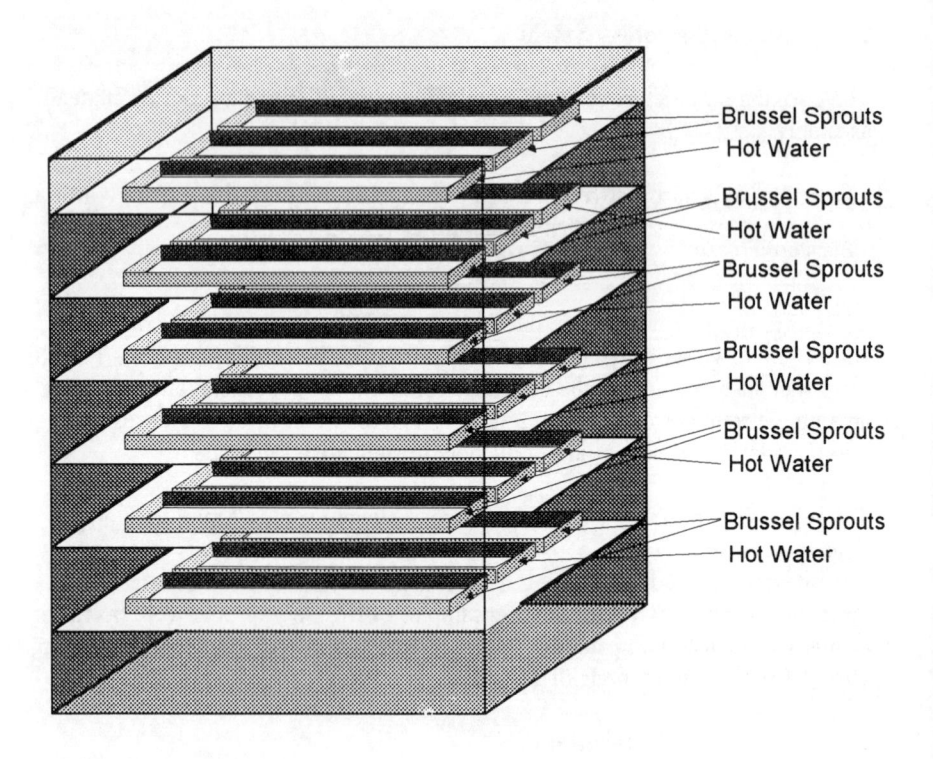

Figure 1. Arrangement of Brussel Sprouts in the Regeneration Trolley

3 Results

Table 1: Average vitamin C losses in Brussels sprouts cooked for 15 minutes and then warmheld for up to 30 minutes in a traditional cook-serve system.

n = 8	Average Vitamin C mg/100g	Standard Deviation	Percentage Loss
Blanched/Frozen	100.2	14.7	0
Cooked	84.2	14.5	16.0
Warmheld 15 mins	82.9	20.8	17.2
Warmheld 30 mins	73.2	18.6	26.9

Table 2: Average vitamin C losses in Brussels sprouts cooked for 30 minutes and then warmheld for up to 90 minutes in a traditional cook-serve system.

n = 8	Average Vitamin C mg/100g	Standard Deviation	Percentage Loss
Blanched/Frozen	67.4	13.5	0
Cooked	52.5	6.1	23.0
Warmheld 15 mins	50.4	4.7	25.3
Warmheld 30 mins	45.6	2.4	32.3
Warmheld 60 mins	38.6	3.2	42.7
Warmheld 90 mins	32.7	4.2	51.4

Table 3: Average vitamin C loss in Brussels sprouts cooked for 38 minutes and then warmheld for up to 30 minutes in a meals assembly system.

n = 8	Average Vitamin C mg/100g	Standard Deviation	Percentage Loss
Blanched/Frozen	96.5	7.7	0
Thawed	85.1	1.4	11.8
Regenerated	69.4	13.2	28.1
Warmheld 15 mins	58.2	6.7	39.7
Warmheld 30 mins	36.1	3.5	62.6

4 Discussion

Cooking accounts for the largest loss of vitamin C in the traditional cook-serve system: 16% after 15 minutes and 23% after 30 minutes. The overall loss of vitamin C in optimum cook-serve processing was 26.9% and in abused cook-serve processing was 51.4%. This compared with a total processing loss of vitamin C in the meals assembly system of 62.6%. (Tables 1, 2 & 3).

These results can be explained by exposure to different time and temperature parameters in heat treatment. The traditionally cooked Brussels sprouts were steamed for 15 or 30 minutes during which the core temperature achieved 85°C whereas the regenerated Brussels sprouts were heated for 38 minutes after which time the average core temperature was 90°C. (Tables 4 & 5).

The Brussels sprouts cooked for 15 and 30 minutes were warmheld at an average air temperature of 86°C, which resulted in average core temperatures of 75° and 69.5°C after 15 and 30 minutes of warmholding respectively. This contrasted with the meals assembly system where an average warmholding air temperature of 102°C was recorded during which the Brussels sprouts achieved a core temperature of 92°C and 75°C after 15 and 30 minutes respectively. This drop in temperature between 15 and 30 minutes warmholding was mostly due to the opening of doors for sampling.

In the meals assembly system the maximum exposure time to heat was 70 minutes. This compared with a maximum exposure time to heat of 45 minutes in the optimum traditional cook-serve system and 120 minutes in the abused one. This demonstrated that the lower warmholding temperatures of the traditional cook-serve system caused a smaller loss of vitamin C than the higher warmholding temperature of the meals assembly system. (Table 4 & 5).

Table 4: Time and temperature parameters and vitamin C loss (%) of Brussels sprouts in a traditional cook-serve system.

15 minute cook			30 minute cook		
Time (Mins)	Temp °C	Loss Vit C %	Time (Mins)	Temp °C	Loss Vit C %
Cook	85	16.0	Cook	85	23.0
Warmheld 15	74	17.2	Warmheld 15	76	25.3
Warmheld 30	69	26.9	Warmheld 30	70	32.3
			Warmheld 60	61	42.7
			Warmheld 90	59	51.4
Average Air Temp. Warmholding 86°C			Average Air Temp. Warmholding 86°C		

Table 5: Time and temperature parameters and vitamin C loss (%) of Brussels sprouts in a meals assembly system.

Time (Mins)	Temp °C	Loss Vit C %
Cook 38+2	90	28.1
Warmheld 15	92	39.7
Warmheld 30	75	62.6
Average Air Temp. On Cooking 144°C		
Average Air Temp. Warmholding 102°C		

5 Conclusion

More vitamin C was lost in a meals assembly system compared with the optimum and abused traditional cook-serve systems. These higher losses were attributed to the exposure time of the Brussels sprouts to higher temperatures. Warmholding Brussels sprouts at lower temperatures preserved more of the vitamin C. This agrees with work conducted by Hill et al[5] and Paulus[8] which showed reducing high temperatures during warmholding could minimise the loss of vitamin C in vegetables. However it is important to acknowledge the possibility of adverse sensory changes which accompany low warmholding temperatures.

Some warmholding is unavoidable in hospital catering, the results of this study suggest the losses of vitamin C are not critical if the warmholding time is minimised. However, when warmholding is used it should be limited to 30 minutes and good menu planning should be observed to avoid poor vitamin C status in the patient. This supports the recommendations of Hill et al[5] and Lachance and Fisher[6] that, in order to minimise sensory and nutritional deterioration of food, 30 minutes should be the maximum allowed period of warmholding.

Despite variable cooking and warmholding times in both systems the vitamin C content is sufficient to comply with the reference nutrient intake standard used by the Health of the Nation Guidelines for Hospital Catering[4] of 40mg a day. Clearly the limiting factor in any system of cooking with a period of warmholding is the sensory quality of the food. If food is not acceptable for consumption then the patient will not benefit from its nutritional value. If caterers observe good catering practice the loss of both nutritional and sensory quality need not be excessive.

6 References

1. Abrams, C. I. The ascorbic acid content of quick frozen Brussels sprouts, *Journal of Food Technology*, 1975, **10**, 203-213.

2. Department of Health and Social Security. *Chilled and Frozen-Guidelines on Cook-Chill and Cook-Freeze Catering Systems*, HMSO, 1989.

3. Harris, R.S. Effects of agricultural practices on the composition of foods, Chapter 6, *Nutritional Evaluation of Food Processing*, ed R.S. Harris & E. Karmas, 2nd Edition, pp 33-57, The AVI Publishing Company, Inc.1975.

4. Health of the Nation. *Nutrition Guidelines for Hospital Catering*, 1995.

5. Hill, M. Baron, M. Skent, J. Glew, G. The effect of hot storage after reheating on the flavour and ascorbic acid of pre-cooked frozen food, Chapter 22, *Catering, Equipment and Systems Design*, ed G. Glew, pp 331-339, Applied Science Publishers Ltd. London, 1977.

6. Lachance, P. & Fisher, B. Effect of food preparation procedures on nutrients in food service practices, Chapter 16, *Nutritional Evaluation of Food Processing*, ed R.S. Harris & E. Karmas, 3rd edition, pp 463-528, The AVI Publishing Company, Inc. 1988.

7. National Audit Office, *National Health Service: Hospital Catering in England*, HMSO, 1994.

8. Paulus, K. *How Ready are Ready to Serve Foods*, S. Karger, 1978.

9. Platt, B.S. Eddy, T.P. & Pellet, P.L. *Food in Hospitals: A Study of Feeding Arrangements and the Nutritional Value of Meals in Hospitals*, Oxford University Press, 1963.

Food and nutrition at environmental extremes - altitude

J.S.A. Edwards

Worshipful Company of Cooks Centre for Culinary Research, Bournemouth University, Poole, Dorset, BH12 5BB, England

Abstract

An increasing number of people are spending time at altitude but often sojourn there too rapidly with the result that they are unacclimatised and in consequence develop Acute Mountain Sickness (AMS). Symptoms of AMS are contributing factors to body weight loss which can have serious consequences. It has been shown that diets high in carbohydrate can lessen the severity of the symptoms of AMS and reduce the period which they exist. This paper considers diet at altitude, reports the results of a study to measure and compare selected aspects of nutrition and food consumption in a group of experienced climbers (n=19) both at sea level and at altitude (5,600 m) following acclimatisation.

Introduction

An increasing number of people are spending time at altitude and, for example, in the United States of America, more than 13 million people visited the Colorado mountains, for business and pleasure in 1990 (Honigman, et al.[1]). During the summer, tourists have traditionally gone in search of the sun, but as the popular destinations have become increasingly crowded and as people have visited them more frequently, many have sought alternative destinations. Summer walking holidays in the European Alps have grown in popularity and often include guided walks, many of which go over 2,500 m (8,200 ft). Winter skiing remains equally as popular but with recent mild winters and low snowfalls in Europe, the attraction of high altitude resorts has increased. Skiing in Europe, in Switzerland, France and Italy, can entail skiing at heights over 3,300 m (10,830 ft) and up to 3,842 m (12,610 ft), although the resorts tend to be at a lower level; typically 1,500-1,800m (4,920-5,910 ft). However, in the United States, whilst skiing is at a similar height, the resorts tend to be much higher; 2,926m (9,600 ft) not being unusual. In addition to those who have traditionally gone to high elevations in search of adventure, the number of people trekking in the Nepalese Himalayas has nearly quadrupled in ten years and each year approximately 10,000 Western Tourists climb over the Thorong Pass, 5,400m, (17,720 ft) (Kayser[2]).

All of these activities involve spending time at high altitude. High altitude has no precise definition but it is generally considered to be elevations in excess of 2,500m (8,200 ft). On rapid ascent to altitude, individuals soon begin to feel sick and nauseous, and experience other symptoms including severe headache and a loss of appetite. They have developed what is generally referred to as Acute Mountain Sickness (AMS). In addition, an increased basal metabolic rate, impaired absorption, and increased energy expenditure all contribute to body weight loss. Cases of AMS have been reported as low as 9,000 ft (2743 m) (Gray[3]), and it is now well established that at altitudes above 3,000 m, AMS is a common phenomenon in unacclimatised individuals (Boyer & Blume,[4] Dinmore, et al.[5]). AMS is dependent upon a number of factors including the speed of ascent and relative fitness but the onset usually occurs within the first six hours. The symptoms usually subside after approximately three days once acclimatisation starts to be achieved and body weight loss can be halted, although these losses tend not to be replaced until return to sea level (Rose, et al.[6] Butterfield, et al.[7]).

Whilst there is a wealth of information concerning ideal diets at altitude, and individuals such as experienced climbers who regularly go there are eager to share their knowledge and experience, much of the information is still anecdotal and unproven. It has been suggested, for example, that climbers deliberately overeat prior to sojourning to altitude in order to counteract the effects of any weight loss; and that they prefer diets high in carbohydrate. A high carbohydrate diet has also been shown to reduce the severity and length of time which the symptoms of AMS prevail (Edwards, et al.,[8] Askew, et al.[9]).

The overall purpose of this study was to measure and compare selected aspects of nutrition, food consumption and absorption in a group of experienced climbers both at sea level and at altitude following acclimatisation. This paper reports part of that research, specifically, whether experienced climbers, either consciously or subconsciously, change their diet prior to sojourning to altitude; determine the adequacy of food and macronutrient intake; compare food acceptability; and measure changes in flavour and taste intensities both at sea level and at altitude.

Methods

Nineteen members, both military and civilian personnel, who had volunteered and been selected to take part in a Joint Services military expedition to the Bolivian Andes, acted as test subjects. All subjects assembled in Poole, Dorset, United Kingdom for a three-day experimental phase at sea level. This regimen was subsequently repeated for a further three days at approximately 5600 m, 28 days later, once subjects had become acclimatised. Results obtained at sea level for all tests were then compared with those obtained at altitude. Data were analysed using appropriate procedures in combination with the SPSS for Windows statistical package. The level of statistical significance selected was 0.05 (p = <0.05).

Baseline measurements taken included body weight, where subjects were weighed in underclothing prior to breakfast using two calibrated, Soehnle digital battery operated scales (accurate to ± 0.2 kg). In addition, body weights were self-reported in a questionnaire administered 12 weeks prior to the start of the study and on the first day once subjects had assembled in Poole. Heights were measured on the first morning of the study ± 0.1 cm.

Diets habitually consumed at sea level were calculated using a Food Frequency Questionnaire developed by Tinuveil Software Systems and modified by Bournemouth University (Diet-Q[10]). These Questionnaires were administered approximately 12 weeks prior to the study and again during the baseline measurements phase in Poole.

All subjects subsisted on standard military 24 hour Arctic Ration Packs which were issued each day and in the same order for both phases, i.e. menu A, B, and C, prepared and consumed under conditions designed to simulate those likely to be experienced at altitude. No other food or drink were permitted. Arctic Rations Packs comprise a nutritionally balanced diet of a breakfast, a main and light meal, snacks and beverages, suitable for consumption over a 24 hour period. The rations are primarily dehydrated food components with an energy value of approximately 4,500 kcal (18.83 MJ).

Dietary intakes were collected using daily 24 hour Dietary Log Books, a technique successfully used on previous field studies under adverse climatic conditions (Edwards, et al[8]). Subjects received instructions on how to record food intakes and completed Log Books were collected at the end of each day by a trained data collector at which time new Log Books were issued for the following 24 hours. Details from the Log Books provided the basis on which to calculate macronutrient intake, using an analysis of Arctic Rations undertaken by the Laboratory of the Government Chemist (LGC[11]).

Food acceptability was recorded in the Dietary Log Books and subjects rated the acceptability of each food item consumed using a 9-point Hedonic scale where 1 = "dislike extremely", 5 = "neither like nor dislike", and 9 = "like extremely".

Flavour intensities of selected food items were recorded in the Dietary Log Books where subjects rated the intensity of flavour of selected food products using a 7-point scale where 1 = "extremely mild", 4 = "neither mild nor strong", and 7 = "extremely strong flavour".

Taste intensities for the four primary tastes (salt, bitter, sweet and sour) were assessed in a similar way utilising a line gradated from "extremely" to "not at all". Subjects placed an ☒ in the appropriate place on the line. These marks were then subsequently graded where 1 = "not at all", 10 = "moderately" and 20 = "extremely". Four liquid items were selected which best represented the taste, thus: soup was selected for salt; coffee was selected for bitter; chocolate drink was selected for sweet; and drink powder was selected for sour.

An Environmental Symptoms Questionnaire (ESQ) was used to assess AMS symptomatology. The ESQ was administered on two occasions, the first during the evening of day two at sea level and the second at the end of day two at altitude. Completed questionnaires were treated and statistically weighted in accordance with the procedures summarised by Sampson et al. 1993[12].

Results

Demographic Data: twenty personnel took part in the expedition, 19 of whom took part in the study; all competed both phases. Seventeen subjects were male and two female. Fourteen subjects (74%) were married, 5 (26%) were single, mean age was 31.9 years (SD ± 8.13; range 23.9 - 53 years).

Body weights: a comparison of mean body weights, prior to, at the start of the study, and at altitude are summarised in Table 1.

Table 1. A Comparison of Mean Body Weights Taken Prior to, at the Start of the Study, and at Altitude

	Height	Self-Reported Weights		Weighed		Changes	
		12 weeks prior	Day 1 of the Study	Phase 1 Sea Level	Phase 2 Altitude		
	m	kg	kg	kg	kg	kg	%
Mean	1.78	73.94	75.61	75.70	*71.35	4.35	5.7
SD	0.06	6.17	5.72	5.38	4.95	1.60	

Notes: * significantly different ($p < 0.05$) - t-tests for paired samples

The mean reported weight gain from 12 weeks prior to the commencement of the study, and on day one, was 1.67 kg. This was not, however, significant. The weight range during the first phase of the study was 67.17 kg - 89.30 kg, whilst the range during the second phase of the study was 62.67 kg - 84.00 kg. All subjects lost weight during the study; minimum 1.93 kg, maximum 7.30 kg, mean 4.35 kg (SD 1.60). Mean weight loss was significant.

Changes in Dietary Intake Prior to Sojourning to Altitude

Nine subjects (47%) reported that they had deliberately changed their diets in the 3 months prior to the study, 10 (53%) had made no change. Mean macronutrient intakes and the percentage contribution to energy intake, three months prior to the study, to the first day of the experimental phase at sea level are presented in Table 2.

Table 2. A Comparison of Food Intake Three Months Prior,
and on Day One of the Study at Sea Level.

Nutrient	Unit	3 Months Prior to the Study		Day 1 of the Period at Sea Level	
		Mean	SD	Mean	SD
Energy	kcal	2787	966	3129	1036
	kJ	11706	405	13139	434
Protein	g	121.56	46.78	129.82	41.62
Fat	g	93.46	48.26	102.83	43.98
Carbohydrate	g	307.41	93.51	350.68	129.30
Alcohol	g	43.50	35.03	52.44	33.76
% of Energy from:					
Protein	%	17.59	2.53	16.96	2.62
Fat	%	29.28	8.56	28.94	6.88
Carbohydrate	%	42.27	7.18	42.20	7.51
Alcohol	%	10.84	7.56	11.90	7.26

Notes: t-tests for paired samples. (No significance)

Macronutrient Intakes

Total macronutrient intakes for each of the six days of the study were calculated from the 24 hour Dietary Log Books. Mean daily intakes and percentage of energy derived from those intakes for the two phases are presented in Table 3.

Table 3. Mean Daily Macronutrient Intake and Percentage Contribution to Energy Intake for the Two Phases of the Study.

Nutrient	Unit	Phase 1		Phase 2	
		Mean	SD	Mean	SD
Energy	kcal	2329	531	2351	568
	kJ	9745	2222	9838	2377
Protein	g	70.06	13.27	72.59	13.96
Fat	g	84.62	20.25	81.07	20.31
Carbohydrate	g	343.28	92.84	355.00	102.29
% of Energy from:					
Protein	%	12.03		12.35	
Fat	%	32.70		* 31.03	
Carbohydrate	%	55.27		* 56.62	

Notes: n = 54 man/days in both phases: * significantly different ($p < 0.05$)

Food Acceptability

An extract from the results for the acceptability of each food item during the two phases of the study, assessed using a 9-point Hedonic scale are presented in Table 4. Although food items in general received a higher preference rating at altitude, (a total of 188.6 at sea level compared with 194.9 at altitude) none of these differences were significant.

Table 4. A Comparison of the Acceptability of Food Items for the Two Phases of the Study.

	Food	Phase 1			Phase 2		
		n	Mean	SD	n	Mean	SD
Breakfast	Porridge	38	6.47	2.70	35	6.57	2.24
Biscuits	Plain Biscuits	38	5.89	1.67	38	6.08	1.50
Chocolates	Milk Chocolate	40	6.78	1.58	35	7.51	1.15
Spreads	Beef	11	6.45	1.75	7	7.00	1.53
Soups	Vegetable	12	6.17	1.19	12	6.58	1.00
	Mushroom	24	6.63	1.10	27	6.81	1.24
Meats	Curried Granules	18	7.22	1.31	18	7.17	1.10
	Mutton Granules	18	6.56	1.54	17	6.29	1.79
Vegetables	Potato Powder	34	6.09	1.78	36	6.50	1.63
Desserts	Apple Flakes	17	6.35	1.77	18	6.56	1.72
Drinks	Chocolate Drink	30	7.07	1.36	34	7.06	0.95
	Coffee	21	5.10	2.39	29	5.38	1.57
	Sugar	23	5.78	1.91	36	6.08	1.48

Notes: 1 = Dislike extremely; 5 = Neither like nor dislike; 9 = Like Extremely.
* t-tests for paired samples. (No significance)

Flavour and Taste Intensities

An extract of the results of flavour intensities rated using a 7-point scale are presented in Table 5. Although there were some differences in the intensity of flavour, none of these differences were significant.

Table 5. A Comparison of Flavour Intensities.

	Food	Phase 1			Phase 2	
		n¶	Mean	SD	Mean	SD
Breakfast	Porridge	15	4.40	0.81	4.27	0.78
Biscuits	Fruit Biscuits	11	4.65	1.37	4.65	0.87
Chocolates	Milk Chocolate	17	5.05	0.66	4.93	0.80
Spreads	All Spreads	12	4.47	1.10	4.04	0.94
Soups	All Soups	15	4.83	0.77	5.04	0.66
Meats	All Meats	18	4.12	1.04	4.47	0.96
Vegetables	Potato Powder	18	3.58	1.14	3.78	0.79
	Dried Peas	17	3.72	0.92	4.03	0.76
Desserts	All Desserts	9	4.80	0.86	4.81	0.80
Drinks	Tea	12	4.56	0.50	4.79	0.66

Notes: 1 = Mild flavour; 4 = Neither mild nor strong; 7 = Strong flavour.
¶ n = number of paired samples: * t-tests for paired samples. (No significance)

Taste intensities were assessed using four 'liquid' items chosen to represent the four primary tastes. Results are presented in Table 6, none of which show a significant difference.

Table 6. A Comparison of Taste Intensities.

Taste	n¶	Phase 1 Mean	SD	Phase 2 Mean	SD
Salt	16	7.29	3.20	7.43	3.17
Bitter	11	11.42	4.05	9.73	4.06
Sweet	6	12.22	2.51	13.06	3.76
Sour	14	11.71	2.69	10.71	2.28

Notes: 1 = Not at all *Attribute*; 10 = Moderately *Attribute*; 20 = Extremely *Attribute*.
¶ n = number of paired samples: * t-tests for paired samples. (No significance)

Environmental Symptoms

In order to determine if levels of carbohydrate consumption influenced AMS, subjects were grouped into one of two groups according to the percentage of energy derived from carbohydrate consumed. Group 1 (n = 9) were subjects who derived less than 56% (range 49.40 - 55.72%) of their energy intake from carbohydrate and group 2 (n = 9) were subjects who derived more than 56% (range 57.36 - 66.21%) of their energy intake from carbohydrate. Intakes and AMS scores were compared utilising Pearson correlation. No significant differences were found between the two groups at altitude.

Discussion

Exposure to altitudes above 3,000 metres is known to cause AMS the symptomatology for which include headache, dizziness nausea and loss of body weight. Body weight loss can be attributed to a number of factors:
  Basal metabolic rates (BMR) which usually increase on exposure to altitude.
  Impaired absorption of both carbohydrate and fat occur at altitude (Boyer & Blume[4]).
  Energy Imbalance.   Probably the most important of all nutritional considerations at altitude is that of energy requirements which are a significant factor in weight loss, much of which could be prevented (Butterfield, et al[7]).
  In this study, subjects reported feeling significantly less hungry at altitude and a significant decrease in the strength of desire to eat. As a consequence, they considered that at altitude there was too much to eat although there were no significant changes over the sea level values. In line with these findings, subjects reported feeling significantly fuller at altitude.
  Dehydration causes amongst other things, drowsiness, impatience, discomfort, weariness and reduced work efficiency. It also predisposes to cold injury and can actually reduce food intake. High altitudes are dry and hyperventilation greatly increases respiratory water losses. In addition, although the environmental temperature may be low, sweat losses can still be appreciable.
  Body weight loss in this study was in line with what would have been anticipated at altitude. The mean loss of 4.35 kg (SD ± 1.65) is acceptable over

a short period of time. However, one subject lost 10% of his body weight and others also lost amounts approaching this figure. If this loss had been sustained over a longer period, as might be found when individuals spend considerable periods at altitude, then this would give cause for concern.

In order to counteract, and in anticipation of weight loss, a number of climbers deliberately gain weight prior to departure. However, high body fat before deployment does not protect against muscle wasting and weight loss hence there is little or no benefit to be gained from increasing body weight prior to ascent (Boyer & Blume[4]). Although over half (53%) of subjects reported that they had not deliberately changed their diet, prior to the study, results from the dietary assessment undertaken three months prior to, and on day one of the study, show a small increase in mean daily energy intakes of 342 kcal from 2787 kcal (SD ± 966) to 3129 kcal (SD ± 1036). This increase was not significant. A small weight gain of 1.67 kg from a mean of 73.94 kg (SD ± 6.17) to 75.70 kg (SD ± 5.38) was observed although again this was not significant.

Energy requirements at altitude are dependant on a number of additional factors including: climate, terrain, work undertaken and loads carried. There is little or no evidence of increased energy costs for work caused by altitude itself (Johnson, et al.[13]) although high elevations are usually cold. Energy expenditure is likely to be increased, not because of the cold, *per se*, unless there is an undue level of shivering, but for reasons that include the hobbling effects needed to traverse the snow and the weight of clothing, and other equipment which is carried.

Mean energy intakes for the first phase of the study were 2329 kcal (SD ± 531) and for the second phase, 2351 kcal (SD ± 568) showing an increase of 22 kcal. This was not significant. The low levels of energy intake were undoubtedly a major factor in the weight loss at altitude indicating that energy intake was well below energy expenditure.

Carbohydrate can improve work performance, and the consumption of a high carbohydrate diet enables individuals to maintain levels of activity for longer periods. It has been shown, for example, in activities such as downhill skiing, that carbohydrate is the primary fuel and that glycogen depletion will occur after several hours activity (Brouns, et al.[14]). Carbohydrate is a good source of anaerobic energy in an environment where oxygen is in short supply. At both 4,000 m (15,000 ft) and 5,200 m (17,000 ft), the ingestion of a high carbohydrate meal, as compared to a high protein meal, has been shown to have a number of benefits and can in effect, lower the physiological altitude by between 300 and 600 m (1,000 and 2,000 ft) (Eckman, et al.[15]). This is due in part to carbohydrate which requires 8-10% less oxygen for metabolism than fat or protein. A high carbohydrate diet has also been shown to reduce the severity and length of time which the symptoms of AMS exist although the exact mechanism as to how this occurs is not known (Consolazio, et al.[16]).

In this study, the increased percentages of energy intake derived from carbohydrate produced no significant change in the symptomatology of AMS.

However, it should be borne in mind that the actual increases in carbohydrate consumption, although significant, were small.

Although there appears to be a greater physiological benefit of a high carbohydrate at altitude, there does not appear to be any natural mechanism for the selection of it. A voluntary shift from fat and protein to carbohydrate has been reported by many authors but other studies demonstrate that, when given a diet *ad libitum*, both unacclimatised (Gray[3]) and acclimatised individuals (Rose, at al.[6]) do not spontaneously choose a high carbohydrate diet. A further suggestion is that high protein meals may be shunned at altitude because of their salt content (Boyer & Blume[4]). The percentage of energy obtained from carbohydrate in this study increased from 55.27% to 56.62% whilst that of fat decreased from 32.70% to 31.03%. Although these amounts are small they are nevertheless significant.

In the climbing literature there are numerous anecdotal reports of changes in the acceptability of individual food components at altitude although little work has been undertaken to quantify how food acceptability changes. This study is unique in that subjects, under 'field' conditions, chose from identical food components offered in the same order at both sea level and altitude following acclimatisation. Food items generally received a higher preference rating at altitude, the total score being 194.9 compared with 188.6 at sea level, but there were no significant differences.

There is considerable weight of opinion and anecdotal evidence reporting changes in taste and flavours at altitude but little hard data to support many of the assertions made. Pugh[17], for example, when considering the rations for the 1953 expedition to Mount Everest comments on the increased demand for sugar but which also 'seems to taste less sweet'. In one of the few studies undertaken, the four basic tastes, sweet, sour, salt and bitter, were studied in climatic chambers at 0, 1,525 and 3,050 m (0, 5,000 and 10,000 ft). Significant differences were found from 0 to 1,525 m (5,000 ft) but not from 1,535 to 3,050 m (5,000 to 10,000 ft) leading the authors to conclude that taste sensitivity can be significantly influenced by altitude (Maga & Lorenz[18]).

In this study, there were no significant changes in flavour intensities. Three flavour intensities decreased, five increased and three remained unaltered. Similarly taste intensities were not significantly different. Salt and sweet both increased slightly whilst bitter and sour decreased slightly. Despite, therefore, the plethora of anecdotal evidence on changes in flavour and taste intensities, these assertions are not supported in this study.

Conclusions

This study was unique in that it directly compared military Arctic rations consumed both at sea level and at altitude using the same subjects.

Although there was a slight increase in energy intake in the 3-month period prior to the start of this study, it was not significant. Energy intakes were not significantly different between sea level and altitude, although energy expenditure at altitude was much higher causing a loss in body weight, the highest of which was 10%. These results are consistent with other studies confirming body weight loss is a common phenomenon at altitude.

There was a small but nevertheless significant increase in the percentage of energy derived from carbohydrate at altitude. There seems, however, to be no increase in the preference for this category of food. Consumption of carbohydrate did not change the symptomatology of AMS, as measured by an Environmental Symptoms Questionnaire.

The acceptability of dehydrated rations, as measured by an Hedonic scale, was not significantly different indicating their suitability for use in the two environments.

Flavour and taste intensities did not change significantly at altitude.

References

[1] Quoted in Honigman, B., Theis, M.K., Koziol-McLain, J., Roach, R., Yip, R., Houston, C.S., & Moore, L.G. Acute Mountain Sickness in a general tourist population at moderate altitudes. *Ann. Int. Med.* 118: 587-592. 1993.

[2] Kayser, B. Acute mountain sickness in western tourists around the Thorong pass *(5400m) in Nepal. J. Wild. Med.* 2: 110-117, 1991.

[3] Gray, E. Le B. Appetite and acclimatization to high altitude. *Mil. Med.* 117: 427 - 431, 1955.

[4] Boyer, S.J. & Blume, F.D. Weight loss and changes in body composition at high altitude. *J. Appl. Physiol.* 57: 1580-1585, 1984.

[5] Dinmore, A.J., Edwards, J.S.A., Menzies, I.S. & Travis, S.P.L. Intestinal carbohydrate absorption and at high altitude (5730m). *J. Appl. Physiol.* 76: 1903-1907, 1994.

[6] Rose, M.S., Houston, C.S., Fulco, C.S., Coates, G., Sutton, J.R. & Cymerman, A. Operation Everest 11 nutrition and body composition. *J. Appl. Physiol.* 65: 2545-2551, 1988.

[7] Butterfield, G.E., Gates, G., Flemming, S., Sutton, J.R. & Reeves, J.T. Increased energy intake minimises weight loss in men at high altitude. *J. Appl. Physiol.* 72: 1741-1748, 1992.

[8] Edwards, J.S.A., Askew, E.W., King, N. and C.S. Fulco. Nutritional intakes and carbohydrate supplementation at high altitude. *J. Wild. Med.* 5: 20-33, 1994.

[9] Askew, E.W., Claybaugh, J.R., Hashiro, G.M., Stokes, W.S., Sato, A. & Cucinell, S.A. Mauna Kea III. *Metabolic effects of dietary carbohydrate supplementation during exercise at 4100m Altitude.* US Army Research Institute of Environmental Medicine Technical Report T12-87, May, 1987. Natick, Massachusetts.

[10] Diet-Q. Tinuviel Software, 2, Penmark Close, Callands, Warrington, Cheshire, WA5 5TG.

[11] Laboratory of the Government Chemist. *Survey of the Nutritional Composition of Operational Ration Packs, 1988-1992.* Analytical Report No. FB2/189/92. Teddington, Middlesex, 1992.

12 Sampson, J.B., Kobrick, & Johnson, R.F. *The Environmental Symptoms Questionnaire (ESQ): Development and Application.* US Army Natick Research, Development and Engineering Center Technical Report NATICK/TR-93/026. March 1993. Natick, Massachusetts.

13 Johnson, H.L, Consolazio, C.F., Daws, T.A. & Krzywicki, H.J. Increased energy requirements of man after abrupt altitude exposure. *Nut. Rep. Int.* **4:** 77-82, 1971.

14 Brouns, F., Saris, W.H.M. & Ten Hoor, F. Nutrition as a factor in the prevention of injuries in recreational and competitive downhill skiing. *J. Sports Med.* **26:** 85-91, 1986.

15 Eckman, M., Barach, B., Fox, C.A., Rumsey, C.C. & Barach, A.L. Effect of diet on altitude tolerance. *Aviat. Med.* **16:** 328-340 +349, 1945.

16 Consolazio, C.F., Matoush, L.O., Johnson, H.L., Krzywicki, H.J., Daws, T.A. & Isaac, G.J. Effects of high-carbohydrate diets on performance and clinical symptomatology after rapid ascent to high altitude. *Fed. Proc.* **28:** 937-943, 1969.

17 Pugh. L.G.C.E. Himalayan Rations with Special Reference to the 1953 Expedition to Mount Everest. *Proc. Nutr. Soc.* **13:** 60-69, 1954.

18 Maga, J. A & Lorenz, K. Effect of Altitude on Taste Thresholds. *Perceptual and Motor Skills.* **34:** 667-670, 1972.

SECTION 3: FOOD HABITS

Section 3.1: Food Habits

The relationship between post-war cultural shift, consumer perspectives and farming policy

S.C. Beer, M.H. Redman
Centre for Land-Based Studies, Bournemouth University, The Lindens, East Lulworth, Wareham, Dorset BH20 5QT, UK

Abstract

A study was undertaken to investigate the relationship between changing popular culture and consumer perspectives with regard to food, and the associated primary production (agricultural and horticultural) systems. Such systems are considered fundamental to food culture and a model is proposed that provides a frame-work for considering: a) the process by which cultural change is communicated to, and influences the activities of, the primary producer; and b) how in turn the primary producer influences culture. This model is tested against various aspects of cultural history and through a series of case studies. Conclusions are drawn as to how those 'players' that are innovative within the food industry can interpret and respond to cultural change, and how they might shape popular culture itself.

Introduction

It might be considered that the true innovator is someone who is performing at the peak of current cultural experience. This may be as a result of innate genius, because the person has "read" current culture and has placed themselves in this position, or because he/she has succeeded in shaping and leading culture itself.

Unfortunately, cultural study is fraught with difficulty as definition is overlaid by interpretation. Culture was defined by Taylor[1] as that "...which includes all capabilities and habits acquired by man as a member of society" and popular culture may therefore be considered as the culture of every day life. Jones[2] highlighted four strands to the "...bourgeois (cultural) ideology present in the western world: the conservative, the liberal-humanist, the technical rationalist, the vulgar Marxist". Other commentators talk about modernism, post-modernism and minimalism, but the concept of culture should not be as exclusive and elitist as these terms suggest. It should, as Williams[3] recommended be "ordinary", and thereby also be understandable and accessible.

A case in point is food: a central part of our experience of popular culture and one which lends itself to cultural analysis. It is, for example, possible to identify changes in national popular culture and the attitude of the consumer to food, as the excellent UK study, *Slice of Life*, by Hardyment[4] (also produced as a BBC television documentary series) illustrated.

Rationale

This study undertook, through a review of relevant literature, to extend such an analysis to the relationship between popular culture, consumers and the food producer. This is presented below as a simple model which is tested against some basic markers in cultural history. By demonstrating a correlation between the popular culture of the consumer and the management decisions taken by the primary producer, the question then arises as to what degree there is a causal relationship. If there is causality then there must be lines of cultural communication.

An understanding of such a relationship should be of interest to innovators in the food industry who are looking to respond to cultural shift, or indeed to shape popular culture itself. Assuming, of course, that the innovator considers the relationship between primary food production and consumer perception to be an important one (it is interesting to note that some of the original work on the uptake of new ideas was done in agricultural communities[5,6]).

A basic premise of this investigation is that consideration of the broad quality characteristics of any food product must start with the original primary production system. It is here that the fundamental nature of the product in qualitative (e.g. consumer perception of production ethics) and quantitative terms (e.g. nutrient content, microbiological contaminants) is laid down. Beyond the "farm-gate" many of these characteristics cannot be altered. This is especially important since consumer perceptions relating to the fundamental nature of the food product, for example in terms of pesticide exposure or animal welfare, are of increasing importance. Indeed, for many consumers these considerations now represent some of the most important factors in product choice, as trends such as the increasing sales of organically-produced foods indicate[7].

This does not mean to say, of course, that other factors (e.g. processing methods, packaging etc.) do not play a part in determining the consumers' perception of food products - since they clearly do! But for those consumers concerned with issues of quality at, or before the "farm-gate", these factors tend to be subordinate to the nature of the primary foodstuff itself.

Note that the words 'agriculture' and 'farmer' are used throughout the text as generic terms to refer to all primary food production systems/managers.

Trends in Post-war Primary Production Systems

The UK agricultural industry has undergone a major revolution since the 1940s and a full analysis of this is well beyond the scope of this paper. Nonetheless, a number of general trends should be noted:

a) unlike most other European countries, the UK had few historical restraints upon imports of agricultural produce and by the early twentieth century had become the world's foremost food importer (e.g. between 1914 and 1927 the area of farmland in England and Wales fell by 300,000 hectares)[8]. However, with the outbreak of war in 1939, maximum output became the official policy with stringent measures against any farmers who refused to co-operate. By 1943, food imports had reduced by 85% and net output increased by 91%;

b) technical progress in animal and crop breeding, combined with the availability of sophisticated inputs and technologies, has helped to raise agricultural productivity and levels of national self-sufficiency significantly since the 1940s;

c) successive government policies since 1947, including adherence to the EU's Common Agricultural Policy (CAP), have provided farmers with stable and secure markets which have helped to create a favourable economic climate in which modern, intensive agriculture has flourished;

d) although farmers have historically been seen as the natural custodians and trustees of the rural environment, public perception of the agricultural industry has profoundly changed over the past 50 years. According to the Organisation for Economic Co-operation and Development (OECD)[9]: "...fundamental perceptions about the farmers...as an economic actor living in symbiotic harmony with the land are beginning to alter as the significant long-term environmental impact of agricultural activities on the landscape, on drinking water, the air we breathe and the food we eat is increasingly recognised";

e) a series of sensational food scares (e.g. salmonella, listeria and BSE) in the late 1980s and early 1990s rocked the public's perception of many traditional foodstuffs as being safe and wholesome, focusing particular attention and criticism upon livestock production;

f) a growing number of consumers have been prepared to pay more for food arising from production methods that are perceived to be 'better' for the environment, animals and people[10]. For those farmers willing and able to modify their production systems accordingly, this has provided an important niche market opportunity.

The Model

A simple model (Figure 1, and subsequently expanded in Figures 2-5) was constructed to illustrate/annotate the relationship between the three areas of Primary Production, Consumer and Popular Culture.

A number of inter-relationships operate within the model. The consumer can obviously be greatly influenced by popular culture, but may also influence it in return. Certain individuals will contribute more to the development of popular culture than others and can be considered at its 'leading edge'. The identification of such people or groups is useful because they are the innovators in society.

In some cases, the consumer may have direct contact with the primary producer, but this will vary depending on the population and time. For example, historically the populations of Britain and France had very good contact with the primary producer. These links were first severed in Britain during the industrial revolution and the urbanisation of large proportions of the population. This might also have happened in France had it not been for the Napoleonic inheritance laws ensuring that land was split evenly between children, thus ensuring that more people retained a link with the land.

The producer may also be influenced by culture directly. This is independent of the consumer, with communication occurring instead via the government, non-government organisations (e.g. pressure groups), or the media. At times the producer has even been a cultural icon. For example, the war-time farmer in the UK was a hero.

The nature of the inter-relationships between Primary Producer, Consumer and Popular Culture is determined by the 'x' factors in the model - these are i) prevailing ideas, ii) major players in the food industry, and iii) important methods of communication (including, of course, food itself).

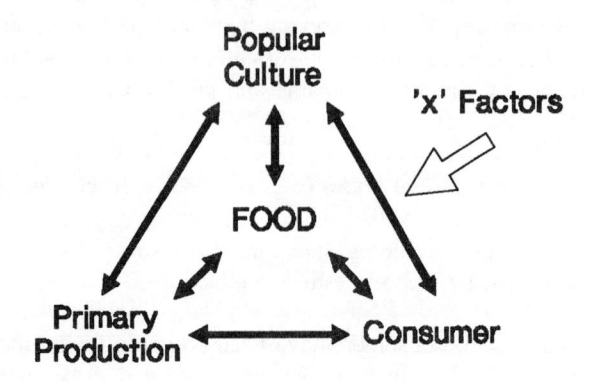

Figure 1: A model of cultural communication linking the areas of Farm, Consumer, Culture and the 'x' Factors.

Post-war Production/Consumption Eras

The proposed model was 'tested' by applying it to a series of four distinctive post-war production/consumption eras: "dig for victory", "food rationing", techno-farming" and "new environmentalism". Numerous information sources were consulted to gain an overview of each era, but the principal texts included Ritson et al.[11] and Hardyment[4].

Figures 2-5 are intended to illustrate for each era:

a) the principal ideas, major players and methods of communication operating;

b) the characteristics of the principal relationships (i.e. cultural dynamics) operating between Primary Producer, Consumer and Popular Culture.

Satisfying the Consumer Through Innovation

The model provides a useful tool for explaining the dynamics of the relationship between the consumer, popular culture and the primary producer. But, what does this tell us about contemporary culture and the opportunities that prevail within it for the innovator?

Firstly, there is an innate value in understanding our cultural systems and the role of food within them since it allows individuals to make their choices in context. This may involve a consumer deciding what goods to purchase or a company deciding what goods to produce.

There is currently considerable debate about the cattle disease bovine spongiform encephalopathy (BSE) and the possibility of its transmission to humans. This is obviously a very serious issue, but it is also an issue that lends itself to sensationalism. For example, it is interesting to compare column inches and air time devoted to the possibility that we might catch the disease, compared to the work of people such as Collinge[12] which indicates that we might not. Bad news tends to be good news for the media, but bad news can also be the product of pressure from parties with a vested interest.

The use of use of artificial hormones, agrochemicals, irradiation[13,14] and genetic engineering in food production systems[15] are other issues for which a more balanced debate might be useful.

Secondly, life cycle analysis is increasingly coming to the fore[16] and looks set to play a role in successful innovation regarding food products[17]. With issues such as animal welfare now at the centre of popular culture[18], it is no longer possible to consider the product in isolation from its primary production system. Food labelling schemes[19,20] may be an answer since they represent a form of direct communication from producer to consumer, although it is debatable whether consumers are informed enough to interpret the schemes and to make appropriate purchasing decisions. Consequently, it is easy to dismiss many (although not all) of the plethora of 'natural', 'traditional' and 'welfare-friendly' labels as little more than 'Green Tokenism'.

'X' FACTORS:

IDEAS	MAJOR PLAYERS	COMMUNICATION
• Food shortage • National isolation • "Food as a munition of War" • Food control • Dig for Victory • Need • National spirit and purpose	• Government* • Regional Agricultural Committees	• Radio; Cinema & Newsreel • Newspapers • Organisations and meetings • Word of mouth • Food = shortages; austere; price

DIG FOR VICTORY

```
              Popular
              Culture
               ↕
      [A] ↗         ↖ [B]
             FOOD
        ↙  ↕   ↕  ↘
   Primary  ←——→  Consumer
  Production
              [C]
```

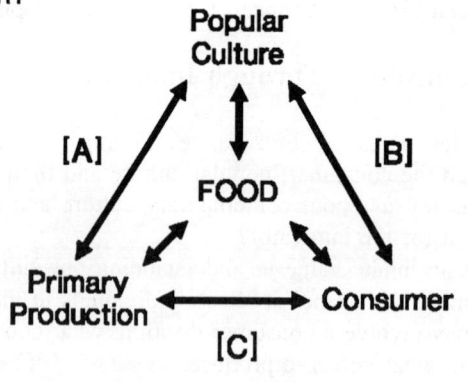

CHARACTERISTICS OF PRINCIPAL RELATIONSHIPS

[A] "Dig for Victory" - Farmers as combatants - Land Army
[B] Consumer as a combatant able to share in War effort
[C] Farmers as heroes and popular cultural icons

NOTES

* Non-interventionist Western Governments can influence food consumption patterns via a variety of policy mechanisms. Contemporary examples include affecting the availability of food supplies, the purchasing power of food consumers and the control of food quality[21,22,23]. Governments can, however, intervene in even more direct ways and during this period especially Government 'influence' was absolute.

Figure 2: A model of cultural communication linking the areas of Farm, Consumer, Culture and the 'x' Factors during the 'Dig for Victory' era.

'X' FACTORS:

IDEAS	MAJOR PLAYERS	COMMUNICATION
• Austerity • Rise of technology • Post-war consumer boom* • Discovery of marketing function • Hope • Food security	• Government • Farming lobby • Rising Agri-Business	• Radio • Newspapers • Organisations and meetings • Word of mouth • Cinema • Food = shortages and austerity, but availability of new foodstuffs which were symbols of hope

FOOD RATIONING

```
                Popular
                Culture
                  ↕
          [A]          [B]
                 FOOD
            ↙    ↕    ↘
      Primary   ←→   Consumer
     Production
              [C]
```

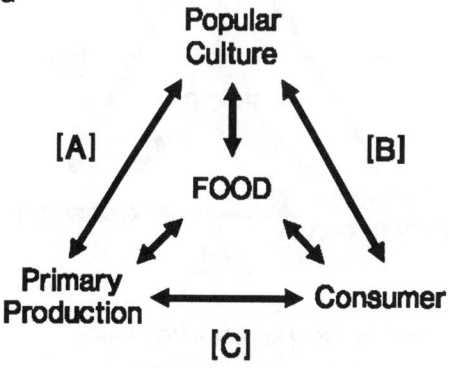

CHARACTERISTICS OF PRINCIPAL RELATIONSHIPS

[A]   Demand for good quality, consistent food at a cheap price was part of the new era

[B]   'Education and purse' period[24]. Better education and more money was supposed to give rise to new, healthier purchasing patterns

[C]   Farmers are still seen as popular heroes but are increasingly taken for granted especially with 'new' imported foods and continued rationing

NOTES

*   Engels Law[25] indicates that with increased personal income, the proportion of income spent on food declines. Although this is open to some discussion[26] the observation appears to be applicable in most situations. It also indicates that with increasing income, consumers are more likely to purchase expensive luxury foodstuffs. These do, however, have to compete with other consumer items.

Figure 3: A model of cultural communication linking the areas of Farm, Consumer, Culture and the 'x' Factors during the 'Food Rationing' era.

358 Culinary Arts and Sciences

'X' FACTORS:		
IDEAS	**MAJOR PLAYERS**	**COMMUNICATION**
• Technological revolution • Cheap and plentiful food • EC Common Agricultural Policy • Conspicuous consumption	• Government • Agribusiness • Increasing dominance of the supermarkets • New pressure groups	• the 'Media' (Television, Radio, Newspapers) • Word of mouth • Food = sign of brave new world; new horizons; advancement

TECHNO-FARMING

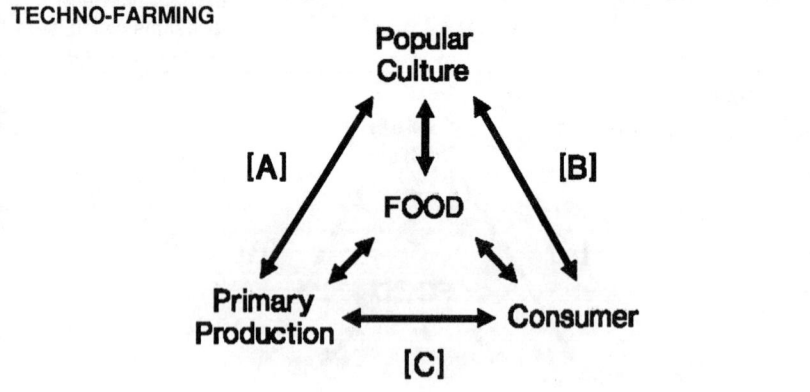

```
              Popular
              Culture
              ↗   ↕   ↖
           [A]         [B]
           ↙   FOOD    ↘
          ↙    ↕ ↕     ↘
      Primary  ←——→  Consumer
     Production
              [C]
```

CHARACTERISTICS OF PRINCIPAL RELATIONSHIPS

[A] Use of new technology viewed as a good thing, with science as a social saviour (e.g. 'Green revolution')
[B] Common acceptance of Science as shaping the future e.g. new food horizons in terms of choice of cuisine
[C] Further distancing of consumer from primary producer leads to an increasing lack of understanding about agriculture amongst the public

Figure 4: A model of cultural communication linking the areas of Farm, Consumer, Culture and the 'x' Factors during the 'Techno Farming' era.

'X' FACTORS:		
IDEAS	**MAJOR PLAYERS**	**COMMUNICATION**
• Decline in consumption • Modernism and systematic attraction to novelty • Selfish 80's • Caring 90's • Care for the environment • Food safety • Change in family structure	• Media • Pressure groups • Supermarkets* • Government • International food • Agribusiness	• the 'Media' (Television, Radio, Newspapers, Cable and Satellite) • Word of mouth • Electronic communication • Food = fuel for mechanics of life; cultural icon in its own right; symbol of family disintegration;

NEW ENVIRONMENTALISM

```
            Popular
            Culture
           ↗  ↕  ↖
        [A]        [B]
       ↙     FOOD     ↘
      ↙    ↗   ↖      ↘
   Primary  ←——→  Consumer
   Production   [C]
```

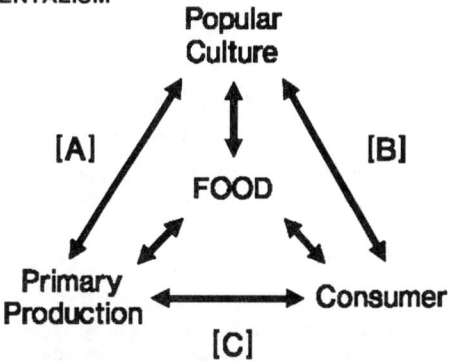

CHARACTERISTICS OF PRINCIPAL RELATIONSHIPS

[A] Farmers under pressure as producers of food of dubious quality (welfare, disease, chemicals, hormones); despoilers of the countryside, and; recipients of large amounts of tax-payers' money

[B] Growth in consumer awareness; conscience with regard to the environment, animal welfare and other 'green' issues, and; desire for new food products which enhance lifestyle

[C] Increasing interest in the origin of food; demands for new products from home and abroad; growing awareness of the concept of quality, and; demand for an immediate response to consumer pressure.

NOTES

* The UK has seen an unprecedented growth in the supermarket sector of British retailing. Not only are supermarkets economically efficient, but with a retail market share approaching 70% they have a huge cultural influence[4,27].

Figure 5: A model of cultural communication linking the areas of Farm, Consumer, Culture and the 'x' Factors during the 'New Environmentalism' era.

An understanding of cultural systems also allows individuals or organisations to exert a direct influence by shaping popular culture. This can be seen in the context of food with the emergent cult of the 'celebrity chef' throughout the 1980s; prominent individuals who became widely-known and celebrated on television and through a series of books. They were, and are cultural figures, innovators who helped extend Britain's food horizons. At the same time many became consultants to national supermarkets. In the evening they would be on the television in our living-rooms as 'friends' encouraging us to try a new dish, then in the morning they would be in the supermarket encouraging us to buy. Such association enabled the supermarkets to latch on to the peak of popular culture, and through their promotion of products and ideas to shape culture itself.

Other organisations use similar associations with cultural figures and while the legitimacy of such an approach can only be considered in the context of individual organisations, its effectiveness cannot be dismissed.

Conclusions

This paper demonstrates the central role of the primary producer and the primary product within our food culture. It has also shown the way in which this cultural system can operate and how it is possible for potential innovators to read and manipulate the system in pursuit of varying objectives.

What is clear, is the interest that the consumer has in the origin of food. If this interest and relationship is to progress in a positive way, then the further development of coherent, transparent and honest links between primary production and the consumer is desirable. A sensible and universal food labelling scheme would be a good starting place, and would be of very great value to all concerned in the supply of our food.

References

1. Taylor, B. *Primitive Culture*, 1871 - cited in Encyclopedia Britannica.

2. Jones, B. *The Politics of Popular Culture*, Sub and Popular Culture Series: SP No. 12, Centre for Contemporary Cultural Studies, University of Birmingham, 1982.

3. Williams, R. Culture is Ordinary, Chapter 1, *Studying Culture: An Introductory Reader*, eds A. Gray and J McGuigan, pp 5-14, Edward Arnold, London, 1993.

4. Hardyment, C. *Slice of Life: The British Way of Eating Since 1945*, BBC Books, London, 1995.

5. Jones, G.E., The diffusion of agricultural innovations, *Journal of Agricultural Economics*, 1963, **15**, pp 387-409.

6. Rogers, E., Categorizing the adopters of agricultural practices, *Rural Sociology*, 1958, **23**, pp 345-354.

7. *Vegetarian and Organic Food*, Mintel International Group Limited, London, 1993.

8. *Agriculture, Fisheries and Forestry*, Aspects of Britain Series, HMSO, London, 1993.

9. *Agricultural and Environmental Policies*, Organisation for Economic Co-operation and Development, Paris, 1989.

10. Anon. Greener Groceries - at a price, *Which? Magazine*, 1992, October, pp 28-31.

11. Ritson, R., Gofton, L. and McKenzie, J. (eds). *The Food Consumer*, John Wiley & Sons, Chichester, 1986.

12. Collinge, J., Palmer, M.S., Sidle, K.C.L., Hill, A.F., Gowland, I., Meads, J., Asante, E., Bradley, R., Doey, L.J. and Lantos, P.L., Unaltered susceptibility to BSE in transgenic mice expressing human prion protein, *Nature*, 1995, **378**, pp 779-783.

13. Jack, F. and Sanderson, C.W., Irradiation of gourmet foods - potential for improving sensory quality and process acceptability?, *British Food Journal*, 1995, **97** (8), pp 29-30.

14. Jack, F. and Sanderson, C.W., Radiophobia: will fear of irradiation impede its future use?, *British Food Journal*, 1995, **97** (5), pp 32-35.

15. Frewer, L.J., Howard, C.and Shepherd, R., Genetic engineering and food: what determines consumer acceptance?, *British Food Journal*, 1995, **97** (8), pp 31-36.

16. Sadgrove, K., *The Green Manager's Handbook*, Gower, 1992.

17. Fojt, M., Precis: step by step to successful innovation, *British Food Journal*, 1995, **97** (3), pp 22-23.

18. Eastwood, P.J., Farm animal welfare: Europe and the meat manufacturer, *British Food Journal*, 1995, **97** (9), pp 4-11.

19. Lowman, B.G. and McClelland, T.H., An overview of quality assurance schemes, pp 55-61, *Proceedings of the British Grassland Society Conference: Quality Milk and Meat from Grassland Systems*, Great Malvern, United Kingdom, 1994, British Grassland Society, 1994.

20. Potter, M.J., The viewpoint of animal welfare organisations, pp 47-54, *Proceedings of the British Grassland Society Conference: Quality Milk and Meat from Grassland Systems*, Great Malvern, United Kingdom, 1994, British Grassland Society, 1994.

21. Josling, T., and Ritson, C., Food and the Nation, Chapter 1, *The Food Consumer*, eds R. Ritson, L. Gofton and J. McKenzie, pp 3-20. John Wiley & Sons, Chichester, 1986.

22. Jones, J., The bad food trap, *The Observer*, 1996, 21 January, p 13.

23. Jones, J., Poverty triggers UK diet crisis, *The Observer*, 21 January 1996, p 1.

24. Mckenzie, J., An integrated approach - with special reference to the study of changing food habits in the United Kingdom, Chapter 8, *The Food Consumer*, eds R. Ritson, L. Gofton and J. McKenzie, pp 115-170. John Wiley & Sons, Chichester, 1986.

25. Burk, M.C., Ramifications of the relationship between income and food, Journal of Farm Economics 1962, XLIV, 1 February.

26. Tangermann, S., Economic factors affecting food choice, Chapter 4, *The Food Consumer*, eds R. Ritson, L. Gofton and J. McKenzie, pp 61-84. John Wiley & Sons, Chichester, 1986.

27. Burns, J., Do retailers really make big profits?, *Farmers Weekly*, 1992, 23 November.

All in the garden is not dim sum. Trends and potential for the development of western style food & beverage operations in the Peoples Republic Of China

J. Sutton
Department of Hotel & Tourism Management, Hong Kong Polytechnic University, Hong Kong

Abstract

The Peoples Republic of China (PRC), with a population of over 3 billion, is fast attracting overseas investors into the hospitality sectors of it's service industries. Recent economic reports demonstrate the comparative wealth and size of the various regional markets within China and their high consumer potential. This paper sets out to examine the changes that are occurring within the arena of China's international joint venture hospitality businesses, and in particular it examines the change in the local market which reflect a developing trend for western and "international" style food and beverage experiences.

1 Introduction:

Economic forecasts have demonstrated a period of sustained growth within service industries over the next decade. In maintaining this growth, most authorities anticipate considerable movement within established labour markets; that is a movement away from the manufacturing sector and into the expanding service sector. With this rapid industrial restructuring among all of the Association of South East Asian Nations (ASEAN) countries there are new social and economic trends emerging throughout all of the South East Asian region. Changing trade patterns show an increase in regional Pacific-Asia interdependence resulting from foreign trade and changing patterns of competitive advantage. Tan[1] considers that this has had a major effect on the economic reforms in China, and has helped encouraged China to develop its trade both on a regional and a global scale. According to Urata[2] much of this regionalization has been influenced by international joint-venture companies (IJV's) and their affiliates operating within the PRC. These companies are keen not only to promote Asian integration, but also to encourage interdependence with North America and Europe. Heenan and Willey[3] propose that capitalism is rapidly beginning to motivate the country's leadership, by giving examples of how Chinese multinationals are increasingly becoming more adept at global reach.

2 Bull In A China Market

In many senses, this move towards international business, commerce and capitalism is a very new concept within the PRC. Balazs[4] comments that during the imperial dynasties, capitalism failed to develop because of the super-abundance of cheap labour and the fact that there was no scope for; individual enterprise, individual freedom, security, nor any guarantees against the predatory actions of the state. Even between the years of 1911 and 1976 the constant state of turmoil within the country hampered any real development. It was only with Deng's modernization policies in the late 1970's that the government slowly began to introduce carefully controlled levels of "capitalism" into the country. Since that time the sustained economic reform in the PRC is producing a consumer market in China that is attracting the attention of a variety of multi-national concerns. Economic reports from the McKinsey Quarterly[5] demonstrate the comparative wealth and size of the various regional markets within China; show the GNP / GDP of these markets and their high consumer potential.

2.1 Social Changes In The PRC

In the same way, measured social and political reform is also affecting the market place within the PRC. One of Hong Kong's leading newspapers, the Eastern Express[6],comments that there are now less than 8% of China's younger generation who remain enamored with the Communist party and its doctrines. Consequently, the concept of the sublimation of one's self to the state; that is, the traditional Chinese Communist idea that; "all is for the People rather than for the individual," would appear to be diminishing. This attitudinal change is being further speeded up by major societal changes adopted by China, such as the "one child" policy. Parents cannot now automatically expect their children to support them in old age and the Chinese people are now having to think about saving and planning for their old age. Such factors appear to be decreasing the reliance on the *"iron rice bowl"* philosophy common in the China of only a decade ago, and are promoting a greater impetus towards intrinsic motives for individual success. This, coupled with increasing exposure to western life styles, and the movement towards a "controlled" market economy is producing a high degree of entrepreneurial spirit within the new generation of Chinese citizens. For example, tipping, seen from an ideological viewpoint, reflects an attitude of servitude rather than one of service, and is still forbidden legally in China. However, tips are universally and graciously accepted, although not, as yet, solicited or expected as too often happens in western societies. Similarly, Coca-Cola, Kentucky Fried Chicken and McDonalds are considered, almost everywhere as barometers of social and economic change within a given society, and whilst the latter has only recently moved into the China market, the former two companies have been established for over a decade and are experiencing a high degree of success with the local market. Similarly, IJV hotels, initially developed to cater for the international traveler are experiencing a tremendous growth from the internal market. Consequently, along with other Multinational Companies (MNC's), many hospitality corporations are turning their attention towards this emerging market of over 3 billion potential consumers.

2.2 Potential And Pitfalls For Hospitality IJV's In China

As with most foreign investment within the PRC, these developments are normally in the form of joint-venture operations with Chinese partners. In an attempt to explain why China wants joint ventures, Shaw and Meier[7] show that consumers in China have established themselves as purchasers of goods requiring substantial discretionary income. They point to a new stage of involvement for the multinationals; that is, the 'strategic investor' stage, characterized by multiple ventures, relationships with decision making authorities and a desire to create a dominant share within the Chinese market. In the hospitality industry this approach is illustrated by the recent joint venture investments of such multi-national hotel corporations as; Sheraton International, Shangri la Hotels & Resorts, Holiday Inn, New World International and others. Shangri la and Sheraton, for example, are not only developing city centre properties in China, but are now involved in resort development which is a very new concept for China's hospitality industry. Similarly, Holiday Inn's flagship hotel the Holiday Inn Lido in Beijing, has, since opening in 1988, adopted the Chinese *"danwei"* principle (self contained work-unit). In addition to the usual hotel facilities, it offers such amenities as; a post office, a supermarket, fast-food restaurants, a bank, a delicatessen, an international school, a 318 unit residential area, a commercial centre and a children's playground. The Lido incidentally was the first of the IJV hotels in Beijing to allow local Chinese to enter the hotel freely and use its facilities. In the same way, the fast-food industry has seized joint-venture opportunities with brand leaders such as McDonalds, Pizza Hut and Kentucky Fried Chicken opening branches within the PRC over the last decade.

The key factors of developing international joint ventures (IJV's) from the PRC's view point are identified by de Bruijn and Jia[8] as being strongly related to the need for an improved foreign exchange balance and for the possibility for full localization of these corporations. They also state that foreign investors are looking for brand introduction in the Chinese market and for profit priorities. There are however problems, de Bruijn and Jia[8] conclude that the results of IJV's are often less effective than originally envisaged. This is a view supported by much current research into IJV's in the PRC. Most writers isolate the problem factors as; insufficient or poor quality local supplies; poor local infra structure; inadequate human resources; difficulties caused by government policies in the initial negotiation and in subsequent administration; and key difficulties in areas of technology transfer. Indeed the McDonalds Corporation delayed their entry into the China market primarily because of the problems of poor quality local supplies and poor local infra structure. In the areas of human resources, and in the associated areas of technology transfer, many IJV companies have experienced operational problems because of cultural differences. These have affected both management style and the evolution of a successful corporate culture. In many Chinese societies, job status and security still have strong traditional values, and the more recent Maoist doctrines intensified the traditional collectivist culture to heights whereby it became expected that the state would provide all. This *"iron rice bowl philosophy"* guaranteed jobs for life with the associated benefits of housing, food, social welfare and support. For many IJV hospitality companies operating in the PRC, there can be problems involved in establishing a corporate culture and identity, as such traditions can cause potential conflict between a worker's perceived job commitment levels, which are allied to cultural values and ideology, and modern corporate expectations. Thomas and Whiteley[9] propose that companies adopt a 'competency' approach to defining management principles, a view supported in an editorial article in the Personnel Journal[10] showing success by major multi-national companies such as McDonald's and PepsiCo who have adopted such a competency approach to help maintain their corporate identity and culture in worldwide operations.

2.3 The Growth Of IJV Hospitality Companies

Despite such intrinsic problems, hospitality development in the PRC has been rapid, Dai[11] shows that from a base of only 203 state-owned hotels with 32,000 rooms in 1981, there are now almost 2,000 hotels offering more than 300,000 rooms. Of the new hotels, some 22%, are owned or managed by foreign companies, and this figure is increasing. Similarly, in the fast food sector,

McDonalds are currently operating 44 outlets in the PRC, propose 61 by the end of 1995 and some 600 by the year 2003. Much of this development is geared to meet the needs of the international traveler in terms of accommodation and food and beverages, however companies report that they are experiencing a concurrent demand, particularly within the scope of their food and beverage operations, from an emerging local market.

China, with a quarter of the world's population and a rapidly-expanding economy, is presenting business opportunities that many western companies can no longer ignore, but as we have seen, success relates to developing an understanding of a very different market, with a vast diversity of cultural structures, complexities and traits. Some of the Chinese cultural traits are common and well documented, for example the issue of face and the use of guanxi. One issue that is often overlooked is the almost national characteristic of '*Guo qing*'. This is a characteristic of the Chinese culture that can inhibit the assimilation of foreign methods and ideas. Yan[12] maintains that foreign companies looking for a share of the market must adapt to such aspects of the Chinese culture. This poses a number of challenges and complications to IJV's in building and managing corporate and brand identities in the rapidly-expanding Asia-Pacific markets, particularly those dominated by Chinese culture. Schmitt and Pan[13] identify the interrelated tasks involved in projecting corporate / brand identity as selecting viable names, establishing the right image and enhancing quality perceptions. IJV hospitality companies operating in China have only a limited history, and face challenges in establishing and maintaining a consistent image. The brand names and corporate identities that are familiar in the west are comparatively unknown to most of the local population, and, if these hospitality groups see their long term future with the local markets, then they need to adapt to this concept of local tradition to achieve sustained growth. Connell[14] argues that hospitality chains are only now realizing the need to offer recognizable and consistent quality service for their customers, and that consequently, the conditions favorable to branding are now present. If this is the case, it is likely that the late 1990's will see the development of pan-Asian hospitality companies, using economies of scale and branding to gain competitive advantage. China's hospitality industry has a huge local market potential, now with rapidly increasing buying power, and by the using multi-brand / multi-concept strategies, coupled with the advantages offered by new technology, could easily adopt and utilize marketing strategies geared to meeting the needs of the internal client in addition to those of the foreign visitor. As already illustrated, product quality and consistency are important considerations to both local and international markets. Sharp[15] CEO of Four Seasons Hotels, defines the key elements for success in the global market as being; a clear purpose and goal, quality service, customer focus, employee commitment and attitude, management integrity, and high performance standards, all which aim at total customer satisfaction. Whilst China's hospitality companies are reaching for such goals they have several problems they must first overcome. Service quality cannot be easily measured, and the service industry naturally has a high cost of quality. Whilst hospitality companies in China are striving to achieve international quality and despite the advantages of relatively cheap land and labour, they are experiencing this "high cost" phenomena. The ongoing costs of almost perpetual training, the provision of an appropriate infra-structure, the need for the introduction of new technology, the problems of sourcing quality operating commodities and supplies locally, and the lack of an adequately trained work-force with modern management potential are all negative factors in achieving this success. It would appear that new initiatives, particularly in the areas education and training, and in human resources policies are amongst the prime needs, and yet again cultural disparity must be considered in establishing these initiatives.

3 Pizza, Pasta And Pie:

The initial research project on which this paper is drawn set out to investigate the potential for and effects of inter-cultural conflict and thus the need for cross cultural education and training in China's IJV hospitality organizations. In the subsequent analysis of the corpus, a significant amount of sequential information emerged about the emerging local market and trends towards changing eating habits amongst the local population.

3.1 Methodology:

Much of the established data was drawn from one area of the original study, that is a series of interviews with expatriate managers working in China's hospitality industry. The interviews with expatriate managers were designed, tested and conducted to obtain the attitudes and opinions of experienced hospitality practitioners about operating in China, and to find out what they felt were the important factors for improving quality within the Chinese hospitality industry, and the problems they encountered in achieving this. Four international hotel management companies operating in China were used as the sample for the survey, and a total of $n=65$ expatriate managers from $n=12$ properties operated by these companies were interviewed. All of the managers interviewed were currently working at the level of department head or above. Of those interviewed some 27% (n=18) were specifically responsible for food and beverage operations, with a further 23% (n=15) being General Managers and Deputy General Managers. The interviews were tape recorded, transcribed and subsequently analyzed.

3.2 Findings:

Several managers commented that IJV hotels were attracting a great deal of interest and inquisitiveness from local sectors of the market, and managers from all of the hotels surveyed commented that the local market was developing rapidly, especially in the food and beverage areas. This was seen as being due to a high level of local curiosity in a new and developing hospitality culture, where new ideas and western concepts were being incorporated and were on public show. By contrast, some managers commented that there is still some resistance from locals towards using a hotel's public facilities and outlets, as many locals still have absolutely no concept of what a hotel is, or of the services and facilities that it can provide. Despite this apparent lack of awareness, many managers commented that, because of the buoyant economy, Chinese nationals have greater disposable income and more purchasing power than ever before, and they had noted substantial recent improvements in their local trade. This was especially marked in food and beverage areas, despite the fact that in some localities there are still restrictions on a Chinese national's use of *"foreign"* trading ventures, and that some properties are required to imposed local currency surcharges on transactions from Chinese nationals. Managers from one property surveyed commented that despite such surcharges 80% of its food and beverage *"walk-in"* trade came from these locals with increased spending power, and that the average food and beverage check from these consumers had doubled over the two year period from 1992-94. It was generally felt that more locals wanted to experiment with western food outlets. Many hoteliers reported a substantial growth in local customers using their western and/or international outlets. International buffet style operations were seen as especially attractive to locals, as they could experiment with western foods in relative safety, knowing that there were also familiar foods available. Similarly, ethnic speciality outlets were also seen as gaining favour with the local population. In one of the properties surveyed, an Italian style trattoria had been recently opened, which was reported to have a substantial local following amounting to over 75% of its lunch time business and around 40% of its evening trade. Other properties reported similar successes with such diverse operations as Irish style bar restaurants, Japanese restaurants, English style pubs, and other such operations. However, most managers also agreed that locals staying in the hotels generally preferred to use the more familiar Chinese restaurant outlets for the majority of their meals, as many of these local users were still not familiar with western food, and many perceived western and international outlets to be more expensive than their Chinese counterparts. One general manager commented that in his hotel around 30% of local Chinese residents used the Chinese outlets as compared to only 15% using western outlets. Several of the managers commented that whilst locals were still not generally accustomed to western foods, they were beginning to use the western outlets for "face", and came to be seen or to entertain in what was considered as a prestigious, and "face gaining" environment. This was seen as something of a new trend, but with a great deal of potential for future development. Banqueting, both in the Chinese and western styles, was also viewed as a big growth area, one which contributed as much as 20% of total food and beverage revenue. This was seen as an expanding avenue for corporate and personal entertaining by local business men, with managers from one Beijing property commented that banqueting represented over 50% of its local food and beverage business.

With this increase in local trade several managers, particularly those from the major cities, commented that this sector of the market were becoming more demanding in their wants and needs. The reasons given were that as a greater number of PRC business men are experiencing overseas travel, are becoming more exposed to western concepts and are better educated. Thus managers felt that their expectations were growing, and that they *"knew what to ask for"* in terms of service quality and provision. By comparison, it was felt that the less experienced local users were not as demanding as either "foreigner visitors" or the local business man, and were far easier to satisfy, as they had less of a base for comparison and lower expectations in terms of service and quality.

China's hotel rates were traditionally cheap, but are no longer so, and this pricing is often reflected in the hotel's food and beverage outlets, as compared to many local restaurant operations which are still relatively inexpensive. Hoteliers were aware that recent rate increases are now putting hotel prices in many of China's major cities on a par with those in other regional and international locations. Consequently their customers are rapidly having to come to terms with increased pricing, but as a result are demanding increased value, increased facilities and consistently high standard of food, beverages and service. The hoteliers interviewed generally agreed that whilst the hotel *"hardware"* packages were often equal to those found elsewhere, the quality standards found in China, particularly as related to food and beverage provision, provision of hi-tech facilities, customer care and general service levels, are much lower than those found in other parts of the region. This was seen as a major problem, and even more so when customer were buying a brand name based on experience with that company in other locations. Despite menu and beverage variety often being limited because of difficulties in accessing commodities, managers also thought that their guests now had greater expectations in food and beverage areas and had high service expectations. In some properties managers thought that their staff were meeting these needs, whilst other still saw a gap in the food and beverage areas. Many locals were also seen to have high expectations in terms of hotel hardware, whilst their tastes and expectations of food and beverage service were still more geared to local taste and expectations.

Most expatriate managers commented on the problems involved in obtaining sufficient resources and supplies of the required standard and quality. Imported products such as food and beverage commodities, electronics, tableware, machine parts and wines and spirits were seen as particular problems. Whilst this problem stemmed from China's national trading policies and import quotas, the procedures necessary to import adequate working stock were seen as unnecessarily complicated and restrictive. The processes involved with obtaining Customs and Excise clearance, and with acquiring work permits and various trading licenses were frequently cited as an example of restrictive practice, where it was noted that to obtain such goods and services, international hotels pay more than double the price charged to local operations.

Local sources of supplies were generally regarded as being of little use, in that they were frequently of a poor standard, that delivery times were excessive, or that in terms of overall costs to the IJV companies such products were either as expensive or more expensive than imported goods. This was attributed to the fact that international standards of quality grading were seldom used in China, with managers quoting that where local products were adopted, the quality would vary significantly from one delivery to the next. Managers reported that suppliers frequently used local measures of quality and quantity, and that these varied from one source to another. Because of the lack of any national standards which could be applied to purchasing decisions, managers consequently saw a great need to educate local suppliers to implement international standards and quality controls, and that to obtain a satisfactory source of local supplies was a struggle. Similarly, taxes, duties and import regulations appeared to vary according to region, for example those IJV properties operating in China's Special Economic Zones (S.E.Z's) obtained preferential rates as compared to other regions.

4 Back To The Future:

Overall managers were confident that the percentage volume of local trade was increasing and felt that such business would continue to expand. This was attributed to the lessening of internal restrictions and the greater exposure of the populace to western concepts and ideas.. The Chinese, like the French have a national pre-occupation with food and eating and are keen to experiment with new tastes and flavours. Thus whilst it is unlikely that national food will ever be totally replaced by western imports, it was seen that there was a great deal of potential for western style products and eating experiences, and that such experience held a great deal of novelty value for many of the local population. In developing and establishing western style food and beverage outlets it was generally felt that the local culture, from the perspectives of both the consumer and the organization, had to be considered. Consideration to the local culture needs to be made in the areas of the product and its development and in developing a viable brand image, and in also strategic management issues such as human resources policies, operating policies and in developing and promoting a strong corporate identity.

International and local hospitality companies in China are currently striving to create an industry which meets with international, yet variable standards expected by culturally disparate categories of customers. In order to develop an industry that can deal with such a divergence of cultural obligations, the industry needs to focus on two significant areas, the consumer and the organization. First, in order to meet the requirements of the consumer, it is essential to become aware of exactly who the customer is, and what exactly are those customers' wants and needs. Second, as the focus of meeting such obligations falls mainly on operational staff, it is necessary for hospitality companies to providing a body of trained staff, who are conversant with both the technical skills, and, the culturally related social skills required by those customers. These skills must also be entirely acceptable to both management and staff in the organization. To do this it would appear that there is a need for hospitality companies to develop both corporate standards and training techniques which capitalize on the use of local cultural norms, and by which the cultural awareness and sensitivity of employees can be increased. Appropriate training is essential in developing a body of skilled hospitality personnel, and consideration must be made in three key areas which are; the nature and level of skills required by the industry in today's market place; the perception of the consumer as to continued provision of service and the maintenance of quality assurance; and, the needs of the work force. Technology is undoubtedly changing both the type of skills required in the industry, and affecting the numbers of personnel employed within its various categories. As the number of international hospitality chains increase within China, both standardized procedures and the use of high-tech equipment are being employed to ensure that customer expectations are constantly maintained, and that there is minimum product variation between one unit and another. Consequently, China's growing hospitality industry must consider its future evolution carefully. Human resource planning, and the associated issues of education and training are of vital importance to economical and environmental success and survival. Thought must be given not only to the procedures required to provide an ample workforce for the industry, but to the task of developing a skilled and technologically knowledgeable body of workers, who are conversant with internationally accepted standards and procedures. For China, expansion in this new market arena of hospitality will predictably produce a growth in the total number of employees required by the industry. With the acceleration of technological dependence, it is ultimately likely that the skills required will be those of the technologist and the specialist. Technical and social skills will become increasingly important, as will the abilities and attitudes needed to adapt to change. This means that flexibility and the acquisition of transferable skills, rather than the development of specific craft skills must be one of the key issues in future education and training provisions. It is in these areas of managing change and developing commitment, that industry and those responsible for educating and training its future personnel, both in the commercial and the educational sectors, must consider in their long term planning.

References:

1. Tan, K.Y. — *"Emerging Economic And Social Realities In East Asia"* Australian Journal of Management (Australia), June 1992 Vol.17, No.1: pp. 67-89
2. Urata, S. — *"Globalization And Regionalization In The Pacific-Asia Region"*. Business & the Contemporary World (Germany), Autumn 1993 Vol.5, No.4: pp. 26 - 46
3. Heenan, D.A. & Willey, D — *"China On The Move"*. Journal of Business Strategy (USA), May/June 93 Vol.15 No.3: pp. 35 - 41
4. Balzar. P. — In Lafayette de Mente, B. *"Chinese Etiquette and Ethics In Business"*. 1994, NTC Business Books, Hong Kong.
5. McKinsey Quarterly — McKinsey Quarterly *Editorial*: 1992, No. 3, pp. 37
6. Eastern Express — Eastern Express; *"Lead Article"*. March 31, 1994
7. Shaw, S.M. & Meier J. — *Second Generation MNCs In China* The McKinsey Quarterly (USA), 1993: No 4, pp. 3 - 7
8. de Bruijn, E.J.& Jia, X — *"Managing Sino-Western Joint Ventures: Product Selection Strategy"*. Management International Review (Germany), 1993, Vol 33 No 4 pp. 335 -361
9. Thomas, I. & Whiteley, A. — *"The Competency Approach To Hong Kong Management Development"*. Human Resources Journal (Hong Kong), Jan 1990 Vol.6; No.1 pp. 17 - 27
10. Personnel Journal: — Personnel Journal: *Editorial:* Aug.1994, p12
11. Dai, Lan — *China Tourist Hotels Towards World Standards* Peoples Daily *(Overseas Edition)* April 8 1991
12. Yan, R. — *"To Reach China's Consumers, Adapt To Guo Qing"* Harvard Business Review (USA), Sep / Oct 1994 Vol.72 No.5: pp. 66 - 74
13. Schmitt, B.H. & Pan, Y. — *"Managing Corporate And Brand Identities In The Asia-Pacific Region"*. California Management Review (USA), Summer 94 Vol.36; No.4: pp. 32 - 49
14. Hansen, C. & Brooks A. — *"A Review Of Cross-Cultural Research On Human Resource Development"*. Human Resource Development Quarterly (USA), Spring 1994 Vol 5 No.1: pp. 55 - 75
15. Sharp, I. — *"Managing For Global Market Leadership"*. Business Quarterly (Canada), Summer 1991 Vol.56; No.1: pp.16-20

The influence of cultural traditions and social change on domestic meals in a New Zealand context

J. Mitchell

Department of Consumer Sciences, University of Otago, P.O. Box 56, Dunedin, New Zealand

Abstract

From 1870-1970 few changes occurred to the New Zealand main meal. It is argued that meals continued to follow British traditions because of the homogeneity of the population, and the presence in the society of an inherent rule system that dictated the form of meals. An absence of intense episodes of social change was also a factor. Substantive changes in meals became noticeable in the 1970s. The factors associated with change at this stage were the movement of women into the work force in the 1960s, technological developments associated with food and the acceptance in the society of new ideas about food and nutrition. The interaction of these factors was important in producing changes.

Introduction

From 1870-1970 few changes occurred in New Zealand meals. The structure of the main meal altered least, although some changes occurred to its elements. In the 1970s more substantial changes were observed in meals. These involved the elements of the meal which were often replaced by new food products, and as the decade progressed 'foreign dishes' also appeared on the menu. In the past historical studies have attempted to explain changes in the New Zealand diet [1], [2] but the authors of these studies, while they have taken some account of cultural traditions — both agree that New Zealanders' food habits were strongly influenced by British immigrants — do not attempt to explain in any detail how tradition operated to preserve food habits or how social change influenced meals. Food consumption figures have also been used to indicate diet change[3]. Consumption figures however, while useful are also limited, because although they indicate food trends over a period of time they do not give information about meal formats or the composition of meals. Consumption figures in New Zealand for example in the period 1914-1986 indicated the continuing dominance of meat and dairy products in the diet and in later years they recorded an increase in the consumption of processed foods, but because they are based on food available their validity depends on accurate recording of local production as well as exports and imports. For this reason household food consumption studies [4] which record information on food consumed daily and weekly in terms of food items and meal format are considered to more informative in terms of domestic meals.

The purpose of this paper is to provide an explanation for continuity and change in domestic meals in New Zealand over a set period (1870-1970) from an historical and a sociological viewpoint. Terms used are defined as follows:

diet: an individual's normal food.

meal: food items prepared and combined in certain ways; it may consist of several courses.

meal format: structure of the meal in relation to the food items/dishes and courses served.

menu: list of food items/dishes.

food item: individual foodstuffs.

dish: food item or items prepared in a particular way.

element: food item or dish that is part of a menu.

An examination of domestic meals and factors influencing their production was carried out using historical writings, newspapers, magazines, advertisements and recipe books. The results of this study suggested that continuity and changes that occurred in meals during this period were associated with cultural traditions, cultural factors and social change. Social change is a product of economic and political factors, as well as cultural factors.

> Culture consists of the values the members of a group hold, the norms they follow, and the material goods they create....Norms represent the 'dos' and 'don'ts' of social life...p 31[5]

The discussion that follows examines the interrelationship in New Zealand society between culture, social change and domestic meals. It is argued that stability was the main feature of New Zesaland meals until the 1970s because episodes of social change were not significant. Significant transformation of New Zealanders' meals occurred in the 1970s because New Zealand had, in the 1960s undergone a period of intense social change. This was associated with the exposure of New Zealanders to more diverse ethnic groups, women's position in the society, economic affluence, rapid advances in domestic technology and new communication methods. These factors acted together at this time to bring about substantive changes in domestic meals.

Domestic meals

The main meal for most New Zealanders in the late 19th century, was served in the middle of the day and was likely to consist of a meat, stewed or roasted, *(haricot stew which included carrots and parsnips)* potatoes and vegetables, boiled or baked *(potatoes and spinach)* and a baked or steamed pudding *(bread and butter pudding)*. This menu was recorded in the diary of Rene a daughter of Mrs Edwin of Wellington who went to England in 1893 and left her three girls to manage the meals for themselves and their father with the help of one servant. The girls father was a retired commander of the Royal Navy but his salary was 'rather niggardly'.[6]

Wealthier families probably included more courses. *The Colonists Guide* (1883) intended for the 'colonial gentry' who immigrated to New Zealand suggested for an ordinary cottage dinner, soup succeeded by fish, roast meat and fowl, game or some savoury pie, followed by pudding, tarts, blancmange, etc. and cheese, concluding with dessert.[7]

The meals served in New Zealand households at this time tended to reflect the income and status of the household. The immigrants to New Zealand who were mainly British bought with them as part of their 'cultural baggage' their ideas about food and their food practices. Some changes were necessary at first because of lack of foodstuffs but as regular supplies were established so were familiar meals.

In September 1962 a survey of 1400 New Zealand households by Market Research (N.Z.) Ltd was conducted to determine the pattern of food and beverage consumption.[4] It found that, for 92% of respondent households, meat (beef or mutton) was still the basis of the main meal, carrots, cauliflower, cabbage, pumpkin and frozen peas were the most frequently served vegetables, and milk puddings, including ice cream and fruit were the most favoured puddings. Only a small percentage of households served baked puddings of which the most frequently served were fruit pies.

This menu suggests that in ordinary households the main meal had changed little in nearly a century. Probably most change had occurred in the technology associated with food production (e.g. frozen peas, ice cream) and meal preparation procedures.

In 1992 a postal survey to determine the style of meal served in New Zealand households was conducted by researchers at the University of Otago.[8] On the day the survey was completed 75% of respondent households served a traditional meat based meal, 6.6 % served a vegetarian meal, 4.6% served a pasta based meal, 4.0% had a rice based meal and 2.6% served a bread based meal. The 'rest' had eaten takeaways or 'other' types of meals. Compared with 1962 when 92% of the respondent households served a meat based meal the fact that only 76% did so in 1992 indicates that the traditional meal style was changing. The 1992 figure however does not necessarily indicate that 16% of New Zealanders have suddenly become vegetarians, although between 1982 and 1992 the proportion of the total budget spent on meat did drop without a compensating amount being spent on fish or poultry. [9] It simply indicates that meal patterns may be changing. Similarily in 1990 in Great Britain a survey carried out by Gallup found that 43% of respondents claimed to be eating less meat (a significant increase in number from 1984 and 1988 when previous surveys were done) but only 3.7% of the population were vegetarians. The figures however do indicate that some changes are occurring in British meals.[10]

After 1970 meals featured more convenience foods, they were less likely to follow the standard format and they included a wider variety of food items and some ethnic type dishes. e.g. in a New Zealand recipe book published in 1972 it was suggested that the following menu would be suitable for entertaining: *Iced cucumber soup, Lamb paprika and noodles, Courgette salad, Peach shortcake, Coffee.*[11]

The influence of cultural tradition on New Zealand meals

Despite some changes in ideas about cultural identity, the notion of an identity associated with British norms and values remained strong in the population until the 1960s and this was reflected in food behaviour. New Zealand was settled by the British from 1840 onwards. "English customs and ways of doing things and attitudes of mind were bought to New Zealand from the home country."[12] Attitudes were slow to change although some adaptations were necessary at first because some foods were not available. Acculturation did not occur. The Maori population was small, was not urbanised, was a hunter-gatherer and horticultural group and consequently had little effect on immigrants from industrialised countries.

Media forces reinforced British values in the population. Recipe books were imported from Britain—Mrs Beeton's cookbooks were widely used—and magazines that included a recipe page were often of British origin. Cooking classes held in the colony for servants and young women also reinforced many British ideas about food. Some aspects of meals did change however. Meat was available for all classes of the population and meat meals became more prominent in the diet than they were in Britain.

Non-British people were also present in the population from a early period but their numbers were small. In 1911 for example they numbered only 1.94% of the population[13] — consequently they had little influence on the values and norms of the mainly British-in-origin inhabitants. After World War Two non-British immigrants increased. Twenty thousand Dutch arrived between 1950 and 1970 but in a population of 3 million their influence was limited. They were however responsible for introducing new foods such as salami to the New Zealand diet. Most working class New Zealanders however continued to eat as they had always done.[3]

British food traditions in New Zealand were preserved for nearly a century partly because of the British background of the inhabitants but also because of New Zealand's isolation from the rest of the world. The fact that the destination of most New Zealand overseas travellers until recently was either Australia or Britain also served to reinforce British food traditions. More flexible air travel in recent years and more local restaurants serving ethnic food has exposed New Zealanders to other food traditions. These new trends have influenced domestic meals[3] which are now less traditional.

Some anthropologists believe there is a rule system inherent in a culture that maintains a stable meal system. The meal formats, but also the appropriateness of food items and dishes served at a meal are dictated by this system. Research on food patterns in English families for example found an

> extraordinarily high degree of structure, corresponding to the time of the day, and the day of the week....Structure appeared as the result of strict rules governing the presentation of food, the varieties permitted at a given occasion and the rules of precedence and combination. p 15 [14]

Leach (1993) also found a rule system associated with New Zealand meals. For example it "took over 50 years for two particular dishes - pasta with meat sauce, and curry and rice - to achieve main course status", not because of lack of equipment or ingredients to reproduce them but because they did not fit the rules of "course composition in New Zealand cuisine". In the case of pasta with meat sauce transition to main meal status (in recipe books before the 1960s it was listed as a tea dish) was dependent on the recognition that pasta was an alternative to potatoes (a concept that took time to accept) and could therefore fit the accepted course structure.[15]

Social change and its influence on New Zealand meals

Social change according to Giddens[5] involves looking at baselines. "Accounts of change involve showing what remains stable, as a baseline against which to measure alterations."

In the period from 1870 -1970 social change was a continuing process in New Zealand society but the pace of change was such that in terms of its effect upon meals it was not noticeable until it 'speeded up', beginning in the 1960s.

Social change in a society is associated with the physical environment, political organisations and cultural factors[5]. In the New Zealand context neither the physical environment nor the political organisation that developed was vastly different from that of Britain. The main catalyst for change in the new society was associated with cultural factors, technology, ideas — especially about food and women's role — and communication methods. Separately before the 1960s they did not bring about substantive changes in the society, interactively from this time they did.

It is difficult to consider the effect of technology on domestic meals aside from the role of women in the household. Consequently these will be discussed together.

In the late 19th century advances in kitchen technology in the colony —mainly the development of an efficient coal range simply meant that baking, for example, replaced roasting as a method of cooking meat, baked pies and puddings became more popular and steamed puddings declined. Technology also improved the distribution of food throughout the country, increasing the availability and variety of food available.

In the 1920s an increase in gas and electrical appliances occurred in the home. For domestic meals this meant a saving in time and less wastage of food. Many houses however did not replace their coal ranges until after World War Two when the supply of electricity had improved and appliances were more affordable. Prosperity in the late 1950s and 1960s meant that New Zealand kitchens were now equipped with refrigerators, deep freezers, automatic ovens, food mixers and processors etc. all of which reduced the time needed for meal preparation. Together they

> combined to provide the technological underpinning for the big changes in New Zealand society and work patterns that gathered momentum through the 1960s and 1970s. p 158 [16]

Before this time, while advances in kitchen technology had 'speeded up' meal preparation and appeared to give women more leisure time, the attitude to women's role in the household meant that the time women saved was simply used on the elaboration of other domestic tasks.

> Womens lives were governed by the doctrine that there existed a so-called women's sphere. They were to bear and rear children and attend to household affairs. p ix [17]

For example, in 1874 only 4% of women under 30 were unmarried and only 20% over the age of 15 worked outside the home.[18] This situation began to change at the end of the 19th century as industrialisation created more jobs for women and the onset of a depression in the 1880s made work a necessity for some. Yet despite enfranchisement in 1893 the ultimate aim for most women was still marriage. Enfranchisement did not place women in the forefront of public life; they failed to compete effectively with men for positions in public life, consequently they remained in the domestic domain. The so-called "cult of domesticity" which confined women's influence to the domestic sphere defined their role and their influence in the society.

The idea in the society of a separate sphere of influence for men and women continued into the 20th century. It was reinforced by new ideas about scientific housekeeping and motherhood and this ideology circumscribed even more rigorously their domestic position. World War One released them to a limited extent. More women joined the work force, they were no longer chaperoned, they had more money and freedom, but expectations for women after marriage did not change. The domestic sphere remained their domain. The 1930s Depression did not change this position — it simply served to emphasise the importance of their domestic skills — World War Two however did. The situation it created meant more jobs were available for women not only in traditional occupation but also in the public service and banks etc. The end of the war however returned them to the domestic sphere. The birth rate rose and women returned to the "prescriptive role of wife and mother."[19] This situation did not last. Experience in the work force during the war, better opportunities for education, a challenge to separate spheres for men and women by the women's movement in the 1960s stimulated a renewed transfer for women to the work force. The move was supported by the development of new technology that changed meal production in the home and new methods of communication that changed ideas about food and nutrition. The interaction of these factors created a situation that allowed a more rapid transformation of New Zealanders' meals. An increase in non-British immigration to this country, overseas travel by New Zealanders, global marketing and the ever changing advertising strategies of food producers have also in the past 25 years contributed to changes in New Zealander's eating habits. Continuity was the main feature of New Zealand eating patterns until the 1970s; change has been a feature in the last 25 years.

New Zealand is a relatively small country and its period of recorded European settlement has been short. This situation provided a manageable context for an historical and sociological explanation of eating habits. The study illustrated the importance of cultural traditions in maintaining food habits; it established a link between changes in meals and more rapid social change and identified the factors associated with the process. In particular the study revealed the strong influence of British food traditions in New Zealand during the period.

References

1. Burton, D. *Two Hundred Years of New Zealand Food and Cookery*, A H. & A.W. Reed Wellington N. Z., 1982.

2. Simpson, T. *An Innocents Delight: The Art of Dining in New Zealand*, Hodder & Stoughton, Auckland N. Z., 1985.

3. Bailey, R. & Earle, M. *Home Cooking to Takeaways: Changes in Food Consumption in New Zealand Between 1880-1980.* Food Technology Department, Massey University N. Z., 1993.

4. Department of Health, New Zealand national food survey, *Journal N. Z. Dietetic Assn.* **18**, 17-23, 1964.

5. Giddens, A. *Sociology,* Polity Press, Cambridge U. K., 1993.

6. Coney, S. *Standing in the Sunshine: A History of New Zealand Women Since They Won the Vote*, Viking, Auckland N. Z., 1993.

7. Leys, T. (ed). *Bretts Colonists Guide,* Capper Press, Christchurch N. Z., 1980.

8. Mitchell, J & McLaughlin, R. The influence of various factors on the main meal format served in New Zealand households, *Journal of Consumer Studies and Home Economics,* in press 1996.

9. *Consumer Expenditure Statistics* New Zealand Department of Statistics Wellington 1992.

10. Beardsworth, A.D. & Keil E.T. Vegetarianism, veganism, and meat avoidance: recent trends and findings. *British Food Journal* **93 (4)** 9-24 1991

11. Flower, T. *Tui Flower's Modern Hostess Cook Book ,* A.H. & A. W. Reed, Wellington N. Z., 1972.

12. Sutch, W. B. *Poverty in New Zealand*, Modern Books, Wellington N. Z.,1941.

13. *N. Z. Official Yearbook,* Govt. Printer 1939.

14. Douglas, M. Standard social uses of food: introduction. In M.Douglas (ed). *Food in the Social Order: Studies of Food and Festivities in Three American Communities,* pp 1-39, Russell Sage Foundation, New York, 1984.

15. Leach, H. Changing diets — a cultural perspective, *Proceedings of the Nutrition Society of New Zealand* **18,** 1-8, 1993.

16. Rennie, N. *Power to the People: 100 Years of Electricity in New Zealand*, Electricity Supply Association of New Zealand, Shoal Bay Press, Wellington N. Z., 1989.

17. Grimshaw, P. *Women's Suffrage in New Zealand,* Auckland University Press, Auckland N.Z., 1972.

18. Dalziel, R. The colonial helpmeet: women's role and the vote in nineteenth century New Zealand, Chapter 4, *Women in History: Essays on European Women in New Zealand,* eds. B. Brookes, C. MacDonald & M. Tennant. pp 55-67, Allen & Unwin, Wellington N. Z., 1986.

19. Bishop, S. *Married with Children: Perceptions of New Zealand Women's Lives in the 1950s : A Social and Literary Study,* B.A. Hons. thesis, Hocken Library, University of Otago N. Z., 1991.

Food habits: concepts and practices of two different age groups

S.S.P. Rodrigues, M.D.V. Almeida
Department of Nutrition, Oporto University, 4200 Porto, Portugal

1 Introduction

Food habits, a result of environmental forces acting within a cultural context, are simultaneously persistent and subject to change. When shifts in cultural patterns and values occur within a society, as a result of urbanization and different working patterns, changes in food habits are likely to occur. Within a society, younger members may have distinct concepts and practices than older ones as they experience a very different youth than their older counterparts did, due to rapid and major social changes. The main objective of this project, was to compare concepts and practices related to food habits of two different age groups of Portuguese women, living in Oporto.

2 Methodology

Using a structured questionnaire with open-ended and closed questions, we interviewed 50 female adolescents (A) aged 13-19 years (mean±sd:15.6±1.4) studying at the secondary school level, and 38 women (W), aged 20-64 years (mean±sd:43.7±13.0), attendants of a public health centre. There was no relationships between the two groups of participants. The following aspects were assessed: concepts of ideal meal, healthy eating and food practices - in relation to one weekday and one Sunday: number of eating occasions, interval between them, place, physical position, commensality and activity. Information about food consumption and detailed comparisons between concepts and practices will be published elsewhere.

Data was analysed using SPSSWIN. The two groups were compared by chi-square test (with Pearson's correction) and student's t-Test. Differences were considered statistically significant at the 0.05 level. Results will be presented as proportions (%) or mean and standard-deviation.

3 Results

3.1 Concepts

We investigated what respondents considered an ideal meal according to: general characteristics, time, place, commensality, activity, preparation time, temperature of consumption, cooking methods, freshness of foods and drinks. In both groups the ideal meal was more often associated with being healthy than to it's taste (Table 1).

Table 1 - Characteristics of the ideal meal

	A(%)	W(%)	p
Healthy	74	58	n.s.
To taste	16	26	
Healthy+to taste	10	8	
Undefined	0	8	

Adults were more likely to think that the ideal meal should be consumed at a fixed time than adolescents (Table 2).

Table 2 - Time of consumption of the ideal meal

	A(%)	W(%)	p
Fixed	62	84	0.02
Flexible	38	16	

Both groups considered that the ideal meal should be eaten at home, with company, at the dinner table, talking or talking and listening to music or watching TV (Table 3).

Table 3 - Place, commensality and activity of the ideal meal

	A (%)	W (%)	p
At home	60	66	n.s.
Indifferent	32	26	
Restaurant	6	8	
Open air	2	0	
With company	88	89.5	n.s.
Indifferent	10	10.5	
Alone	2	0	
At table	90	95	n.s.
Standing, on sofa or on the bed	10	5	
Talking (T)	50	45	n.s.
T + listening to music (LM)	28	16	
T + watching TV (WTV)	10	26	
WTV + LM	10	10	
"Nothing"	2	3	

It was found that these two groups had very different concepts related to time of preparation and temperature of consumption: adults considered that the ideal meal required more than 30 minutes to prepare in contrast to adolescents who believed that it should take a shorter period of time (Table 4). For most women (82%) the ideal meal should be consumed hot, whereas for teenagers it was indifferent (Table 4).

Table 4 - Time of preparation and temperature of consumption of the ideal meal

	A (%)	W (%)	p
<=30 minutes	67	36	<0.01
> 30 minutes	33	64	
Hot	41	82	<0.001
Hot or cold	59	18	

Concerning cooking methods there were no differences between the two groups in relation to the number of those who favoured the inclusion of broiled (A 90%, W 90%), boiled (A 76%, W 90%), stewed (A 48%, W 68%), fried (A 38%, W 55%) and raw foods (A 36%, W 18%) but adults mentioned roasted more often than adolescents (A 62%, W 82%, $p<0.05$). All respondents preferred fresh foods (A 100%, W 100%) and to a lesser extent frozen (A 58%, W 63%), canned (A 30%, W 16%) and smoked (A 26%, W 26%) ones.

Drinks to be included were differently indicated by the two groups: adolescents indicated natural fruit juices (A 92%, W 61%, $p<0.001$) and water (A 90%, W 66%, $p<0.01$), whereas adults chose wine (A 10%, W 58%, $p<0.001$) and beer (A 10%, W 29%, $p<0.05$).

Respondents expressed many and varied ideas related to healthy eating, with marked differences between the two groups: adolescents were more likely to mention the need to reduce some dietary constituents (sugar, sweets, soft drinks, fat, alcohol and salt) than adults (A 52%, W 21%, $p<0.01$). Although differences were not significant, they tended to use more "technical vocabulary", referring to macronutrients, Kcal, fibre, vitamins and minerals than their adults counterparts (A 12%,W 5%). In addition, a much higher proportion of adolescents than women mentioned the need to eat the "right amount", according to the "food circle", balanced, varied and with several meals per day (A 68%,W 16%, $p<0.001$). They also pointed out the need to drink water (A 22%,W 5%, $p<0.05$). In contrast, women associated healthy eating to freshness, cooking methods (boiled and broiled), the importance of being home made and cooked properly; they also valued to have time to eat (A 18%, W 53%, $p<0.001$). It is interesting to note that more adult women than adolescents mentioned that healthy eating should include consumption of milk and milk products (A 4%, W 32% $p<0.001$).

3.2 Practices

No differences were found between A and W in the total number of eating occasions or the interval between them, either during the week or on Sunday. On the other hand, within each group there was a decreased number of eating occasions on Sunday (Table 5). However this would not cause a longer interval between eating occasions due to the fact that, on Sundays the number of awakening hours was shorter in both groups.

Table 5 - Number of eating occasions, interval and number of awakening hours

	Week day			Sunday		
	A	W	p	A	W	p
Total # eating occasions	5.0±0.9	4.7±1.1	n.s.	4.2±0.9	3.9±1.0	n.s.
Time between eating occasions (hours)	3.1±0.8	3.4±0.8	n.s.	3.2±0.1	3.8±1.0	n.s.
# hours awaken	15.0±1.4	15.3±1.5	n.s.	12.8±1.9	14.4±2.0	n.s.

Considering the places where eating occasions occurred no differences were found between the 2 groups in both days: the most common place was home, followed by school or work cafeterias on week days or, on Sundays, by restaurants or a relative's house (Table 6).

Table 6 - Place of eating occasions

	Week day		Sunday	
	A (%)	W (%)	A (%)	W (%)
At home	71	79	82	77
School/work	19	11	0	5
Restaurant	7	6	9	7
Friend's/relative's house	3	4	6	10
In the street	0	0	3	1
p	n.s.		n.s.	

When asked to indicate a precise location, women ate in the kitchen, on weekday and Sunday, more often than adolescents (Table 7).

Table 7- Exact location of eating occasions

	Week day		Sunday	
	A (%)	W (%)	A (%)	W (%)
Kitchen	37	65	38	53
Dining/living room	33	30	54	45
Canteen/school bar	19	1	0	0
Bedroom	11	4	8	2
p	< 0.001		< 0.001	

Most respondents sat at the dinner table during the week but on Sunday, curiously a higher proportion of W had their meals while standing (Table 8).

Table 8 - Physical position of eating occasions

	Week day		Sunday	
	A (%)	W (%)	A (%)	W (%)
At table	60.5	65	74	69.5
Standing	25	26	11	22
Sofa/bed	14.5	9	15	8.5
p	n.s.		< 0.01	

During the week there was a higher proportion of A who ate with relatives/friends than the W but the 2 groups shared their eating occasions in a similar way on Sunday (Table 9).

Table 9 - Commensality of eating occasions

	Week day		Sunday	
	A (%)	W (%)	A (%)	W (%)
Family	42	52	71	62
Alone	33.5	40	18	27
Family and/or friends	24.5	8	11	11
p	< 0.001		n.s.	

When asked to indicate what they were doing while eating, more A were engaged in conversation and watching TV/listening to music (WTV/LM) than W, in either day. Interestingly a higher proportion of W referred that they just ate (they did "nothing else") (Table 10).

Table 10 - Activity during eating occasions

	Week day		Sunday	
	A (%)	W (%)	A (%)	W (%)
Talking (T)	40	25	45	33
"Doing nothing"	19	25	8	18
T+WTV/LM	16	13	28	25
WTV	14	20	11	18
Studying/cleaning	5.5	11	7	5
LM/radio	5.5	6	0	0
p	< 0.01		< 0.01	

4 Discussion and Conclusion

Researchers in the area of the socio-cultural aspects of food habits have suggested the importance of analyzing food consumption in the broader framework of foodways. This term which includes methods of eating, preparation, time of eating, number of meals per day as well as food choice, permits the description of food habits as an integrated part of a cultural pattern (Fieldhouse[1]). Industrialization and urbanization have been associated with fundamental changes in various aspects of lifestyle, including eating patterns (Bogin[2], Thomas[3],Prattalla[4]). Food selection is influenced by increased affluence and these changes seem to be affecting the nutritional status and health of populations living in largely urbanised societies (Thomas[5]).

The main objective of this project was to identify what concepts and practices related to food habits, were shared by two different age groups of female participants.

Adolescents and adults had similar concepts in relation to an ideal meal being a healthy one, which should be consumed in the company of others, taken at the dinner table, at home and an occasion to socialize. However younger respondents showed more flexible ideas in relation to time of preparation, time and temperature of consumption. This may indicate a desire for convenience, a growing demand for ready-made meals, light foods and salads, consumed within more informal time schedules. The two groups also exhibit different concepts of healthy eating: adolescents not only use the technical terminology of health workers and teachers but also quote the messages of balance, variety, adequacy, eating according to the "food wheel" and the need to reduce some dietary components (like salt, alcohol, fat). On the other hand, when adults talk about healthy eating , they mention foods, freshness, cooking methods and they attach importance to being home made and properly cooked. We can say that these two groups have concepts at distinct levels: women at the level of food and adolescents at the nutritional one.

In relation to practices, similarities were found in relation to number of eating occasions and the interval between them. There were no differences in the proportion of those respondents who ate at home but more adults were alone and in the kitchen than adolescents. These, more often had their meals with relatives/friends, in the dining room. A significant percentage of adults ate standing, when doing their household taks.

When practices and concepts were compared it was found that, both for adolescents and adult women, the "real"meals were different from the ideal one in relation to commensality (more respondents ate on their own), precise location (fewer sat at the dinner table) and activity (fewer had the opportunity to socialize during meals).

Results of this study, although limited to the population groups studied, emphasize the need to investigate elements of continuity and change in concepts and practices related to food habits.

References
1. Fieldhouse, P. *Food and nutrition: customs and culture*, Croom Helm, London and Sydney, 1986.
2. Bogin, B. Biological aspects of human migration, Chapter 5, *Rural-to-urban migration*, ed C.G.N. Mascie-Taylor & G.W. Lasker, pp 90-129, Cambridge University Press, Cambridge and New York, 1988.
3. Thomas, J.E. Food habits of the majority: evolution of the current UK pattern, *Proceedings of the Nutrition Society*, 1982, **41**, 211-227.
4. Prattalla, R., Pelto, G., Pelto, P., Ahola, M. & Rasanen, L. Continuity and change in meal patterns: the case of urban Finland, *Ecology of Food and Nutrition*, 1993, **31**, 87-100.
5. Thomas, J.E. A balanced diet?, Chapter 8, *Changing lifestyles: the effects on a balanced diet*, ed J. Dobbing, pp 143-168, Springer-Verlag, Berlin and New York, 1988.

Application of Kosher dietary laws by the food industry

N.A.M. Eskin,[1] C. Grysman,[2] G. Brojges[3]

[1] Department of Foods & Nutrition, University of Manitoba, Winnipeg, Manitoba R3T 2N2, Canada
[2] Rabbi, Toronto, Ontario
[3] Kashruth Coordinator, Vaad Ha'Ir of Winnipeg, Manitoba, Canada

Abstract

The popularity of kosher foods in North America has increased to where there are over 20,000 products kosher supervised today. This increase may be due to the perception that additional quality control steps beyond those normally required by the regulatory agencies are being implemented. This paper will discuss how kosher dietary laws are applied in the food industry in Europe and North America and how rabbinic law is dealing with new developments in food production such as genetic engineering.

Introduction

The increased demand for kosher products has encouraged many food companies to seek kosher endorsement from one of a number of recognized kosher certifying councils. The word kosher is a hebrew word meaning "fit" or "proper" as prescribed by Jewish Dietary Laws. Permitted foods generally fall into three categories; meat, milk and pareve or neutral (neither meat or milk). Approval of the manufacture of any of these types of products is far from simple requiring both a detailed understanding of food technology and halacha (Jewish law). This paper will attempt to review Jewish law as it pertains to the production of certified kosher products.

Endorsement of a Food Processing Plant

This process involves the willing participation of a particular company to do the following:

1. Make a formal application to a recognized rabbinical authority.
2. Provide a detailed list of all products made as well as a list of ingredients and suppliers.
3. Permit on-site inspections of the processing plant plus provision for follow-up visits as needed.

Either a local or national Kosher Certifying Council will approach a company due to community interest in a local product or, as is generally the case, the company seeks kosher endorsement of its products. In most cases the company has little understanding of what kosher involves so that a simple yet comprehensive qualitative description should be provided. It is insufficient to merely state that the product must comply with the requirements set out by Jewish law or custom. A practical description should include:

Kosher as described in Deuteronomy xiv, 3 indicates that certain things are prohibited as food because they are considered unclean. This can be summarized as follows:

1. Nothing is forbidden of minerals unless they are known to be dangerous or injurious to health. Modern technology, however, has produced a range of diverse secondary/adjunct ingredients used in the Production Plants which has brought this assumption into question. Kosher Certifying Councils now even scrutinize plastic packaging, water treatment chemicals and food belt lubricants used in a secondary or mechanical aspect.

2. All vegetables or fruits are permitted except for:

    (a) Some fruits and vegetables such as apples and cucumbers may be waxed. The material used should be acceptable.

3. With respect to meat foods the following applies:

    (a) The rule for animals is "Whatsoever parteth the hoof and is clovenfooted and cheweth the cud, that you may eat (Leviticus xi, 3). Thus having cloven hooves and bringing up the cud are key indicators of kosher domestic animals. Thus cows, sheep and deer would be considered acceptable. The bible lists the kosher species as well as providing a warning not to consider the camel, hyrax, hare and pig kosher. The latter are the only four mammals exhibiting one of two requirements for kosher. On this basis the giraffe would be considered kosher, however, because of its long neck we are not sure how to slaughter it. For meat to be accepted as kosher, however, the animals must be slaughtered according to Jewish law as stunning or shooting is forbidden. The meat must be handled through its distribution system in a carefully controlled manner. Animal by-products such as shortenings, emulsifiers, and stabilisers must also be closely supervised.

    (b) The number of birds forbidden are enumerated in Leviticus xi, 13. Twenty-four species forbidden all have a common trait in that they are birds of prey. Fowl, which have one of the following three signs, an additional finger or claw, a crop, or a gizzard were considered acceptable members of the kosher species. Such birds include hens and geese. Other species permitted are based on the tradition of use in a particular country. For example turkeys, not used widely in Europe, were eaten in other countries including the United States. Ducks, on the other hand, were eaten in Europe but not in the United States. A further example of this is evident from the extensive debate among the different rabbinical experts as to whether pheasant was permissible or not. Authorities in Europe accept the kashrut of pheasant as they were permitted in London and Paris but not in North America. Birds, like animals, must also be slaughtered according to Jewish law. The basis of ritual slaughter is to minimise any pain to the animal. The actual act involves cutting the animal's throat in a smooth backward and forward motion using the finest tempered and extremely sharp steel knife. The individual performing this act, referred to as *shochet*, must be highly educated in all the laws pertaining to ritual slaughter. Following slaughter, the blood must immediately be drained to avoid eating the blood of beasts and birds (Leviticus xvii, 12, 14). In addition to removing as much blood as possible, the meat undergoes a kashering process to further remove any blood remaining. This involves soaking the cut-up carcass in water for half an hour then covering with coarse salt for one hour, although 18 minutes would still be acceptable. Afterwards the salt is removed by rinsing.

4. With respect to fish the following applies:

   (a) Permitted fish are those with fins and scales (Leviticus xi, 9). Determining whether fish have fins and scales is not an easy task as some scales in fish are similar to rough skin while others act as a covering over the skin. Kosher fish, according to Talmud must have at least two scales that are removable. As a result, there has been considerable controversy over such fish as the swordfish. The adult swordfish does not have any scales while the juvenile fish has scales as classified by biologists. These however, do not meet those described by Jewish law. Fish is considered parve but must not be mixed with meat or poultry according to some traditions although it may be eaten at the

same meal with meat or with dairy foods (Blucher, 1991; Regenstein and Regenstein, 1979).
5. Miscellaneous:
   (a) There are eight kosher species of grasshoppers each having four legs, four wings and jumping legs. However, it became the established custom that even if these were identifiable they were no longer permissible. Certain Moroccan and Yemenite communities, however, still maintain a tradition of eating them.
   (b) No insects are permitted for food. This includes insects, worms or any tiny beings. These are referred to as *sherets* any small creature that breeds on the land. This necessitates taking great care to ensure no insects or worms are present in foods such as romaine lettuce. Some Kosher Certifying Councils very scrupulous about insects will avoid vegetables such as broccoli, cauliflower and tight lettuces such as romaine, endive etc. Others require vegetables, canned or frozen, be examined carefully to ensure none are present (Schwab, 1986).
   (c) Alcohol produced from any fruit except grapes is permissible. Grape alcohol, however, requires rabbinical supervision from the time the grapes are harvested until fermented into wine. This creates additional restrictions with respect to wine vinegar and grape products such as tartaric acid. Even grape juice requires kosher endorsement (Regenstein and Regenstein, 1988).

In order to comply with kosher requirements the following criteria must be established:

1. That the equipment being used in the manufacture of the product(s) must not have been used for non-kosher products. If it has been, then it must undergo a rigorous cleaning or kashering process; or the non-kosher product might be farmed out to a secondary supplier. If this is not possible then new equipment may have to be acquired dedicated only to the production of the kosher product line. It is also important to ascertain whether this production line is next to one producing non-kosher products. If so, will it be separate and quite distinct or incompatible; and are there any risks of non-kosher ingredients mixing?
2. That all ingredients being used are obtained from only approved suppliers containing no non-kosher sources. Some kashrut councils have exhaustive data bases that are constantly updated

listing all the ingredients permitted. One such list by the Orthodox Union of Rabbis has 50,000 entries. Any changes in a single ingredient or supplier must be pre-checked to ensure the kosher endorsement has not been compromised.
3. With the exception of meat plants and wineries where the presence of full-time authorized kosher inspectors or mashgichim are mandatory, other processing plants may only require occasional on-site inspections. The latter is only possible where there is a high level of confidence with the plant management who have contractually agreed to notify any future changes in ingredient suppliers or processing operations. Unannounced spot checks are also agreed to.

A Kosher Certifying Council, after inspecting the processing equipment and ingredients, then decides whether to grant kosher endorsement or not. If the decision is positive, the product in North America will be required to carry a symbol to show the consumer that this product has been kosher endorsed. Jewish consumers in North America are conditioned to look for these symbols which enables them to shop in major supermarkets throughout the country and find a whole range of products with different kosher endorsing symbols. An example of some of the symbols used in North America are listed in Table 1. Jewish consumers are legally

Table 1. A List of Some Kosher Endorsing Symbols

Symbol	Organization
Ⓤ	The Union of Orthodox Rabbis (U.S.A.)
Ⓚ	The Organized Kashrut Laboratories (U.S.A.)
כ	Kosher Supervision Services (U.S.A.)
MK	Montreal Kosher, Canada
COR	Council of Orthodox Rabbis Toronto, Canada
ⓌⓀ	Winnipeg Kosher, Canada

protected from misrepresentation and violations of kosher in both Canada and in the United States. In Canada there is an Act under the Department of Agriculture with similar types of legislation in several states in the United States with large Jewish populations. The Europeans tend to have a less structured system compared to North America. For example, the

Chief Rabbi's Office in London provides up-dated information on the status of a wide range of products. This is included in an *Annual Guide* with quarterly updates. The major difference from the North American system is an additional classification of products found to be acceptable for use after research even though there is no formal agreement or relationship between the manufacturer and the Rabbinical Council. The general categories in this guide are as follows:

1. **Products manufactured under Rabbinical Supervision.**

    (a) Products produced under supervision of the London Rabbinical Court have a symbol *BDL* on the left hand margin.

    (b) Products made under the supervision of other recognized Rabbinical Supervisory have a symbol *S* on the left hand margin.

2. **Products not manufactured under Rabbinical Supervision**

    Products approved as kosher based on information provided through correspondence, with factory visits in some instances. These products do not have any symbols.

An additional requirement on kosher symbols states whether the product is dairy or pareve. Based on the phrase "Do not cook a kid in its mothers milk" (Exodus xxiii, 19) and further expounded in the Talmud, a separation of dairy and meat products is required. All utensils used for the production of dairy products must be separate from those used in the preparation of meat products. All dishes and cutlery used for meat and dairy meals must also be separated. Consequently, a kosher household has a minimum of two sets of dishes. A similar arrangement is required in a processing plant where dairy, meat and pareve operations must be operated with separate utensils.

3. **Milk, Meat and Dairy Products**

    (a) These include products listed under 2 that contain milk ingredients, indicated by a symbol *M*, not manufactured under any Rabbinical Supervisory Authority.

    (b) These include products listed under category 1 that contain milk, indicated by a symbol *MS*, manufactured under a Recognized Rabbinical Supervisory Authority.

(c)   All products containing meat have the symbol *(MEATY)* after the product name.

4.   Non kosher products.

Any products found to be non-kosher is indicated by the symbol *NK*.

5.   Wines and Spirits

This section provides a list of kosher wines and spirits (including grape juices) plus non-supervised spirits and liqueurs.

Consequently in England there are many products which are acceptable as kosher which do not have a symbol but are listed in the guidelines under category 2. This contrasts with the North American system where, if it is under a Recognized Kosher Cetifying Council, it is only kosher if it has a symbol. The latter does have the convenience of identifying the product immediately without having to refer to an Annual Directory or Guide. However, individual Rabbis in North America do give approval to products, such as those listed in category 2, on an informal basis.

Genetic Engineering - A New Panacea for Kosher Foods

The development of new and improved food products through genetic engineering provides new opportunities and challenges for kosher food products, ingredients and processing aids. One such aid is calf rennet, an enzyme used for the production of cheeses (Eskin, 1991). Calf rennet is a unique enzymes that coagulates milk protein during the cheese process. A serious shortage of calf rennet has resulted in it being blended with porcine pepsin for use in cheese manufacture. This represents a serious problem for the production of kosher cheeses as:

1.   Rennet should be obtained from calves slaughtered according to Jewish ritual practices as discussed earlier.
2.   Porcine pepsin is unacceptable as it is not kosher.

To overcome these problems a number of microbial rennets have been approved for use in the production of kosher cheeses. Development of genetic engineering techniques, however, has provided a new source of rennet that may be acceptable to rabbinical authorities. This involves transferring the gene responsible for rennet from the calf into a microorganism (Eskin, 1991). The genetically engineered organism now produces a rennet that is indistinguishable from calf rennet. Because of the rapid growth of microorganisms the shortage of calf rennet should be solved. Such developments provide unique and new challenges for rabbinic authorities to deal with.

Conclusion

Kosher endorsement of food processing operations requires careful examination of the processing equipment as well as individual ingredients. Adequate inspection systems must involve a constant flow of information between the processor and rabbinical authority. The growth of consumerism as well as constant changes in new ingredients and processing demand constant vigil, review, self education and good communication between both sides to ensure a high confidence level is maintained in the food kosher endorsed.

References

1. Blucher, D.K. 1991. Kosher foods and food processing. Encyclopedia of Food Science and Technology, (Y. Hui editor-in-chief), John Wiley & Sons Ltd. Vol. 3, p. 1585.

2. Eskin, N.A.M. 1991. Biochemistry of Foods. Second edition. Academic Press, New York. p. 524.

3. Regenstein, J.M. and Regenstein, C.E. 1978. An introduction to the kosher dietary laws for food scientists and food processors. Food Technol. 33(1): 89.

4. Regenstein, J.M. and Regenstein, C.E. 1988. The kosher dietary laws and their implementation in the food industry. Food Technol. 42(6): 86.

5. Schwab, S. 1986. Inspection of vegetables. Kashrus. 6(5): 22.

Supermarket shopping and the older consumer
C. Leighton, C. Seaman, M. McGlade
Centre For Food Research, Queen Margaret College, Clerwood Terrace, Edinburgh, EH12 8TS

Abstract

The increasing proportion of older consumers affects Food Retailers and through them the whole supply chain of the Food Industry. Food Retailers have a dominant position over suppliers in the food chain and are able to exert a powerful influence over the development of both branded and own-label food products.

This paper will present the main findings of current qualitative and quantitative research being undertaken at the Centre For Food Research (CFR) into food choice amongst elderly people. The paper will explore factors in the food retailing environment which affect food choice amongst consumers aged 60 years plus in a city, town and rural location in Scotland and relates factors in the retailing environment to current nutritional research.

The factors affecting the decision-making process in terms of location of shopping and consequently the choice, range, price and quality of food available to the consumer will be discussed. Positive and negative factors will be analysed, giving an indication of the elements of the retail offering that must be developed to meet the future needs of the elderly.

Introduction

The numbers of people aged over 60 and the income which they have at their disposal is increasing (OPCS[1]). In Scotland, there were 776,000 people over the age of 65 in 1993, accounting for 15.1% of the Scottish population. It is projected that by 2018 this will have risen to 960,000 and 19% respectively (Scottish Office[2]).

Initiatives led by Age Concern such as 'Through Other Eyes' (sponsored by Tetrapack and IGD) have contributed to the current understanding of the impact this will have on food retailing, but there is an urgent need for information and food research to establish the needs and preferences of independently living elderly people and those whose ageing opens up a new market.

Recent research which examined the diets of 300 elderly people living in sheltered housing in Scotland highlighted a number of factors which are of critical importance when shopping for food. The proximity of the shops to the sheltered housing scheme, the quality and variety of the foods available and food costs affect food choice (Caughey et al[3][4][5][6][7]). It seems likely that these factors and many others will affect food choice.

This paper presents the main findings of current qualitative and quantitative research being undertaken by the CFR exploring the affects of the Food Retailing environment on elderly consumer food choice in city, town and rural locations in Scotland.

Nutritional Status Of The Elderly

The rise of health and nutritional awareness amongst the Scottish population (and the U.K. as a whole) has influenced consumer and retailer behaviour.

The field of nutrition and health in the elderly population attracts much research from a breadth of fields. Health and social services, political and health organisations and charities are all concerned with the issue (Tilston et al[8]; McMahon[9]). Milestones of health awareness development have included the Health Of The Nation report (Dept. of Health[10]) and Nutrition of Elderly People (Dept. of Health[11]) have been born from the general poor nutritional status of the U.K. population. However Scotland has a particular notorious health record with diet-related diseases such as Heart Disease causing high numbers of deaths (Scottish Office[2]).

The ageing process itself can be influenced by diet. Disease and illness characteristic in the elderly population can be prevented, managed or minimised by an appropriate diet (Brown[12]; Scottish Office[2]; Whitney et al[13]; Harvey[14]).

Research has found the elderly population to be a risk group, for many reasons, for poor nutrition in residential care (Herne[15]; Dept. of Health[11]) and in supported accommodation such as Sheltered Housing or with support from social services (Caughey et al[3]; Tilston et al[8]). However elderly people living independently in the community are also deemed a risk group (Wells[16]) Most elderly people in Scotland live in their own homes and most live alone (Scottish Office[2]). It should be considered at the outset that elderly people living in the community independently is difficult to define with many being physically and psychologically dependent on others. Research into the dependence elderly people have on others found that 80% were dependent on others for activities such as shopping (Moane[17]).

Many factors affect this nutritional risk group (Shepherd[18]; Conner[19][20]; Nutrition & Food Science[21]):
· Food e.g. sensory properties
· Individual e.g. mental illness, lonliness, bereavement, physical health, income, dependence (Herne[15][22]; Wells[16]; Pender[23]; Schewe & Balazs[24])
· Environment.
It is the shopping environment which is under-researched and neglected in the light of a recent survey that showed 31% of respondents aged 65 years plus had health as a priority in food choice (Key Note[25]).

Current Trends

It is first necessary to examine the key trends of the retailing environment to understand the environment in which consumers food choice is made and assess the positive and negative impact this has on elderly consumers food choice.

Firstly, the market structure must be considered. The retail environment is dynamic, responding to macro-environmental changes by monitoring consumer characteristics and competition and pursuing opportunities in the industry, through strategy (Cox & Brittain[26]; Johnson & Scholes[27]; Johnson[28]).

Porter's[29] framework for analysis of the environment has often been applied to the Food Retailing sector and found that the industry is dominated by the six multiples: Tesco; Asda; Isoceles Group (Gateway and Somerfield); Argyll Group (Safeway) and Kwik-Save. This concentration of power has produced intense competition for market share and high barriers to entry because the multiples benefit from economies of scale and have the capital, experience and technology to limit new entrants. Consumers are deemed to have choice, but suppliers are dominated, particularly as own-label goods increase (Duke[30][31]; EIU[32]; McHugh et al[33]). This market structure is at the expense of small independent and multiple retailers, who would have the flexibility for prime location.

Secondly consumer mobility has had a significant impact on Grocery Retailing. In 1993, 45% of households in Great Britain had one car/van, which has been a relatively stable figure over the past 20 years, however the rise of 2 car/van households has risen more sharply to being 20% in 1993. Overall, 69% of households are estimated to have a car/van, although car ownership in Scotland is lower than the rest of the U.K.. Retired households who are dependent on state pensions is low at only 9.9%. (General Register Office Scotland[34]; CSO[35]).

This increase in mobility for consumers in general has led to the trend of out-of-town shopping and the one-stop shopping concept, although this trend is not as prominent in Scotland than the rest of the U.K.. Generally however the population do depend on a car for grocery shopping. (Sainsbury[36]; IGD[37]) Elderly people have less access to cars and so are relying on retailers that can be reached by public transport or by foot.

Thirdly, specific location and structural trends have taken place. The move to out-of-town shopping sites has been synonymous with the increase in store size, at the expense of other store formats.

Retailers have widened their product range and depth and can offer service features which allow for one-stop shopping. This also facilitated the use of new technology which improved stock control and service-levels and ultimately improving the freshness of produce. The economies of scale derived from such store formats also allows for competitive pricing. Such retail offerings have become the competitive arena (Ogbonna & Wilkinson[38][39];EIU[32];IGD[37]). This trend has been slowed down by planning regulations because of the negative effect on town and local retail activity (Larkham[40]; IGD[37]; Fernie[41], with some of the large multiples now seeking to return to city centres with Tesco Metro and Sainsbury Central, however superstores will remain as the dominant feature of grocery retailing.

Therefore quality and price in food choice is at the large superstore/supermarket multiples and such factors influence store choice (Nutrition & Food Science[42]however, the elderly do not have as great access to appropriate transport to allow them this choice.

Fourthly, the growth of own-label products have allowed retailers to differentiate themselves from the competition and brand their retail offering with the aim of creating loyalty to the retailer and store (McGoldrick[43]).

From a consumer's perspective, the own-label product allows for greater choice and in many instances a quality and cost alternative. Consumers, like the elderly, who are unable to access the superstore/supermarkets because of their location, do not have the choice of foods other segments of the population have.

Method

People aged 60 years plus were interviewed using a structured questionnaire. Topics covered included: accessibility and transport to shops; dependency on others; perceptions of in-store environment and use of service features. People were recruited through Community Centres in Edinburgh (city), Dunfermline (town) and Fallin (rural) locations.

Consumer location was important to the study as Scotland's population is concentrated in the Central Belt, but has a large scattered population in rural areas (although Islands were not included in the study).

There were 63 quality responses with even distribution amongst locations and population characteristics were:

Table 1   Sample Characteristics

Population Characteristic	Percentage and base numbers
Receive meals on wheels	2% n=60
Wear glasses	89% n=61
Live in Sheltered Housing	3% n=62
Suffer from Arthritis	49% n=60
Living alone	63% n=59

Thus the sample were independently living with sight and mobility problems of varying levels and living alone. One characteristic that was not measured but indicated in the study qualitatively was that most were dependent on state pension and so of a lower income group.

Results And Discussion

The Elderly Market

The increasing elderly segment of the market is indeed a viable market for food retailers to penetrate.

Results showed that elderly people are frequent shoppers, mostly once a week or more confirming research that elderly people shop over time (Schmidt et al[44]), spending up to £20.00 on their main grocery shop. There were however a significant number spending more than this.

This indicates that it is not just the volume of this market but also their shopping characteristics that make them a potentially loyal market. Consumer loyalty is gained by meeting and exceeding customer needs, wants and expectations and once loyal, consumers will shop more frequently and spend more as well as choose one store over other stores (Walters & White[45]; Corstjen & Corstjen[46]; Lewison[47]).

From a health and nutritional perspective, the retail environment in which elderly people make food choices, must be researched, particularly as they become a target for retail marketing activity.

Transport And Store Accessibility

Results would indicate that elderly people shop at supermarkets and so are not disadvantaged in terms of food choice, particularly as the most important reason for store choice was given as convenience. However, those in a rural location were less dependent on supermarkets and more dependent on small independents than the city and town locations.

This would suggest that food retailers are largely meeting the locational needs of the elderly. However there are several points to note in order for these results to be interpreted in their true context.

Firstly, when questioned about their method of transport for grocery shopping, 43% *(n=62)* of respondents stated they went by car, however only 18% *(n=27)* actually drove themselves. Those in a rural location had the greatest dependency on a car.

A relatively high percentage also stated that they walked or used public transport. Of these people, 94% *(n=17)* and 71% *(n=7)* respectively said it limited the amount they bought and indeed what they bought, as they could only buy amounts and types of food they could carry.

Secondly, 57% *(n=63)* of elderly people questioned received help with their shopping, particularly from family. This was largely in the form of car transport. Again, town and rural locations had greatest dependency for help.

The high proportion of home-helps helping with or doing shopping would indicate that the elderly population have many influences over their food choice and store choice. Such high levels of dependency would suggest that Scotland is similar to previous research finding elderly people to be highly dependent, even living independently in the community (Moane[17]).

Therefore, elderly people may largely shop at supermarkets because mobility is facilitated by others. If this informal support network is unavailable, then elderly people are indeed disadvantaged. Only 6% *(n=33)* of respondents used a special bus, delivery or taxi service, which indicates that there is very limited formal community support with geographical variations.

Locational trends and consumer mobility have to some degree disadvantaged the elderly. Elderly people who have access to supermarkets are dependent on a network of support, without which their expenditure and food choice would be limited, either by having to shop alone or depend on local retailers, who have sometimes been found to have less choice, higher prices and less fresh produce (Wells[16]; Consumer Affairs[48]; Keane & Willetts[49]; Stitt et al[50]). This could only compound the risk of poor nutrition amongst this group.

However, given the large proportion of the sample who shopped at supermarkets, it has given insight into the experience of the elderly supermarket shopper and the possible influences on food choice.

Shopping Characteristics And Retail Offering

The retail offering is the composite offering provided by the retailer to the consumer. Three main elements are considered: product offering; customer services / facilities and in-store design and environment.

Firstly, product offering. The growth of own-label products is an important feature that can be used with the elderly.

Results show that 85% *(n=59)* of elderly people purchase mostly own-label or at least a mixture of own-label and branded goods. However, 8% reported that own-label goods were unavailable. Elderly people therefore, are not adverse to own-label goods, although previous research has cautioned that elderly people are apprehensive about the quality of own-label products (Schmidt et al[44]).

Appropriate marketing activity can build upon this positive element and closely related to the development of own-label, are the following features.

Firstly, elderly people were found to be actively trying new products when introduced into stores.

Retailers therefore can influence product development to the market and particularly so through own-label. Again, however, this is relative to store accessibility. Product development could also be communicated in-store, as when questioned about promotions such as tasting in-store, 40% *(n=36)* reported that promotions were available, but less than half used them. Such methods of in-store marketing may help promote product development and encourage nutrition.

Secondly, the ready-made meals is an area to be developed, with 60% *(n=62)* actively purchasing them. This, with appropriate marketing, may be a product sector to be expanded, given the large amount of people living alone.

Thirdly, a point of note that only 56% *(n=61)* found the quantities of packaged foods to be suitable to their needs. Therefore the demographics of this market must be considered in product development.

Therefore, supermarkets have the power through own-label and through their dominance of suppliers to meet elderly customers needs both through product development and marketing.

The second part of the retail offering studied was customer services and facilities. This element influences the elderly shoppers experience, which are generally available at the multiple retailer's supermarkets and superstores. This area was largely qualitative research.

It was established that many customer service facilities were unimportant to the elderly, such as credit paying facilities, store loyalty cards etc.

However, certain features were used well by consumers e.g. some reported enjoying coffee shop facilities and many purchased from specialist counters such as fresh fish. This allows the elderly to buy fresh produce and in appropriate quantities. However, again this was dependent on availability of these services. Such customer service and store features are a key element in providing the consumer with quality food and in enhancing the shopping experience.

The third and final area to be considered was in-store design and environment. This provided an indication of how well consumers functioned in the store environment. Respondents rated their experience to give an accurate indication of ease and difficulty.

It was found that reaching the highest shelf, loading and unloading trolleys and reaching into less than full freezers were most difficult for elderly people.

Therefore, store design and the environment are important in meeting elderly peoples needs. This has already been recognised to some extent by some retailers who are sponsoring design programmes and contributing to the development of appropriate design and incorporating recommendations into their store design (Turner[51]; Shannon[52]).

Conclusion

This paper has investigated the retail factors affecting food choice of the elderly, taking account of consumer location.

It has been established that retail location has a major influence on elderly peoples food choice, as the store shopped at will determine the range, quality, price and freshness of food. However, there is a wider retail offering including customer services and facilities and the in-store environment. All these elements which can be categorised as the retail offering can be controlled to varying degrees by the retailer.

At present, the industry's strengths appear to be that quality, choice and value do exist within the large multiple sector. However, such locational trends have been at the expense of alternative store formats and local retail activity. Results have indicated that this may not be having as detrimental an effect on elderly consumers as first thought, but the informal support network required by elderly shows that indeed, elderly people who are not mobile and independent are disadvantaged. Location is very important, as convenience was identified as the most important factor influencing store choice. If location is right, then other factors will become more influential in store choice, which may improve store/retailer loyalty. Locational strategy therefore, must be on the agenda of retailers seeking to gain consumer loyalty in the future.

The growth of own-label products is an area for development with the elderly. Particularly with product development, packaged goods and promotional activity. These factors are also applicable to brand manufactures.

Generally, store design is improving, but will require more development.
Weaker areas, other than location, would be customer services/facilities. These were not being used significantly and the reasons for this, whether because they were unavailable or not appropriate, needs to be investigated.

Retailers must be aware of their influence on elderly peoples food choice and indirectly their nutritional state. Health and business disciplines have different motives for ensuring a healthy elderly population, but retailers must not be isolated in the pursuit of the goal. Their influence is significant on the elderly consumers food choice and appropriate strategies and contribution can enhance the food choice available to elderly and their shopping experience.

References

1. OPCS, *General household survey*, 1993, **24,** HMSO, London.
2. Scottish Office Home & Health Department, *Health in Scotland*, 1993, HMSO, Edinburgh.
3. Caughey, P. et al *Nutritional status of the elderly in sheltered housing*, The Joseph Rowntree Foundation, 1994.
4. Caughey, P. et al Factors affecting dietary intake and nutritional status of tenants in sheltered housing, *Journal of Human Nutrition and Dietetics*, 1994, 7, 269- 273.
5. Caughey, P. et al Nutrition of old people in sheltered housing, *Journal of Human Nutrition and Dietetics*, 1994, 7, 263-268.
6. Caughey, P. et al What do elderly people eat?, *British Food Journal*, 1995.
7. Caughey, P. et al , Weight and skeletal size, *Journal of Human Nutrition and Dietetics, 1994*
8. Tilston, C., Neale, R., Gregson, K., & Price, A., Study of food preferences of an elderly population receiving meals on wheels in Nottinghamshire, *Journal of Consumer Studies and Home Economics*, 1994, 18, 31-43.
9. McMahon, K., Consumer nutrition and food safety trends, *Nutrition Today*, 1995, 4, 152-156.
10. Department of Health, *The health of the nation: a strategy for health in England*, 1991, HMSO, London.
11. Department of Health , *The nutrition of elderly people: a report of the working group on the Nutrition Of Elderly People of the committee on medical aspects of food policy, 1992*, HMSO, London.
12. Brown, J. *Nutrition now.* West Publishing, Minneapolis, 1995.
13. Whitney, E., Cataldo, C., Debruyne, L. & Rolfes, S, *Nutrition for health and healthcare*, West Publishing Company, Minneapolis, 1996.
14. Harvey, J., Healthy eating in later life, *Elderly Care*, 1993, **1**, 35-36.
15. Herne, S., Healthy eating in old age, ,*British Food Journal*, 1993, 5,, 36-39.
16. Wells, D., Nutrition, social status and health, *Nutrition & Food Science*, 1992, 2, 16-20 Mar/Apr.
17. Moane, G. Dependency and caring needs among the elderly. Special issue: psychological aspects of ageing: well-being and vulnerability, *Irish Journal of Psychology*, 1993, 1,189-203.
18. Shepherd, R., Overview of factors influencing food choice from Why People Eat What They Eat (ed. by M. Ashwell) *in Conference Proceedings of the 12th British Nutrition Foundation Annual*, 1990, British Nutrition Foundation, London.

19. Conner, M. Individualized measurement of attitudes towards foods. *Appetite*, 1993, **20**, 235-238.
20. Conner, M. Accounting for gender, age and socioeconomic differences in food choice, *Appetite*, **23**, 195.
21. Nutrition & Food Science, Food Choice, *Nutrition & Food Science*, 1993, **1**, July/August.
22. Herne, S., Research on food choice and nutritional status in elderly people: a review, *British Food Journal*, 1995, **9**, 12-29.
23. Pender, F. (ed), *Nutrition and dietetics*, Campion Press, Edinburgh, 1993, 92-95.
24. Schewe, C. & Balasz, A., Role transitions in older adults: a marketing opportunity, *Psychology & Marketing*, 1992, **2**, 85-99.
25. Keynote , *UK food market*, 1993.
26. Cox, R. & Brittain, P., *Retail management*, Macdonald and Evans, London.
27. Johnson, G. & Scholes, K., *Exploring corporate strategy : text and cases*, Prentice-Hall, London, 1993.
28. Johnson, G. (ed), *Business strategy and retailing*, John Wiley & Sons, Chichester.
29. Porter, M., *Competitive strategy*, Free Press / MacMillan, London, 1980.
30. Duke, R., A structural analysis fo the UK grocery retail market, *British Food Journal*, 1989, **5**, 17-22.
31. Duke, R, Structural changes in the UK grocery retail market, *British Food Journal*, **2**, 1992, 18-23.
32. EIU Grocery retailers, *Retail Trade Review*, 1993, **28,** December.
33. McHugh, M., Greenan, K. & O'Rourke, B., Food retailers: stuck in the middle? *British Food Journal*, 1993, **3**, 32-37.
34. General Register Office Scotland (GROS), Census 1991 Report for Scotland Part   2 of 2, HMSO : Edinburgh.
35. CSO *Family spending : a report on the 1994-95 family expenditure survey*, 1995, HMSO, London.
36. Sainsbury, D., Food retailing in a new world, *in Transcript from Convention '94 Competition: The Winds Of Change*, IGD.
37. IGD (ed. Anna Dawson), *Grocery retailing 1995: The market review*, 1995, IGD Business Publications.
38. Ogbonna, E. & Wilkinson, B., Corporate strategy and corporate culture: the management of change in the UK supermarket industry, 1988, *Personnel Review*, **6**, 10-14.
39. Ogbonna, E & W, Barry, Corporate strategy and corporate culture: the view from the checkout, *Personnel Review*, 1990,4, 9-15.
40. Larkham, P., The style of superstores: the response of J. Sainsbury plc to a planning problem, *International Journal of Retailing*, 1988, **1**, 44-58.
41. Fernie, J., The coming of the fourth wave: new forms of retail out-of-town development, *International Journal of Retail and Distribution Management*, 1995, **1**, 4-11.
42. Nutrition & Food Science, Food retailing, 1993, *Nutrition & Food Science*, **2**, Mar/April.
43. McGoldrick, P., *Retail marketing*, 1990, McGraw-Hill Book Company, London.
44. Schmidt, R., Segal, R, & Cartwright, C., Two-stop shopping or polarization. Whither UK grocery shopping? *International Journal of Retail and Distribution Management*, 1994 , 12-19.

45. Walters, D. & White, D., *Retail marketing management*, MacMillan Press Ltd, Basingstoke, 1987.
46. Corstjens, J. & Corstjens, M., *Store wars the battle for mindspace and shelfspace*, Wiley & Sons, Chichester, 1995.
47. Lewison, D., *Retailing*, MacMillan College Publishing Company, New York, 1995.
48. Consumer Affairs, The challenge of healthy eating, 1994, July/August, 130.
49. Keane, A. & Willetts, A., Factors that affect food choice, *Nutrition & Food Science.*, 1994, 4, July/Aug 15-17.
50. Stitt, S., O'Connell, C. & Grant, D., Old, poor and malnourished, *Nutrition & Health*, 1995, 10, 135-154.
51. Shannon, B., Healthy eating: a retail perspective, *British Food Journal*, 3, 26 -28.
52. Turner, B, When the problem of age is gripping, *The Times*, 18 November, 1994, 40.

Class, income and gender in cooking: results from an English survey

T. Lang,[1] M. Caraher,[1] P. Dixon,[2] R. Carr-Hill[2]

[1]*Centre for Food Policy, Wolfson School of Health Sciences, Thames Valley University, 32-38 Uxbridge Road, Ealing, London W5 2BS, UK*
[2]*Centre for Health Economics, University of York, Heslington, York YO1 3DD, UK*

Abstract

In recent years, there has been an upsurge of interest in what explains differences in people's diets and whether these play a part in explaining variations in health. That there are considerable variations in what people eat and that diet affects health is not in doubt. The issue is how to explain these variations. There are a number of possible explanations: biophysical, social, cultural, financial, attitudinal and behavioural. One possible source of variation in behaviour is cooking skills. This paper investigates evidence on the English public's attitudes and behaviour with regard to cooking. It analyses data from the UK Health Education Authority's 1993 Health and Lifestyles Survey. It reports evidence of considerable differences associated with income, gender and social class. It is not clear whether these differences are transient or signs of longer term cultural shifts in skills associated with purchasing, cooking and eating food.

1 Background

In Britain, there is growing recognition that consumers' cooking skills cannot be taken for granted. Skills acquisition has considerable implication for food industries (e.g. for cooking instructions), caterers (for training), health promoters (for translating dietary advice into everyday terms) and domestic life (the pleasure of eating home-cooked food). Until recently, it was assumed that the growth of pre-processed foods symbolised positive changes in domestic labour, particularly for women (e.g. Mennell, Murcott & van Otterloo[1]). However, an argument has now been voiced, by Government and other interest groups, including caterers, that this decline - or restructuring - of cooking skills could have an impact on consumers' capacity to control their diet (NFA[2]). Another argument, propounded by chefs such as the Académie

Culinaire, is that the erosion of cooking skills is part of the decline in taste - the result of a supposed rise of bland, sweet, 'pappy' foods. First in France in the late 1980s, and since 1994 in the UK, a system of 'Days of Taste' has been instituted when chefs go into schools to induct children into the joys and peculiarities of different tastes.

The argument about place of cooking skills has reached particular intensity because of evidence about health differences between social classes and social groups (Wilkinson[3]). Variations in health have long been associated with differences in income and behaviour. For many years, there has been a debate about what to do about these differences, and indeed whether there is anything that can be done about them by health education or promotion (Naidoo[4], Beattie[5]). There has been a parallel debate in food policy about the role of diet as a risk factor in significant diseases, and the role of health education in promoting improving diet (DHSS[6], NACNE[7], DHSS[8], H.M.Government[9]) Since early this century, the impact of material circumstances such as cooking facilities, income, available foods, on diet has been recognised (Rowntree[10], Rowntree[11], Boyd Orr[12], Burnett & Oddy[13]).

In the late 19th century and early 20th century, social reformers were concerned about the poor lacking cooking facilities (e.g. Spring Rice[14]). Today, people on low incomes and in the lower social classes have equipment their predecessors might have dreamed of, but there is still a class and income gradient in the distribution of facilities, and also a gender gap in health (Graham[15]). The issue of access to facilities was raised in the 1980s, but only for socially excluded groups such as the homeless (Conway[16]).

It could be argued that this 'old' debate is irrelevant or inappropriate today when the vast majority of the population have kitchens, cookers, access to decent shops, and so on. The need to explain variations in health is, however, encouraging researchers to reconsider the impact of social, rather than biophysical factors in health (Nestle[17]) and to begin to develop a different analysis of resources and circumstances that affect health (Blane, Brunner & Wilkinson[18]).

One avenue of work currently being considered is the role of differences in consumption of foods, for instance fruit and vegetables (National Heart Forum[19]). According to National Food Survey data, in the last three decades, there has been a shift from fresh items (unprocessed, uncooked materials) to frozen or pre-cooked foods. Illustrative of this trend is the declining consumption of fresh potatoes and a rapidly expanding market for value-added potato products such as frozen chips (MAFF[20]). The British eat less vegetables than any other European country, at least half the amount of France, Spain and Italy. This could be a significant factor in the UK record of food-related ill-health. Vegetables (and fruits) contain many of the protective factors such as anti-oxidants, for coronary heart disease and some cancers which are Britain's main sources of premature death (WHO[21]). Research which has investigated the diet of low income consumers has recently highlighted the constraints of money on food purchasing patterns (Leather[22], National Consumer Council[23]). There has also been some interest in the issue of access to shops, and that there might be 'shopping deserts' with low provision (Beaumont et al[24]). The range of foods on sale varies considerably between more and less affluent areas (Lewis[25], Mooney[26]).

Another focus for attention has been the impact of gender and age on division of food within the household, in particular the role of women and mothers in affecting what is purchased, cooked and consumed (Charles & Kerr[27], Brannen et al[28]). Some research has looked at the particular constraints of special groups, such as single women and mothers (Dowler[29]). Other research by the present authors has explored the impact of gender on a wide range of food attitudes and behaviour (Caraher, Dixon & Lang[30]).

In 1994, a new system of a national curriculum was introduced in Britain. This entailed the removal of what previously was known as Home Economics in which practical, hands-on cooking was taught. In its place a more theoretical, technology-oriented approach to food was introduced. There was considerable concern expressed, even by other Government Departments, about the impact of changing cooking skills upon diet (NFA[31]). Unless, it has been argued, people actually cook from relatively unprocessed ingredients, how can they control their dietary intake and make use of the nutritional advice given to them by health educators? In reaction to the curriculum change, across the UK, hundreds of schemes and projects to teach basic cooking skills have been set up (e.g. Hunton[32], Health Promotion Agency for Northern Ireland[33], Caraher & Lang[34]). A number of pilot intervention studies in the UK and elsewhere have suggested that re-designed cooking and food classes for young people can change, not just their own diets, but their family's too (Demas[35]). From previous studies (MORI[36], DoH[37]), it was unclear whether public skills and attitudes towards cooking are affected by income or social class, and whether these could be a material factor in health variations.

2 Methodology

The Health and Lifestlyes Survey is an occasional survey conducted by the English Health Education Authority (HEA) with MORI. It provides baseline data on behaviour and attitudes of relevance to public policy concerns in health education. The 1993 sample had 5,553 observations, reduced to 4,438 in usable form. The survey asked approximately two hundred questions across a wide variety of health issues, and gave multiple options for responses. The present paper draws upon the questions which refer to cooking.

The data reviewed here is analysed only for gender, social class and per capita household income effects. A full explanation of the sample and rationale for income and class indicators is given elsewhere (Caraher, Dixon & Lang[38]).

The Health and Lifestyles Survey data is self-reported, with all the advantages and disadvantages that this entails.

Surveys adopt several standard approaches to defining class. The two most readily available from the Health and Lifestyles Survey are socio-economic group (SEG) and the Registrar General's 6 point scale of social class. The 6 point Registrar General's definition of class was used as it has a more consistent association with both income and education (two traditional markers of social status) and is therefore more suitable for detecting trends relating to status. A major and related issue is whether to base the analysis on the class and income of each respondent, or on measures and class and income related to an entire household. The argument for variables limited to the individual respondent is that it will be the individual's education, culture and knowledge that influences his or her eating and cooking behaviour. But it can also be argued that these will be constrained by the overall material circumstances of the household, in this case represented by per capita income. They may also be constrained by the cultural inheritance of the household, here represented by the class of the 'head of household' - currently defined as the person with legal responsibility for the accommodation. The use of the social class of the head of household is justified for the purposes of this study.

In this study, per capita income is computed as total annual household income divided by number of persons in household - according to following formula: the first adult is weighted as 1.0, subsequent adults weighted 0.75, children and adolescents aged 1-16 weighted 0.5, and babies weighted 0.33. The tables use a variable that re-codes this income into four bands (£3000 or less, £3001-7000, £7001-14000, over £14000).

3 Findings

The survey asked about where people ate their food. The majority (55%) ate at least 7 main meals at home per week, and of meals eaten at home, three quarters (73.7%) were not described as ready-prepared. There were few class, income or gender differences. Three quarters (74.4%) of respondents also said that they did not eat one take-away meal in the last week, but when asked how often they cooked a meal, i.e. any meal, less than half said they did every day.

There was a clear class and income effect on how likely a respondent was to have a main meal purchased within the previous week. 80.1% of the least affluent had purchased no ready-prepared main meal within the last week, compared to 69.5% of the most affluent. 73.3% of social classes I and II had had none, compared with 83% of social classes V and VI.

Marked effects appeared with regard to actual cooking. When asked how often they cook a meal, income and social class produced few variations but gender did. For instance, 92.6% of females, compared to 77.1% of males, felt either fairly or very confident about being able to cook from basic ingredients.

66.5% of females, for instance, compared to 40.1% of males, were confident about using steaming. Females were more confident also about stewing, braising and casseroling. Table 1 and 2 suggest that confidence in general is quite high, but varies between techniques, and is income and class related.

TABLE 1 Confidence in using cooking techniques, by per capita income

Cooking technique	Percentage in each income group who are confident with the technique			
	£3000 or less	£3001-7000	£7001-14000	£14001 and over
Boiling	87.6	91.0	93.4	93.3
Steaming	41.6	56.6	58.7	65.8
Shallow-frying	66.8	75.6	80.3	80.9
Deep frying	61.5	68.1	65.2	60.5
Grilling	88.9	91.9	92.6	95.0
Poaching	45.7	63.9	67.9	70.2
Oven baking or roasting	77.4	84.5	86.4	86.1
Stewing/braising/casserolling	52.8	70.4	75.8	77.7
Microwaving	65.4	56.3	65.9	65.6
Stir frying	51.2	51.5	63.7	68.1
None of these	1.4	1.8	1.1	0.5
Don't know/non response	0.1	0.1	0.2	0.2
Number in income group	1148	1326	1309	655

If knowing about cooking techniques is one thing, applying them is the ultimate test. There are gender, class and income variations in these applied skills, but the most marked are for gender. For instance, 81.9% of females are confident in cooking pasta, but only 60.9% of males. Even in cooking fresh green vegetables, there is a gender divide, with 78.6% of males confident and 94.4% of females, and also an income divide with 93% of the most affluent being confident, but 79% of the least affluence reporting confidence. As Table 3 shows, confidence is also related to income, with the more affluent invariably more confident about cooking named foods.

The present survey found 98.6% of people reported that they had fairly or very easy access to cooking facilities. There were almost no class, gender or income variations, except in the case of steamers, food processors and liquidisers, when higher incomes and classes are more likely to have them. 54.8% of classes I and II had a food processor, for instance, compared to only 27.2% of classes V and VI. The chip pan or deep fat fryer is the only piece of equipment which the lower social classes are more likely to have than the higher. 65.3% of classes V and VI had one, compared to 45.1% of classes I and II. There were fairly strong requests for more information on food preparation. For instance, 55% of respondents wanted more information on preparing meat and poultry, and 52.5% on fruit and vegetables. There were few class or income variations on this point.

TABLE 2   Confidence in using cooking techniques, by social class

Cooking technique	Percentage in these social classes who are confident with this technique			
	I x II	IIIN	IIIM	IV x V
Boiling	94.6	93.2	91.0	89.5
Steaming	66.8	62.3	55.8	55.3
Shallow-frying	82.3	77.7	74.8	73.4
Deep frying	64.0	65.3	69.7	68.7
Grilling	93.4	92.9	91.8	87.6
Poaching	73.3	67.8	63.9	60.7
Oven baking or roasting	87.6	86.5	84.4	81.6
Stewing/braising/casserolling	80.7	75.7	70.4	68.7
Microwaving	67.7	62.3	56.2	49.9
Stir frying	67.7	59.1	54.3	46.4
None of these	0.8	0.6	2.0	2.7
Don't know/no response	0.2	0.1	0.3	0.7
No. in class group	1535	717	1323	854

TABLE 3   Confidence in cooking particular foods, by income

Food/food type	Percentage in each income group who are confident they can cook this food			
	£3000 or less	£3001-7000	£7001-14000	£14001 and over
Red meat	66.9	81.7	84.2	83.2
Chicken	76.2	87.6	88.3	87.5
White fish	56.8	75.3	76.2	78.6
Oily fish	30.5	49.0	55.0	57.3
Pulses	44.4	53.3	57.5	64.4
Pasta	71.6	69.5	75.7	82.0
Rice (not rice pudding)	75.7	77.4	84.3	88.1
Potatoes (not chips)	88.8	92.3	94.8	94.7
Fresh green vegetables	79.0	89.4	92.0	93.0
Root vegetables	74.5	86.7	89.6	90.2
None of these	3.1	3.1	2.2	1.1
Don't know/no answer	0.3	0.2	0.1	0.2
No. of respondents	1148	1326	1309	655

There was a fairly strong desire in the sample to learn more about cooking in general, with 52.8% saying they would like to learn more, and 44.3% saying they would not. Of the latter group, the majority felt they knew enough already. The largest difference was between the sexes, with 65.1% of females and 44.3% of males saying they knew enough already, and more males (36.2%) than females (22%) expressing no interest in cooking (see Table 4).

TABLE 4   Reasons cited for not wanting to learn more about cooking, by gender

	Percentages giving these reasons	
	Women	Men
Know enough already	65.1	44.3
No time to cook	6.3	8.9
Prefer to: eat out; eat takeaways; eat cook-chilled foods	0.2	1.0
Not interested in cooking	22.0	36.2
Other	9.0	14.2
Don't know / non response	0.9	1.4
No of respondents	1374	1087

97.2% thought it fairly or very important to teach boys to cook and 98.6% to teach girls to cook. There were no class or income differences shown.

Asked where people first learned to cook from, mothers were by far the most significant source, with cookery classes at school the next, and cookery books third most cited, but there were gender differences. Also, cookery books were more important for the higher social classes, whilst cookery classes at school were more important for lower social classes (see Table 5), whereas mothers seem to rise above class differences.

Later in life, cookery books appear to be by far the most cited source for learning about cooking (see Table 6). With many of these potential sources of learning, there were clear differences between the sexes and some less marked differences in income groups and social classes. For instance, 54.7% of females cited cookery books, and 30.5% of males. 38.3% of lower income groups cited cookery books, compared to 49.6% of the higher groups. Cookery books were cited by 54.1% of social classes l and ll, and by 37.6% of classes V and Vl.

4   Conclusion

The survey data supports the view that differences in food behaviour are subject to pressures of class, income and gender. These differences are sufficiently marked to warrant more investigation than they have been accorded before now. Compared to the 19th century or early 20th, almost all people in this English survey had well equipped kitchens, but the use to which this

TABLE 5 'When you first started learning to cook, which if any of these did you learn from?', by social class

	Percentage in this class group learning from these sources			
	I x II	IIIn	IIIM	IV x V
Mother	67.4	70.7	67.7	65.2
Father	7.0	7.4	8.0	8.2
Grandmother	10.3	9.9	10.3	9.4
Wife/Husband/Partner	12.0	6.7	9.3	7.8
Other relatives	6.6	5.7	6.5	6.4
Friends	10.2	7.1	5.4	5.7
Childminder	0.3	0.6	0.2	0.1
Cookery classes at school	29.1	35.0	34.0	36.3
Other cookery classes	3.5	3.9	2.6	2.7
Cookery books	29.5	23.8	18.1	13.7
Cookery programmes on media	6.6	5.6	3.7	2.6
Specialist cookery	3.2	2.1	1.4	0.8
Articles in magazines	6.5	5.3	3.4	2.3
Booklets from supermarkets	3.2	1.7	1.4	0.5
Booklets from food producers	2.3	1.1	0.8	0.7
Health centre/doctor	0.3	0.1	0.2	0.0
None of these	2.7	2.0	3.3	3.4
Haven't learnt to cook	2.5	2.4	4.0	4.2
Don't know/non response	0.8	1.0	0.7	1.2
No. of respondents	1535	717	1323	854

equipment is put may be affected by factors such as knowledge, confidence and skills. Food culture, not just in England, is subject to considerable change and has led change in wider culture (Ritzer[39], Gabriel & Lang[40]). These are global, not just national changes (Barnet & Cavanagh[41]). Debates about the role of cooking skills are not new in Britain (Smith & Nicolson[42]), but a serious question has been posed by this survey as to whether late 20th century homes are now full of equipment which is predominantly used to serve pre-processed ingredients and meals, over which consumers have little nutritional control. This has implications both for health promotion, catering and education policies. Without cooking skills, consumers cannot act on the educational advice that research suggests they already have (Sheiham et al[43,44]). This research is a reminder that, while emphasis in national economic policy is understandably put on the economic value of hi-tech skills such as computing and educational flexibility, public policy also needs to recognise that some of the older, and more low-tech skills also warrant urgent attention.

TABLE 6 'And later on, which if any of these were useful to you in learning more about cooking?'

Sources	People who said these sources were useful 'later on'	
	Number	Percentage
Mother	748	13.5
Father	122	2.2
Grandmother	113	2.0
Wife/husband/partner	701	12.6
Other relatives	309	5.6
Friends	633	11.4
Childminder	7	.1
Cookery classes at school	612	11.0
Other cookery classes or courses	314	5.7
Cookery books	2475	44.6
Cookery programmes on TV	1115	20.1
Specialist cookery/food magazines	392	7.1
Articles in magazines/newspapers	915	16.5
Booklets/leaflets from supermarkets	469	8.4
Booklets/leaflets from food producers	221	4.0
Health centre/doctor	69	1.2
None of these	524	9.4
Haven't learnt to cook	182	3.3
Don't know/no answer	120	2.2
No. of respondents	5553	180.8 *

* The total is more than 100% because it was a multiple response question

Acknowledgments

This paper draws upon a more extensive study for the Health Education Authority reviewing variations in health. The authors wish to thank the HEA Research Unit for permission to quote from this work.

References

[1] Mennell, S., Murcott, A. & van Otterloo, A. *The Sociology of Food*, Sage, London, 1992.
[2] National Food Alliance *Get Cooking!*, National Food Alliance, Department of Health, BBC Good Food, London, 1993.
[3] Wilkinson, R. Income Distribution and life expectancy, *British Medical Journal*, 1992, 304, 165-8.

[4] Naidoo, J. Limits to Individualism. *The Politics of Health Education: Raising the Issues*, ed S. Rodmell and A. Watt, Routledge and Kegan Paul, London, 1986.
[5] Beattie, A. Knowledge and Control in Health Promotion: A test case for social policy and social theory, *The Sociology of the Health Service*, ed J. Gabe, M. Calnan & M. Bury, Routledge, London, 1991.
[6] DHSS *Diet and Coronary Heart Disease*, Report on Health and Social Subjects no. 7, HMSO, London, 1974.
[7] NACNE *A Discussion Paper on Proposals for Nutritional Guidelines for Health Education in Britain by the National Advisory Committee on Nutrition Education*, Health Education Council, London, 1983.
[8] DHSS *Diet and Cardiovascular Disease, Committee on Medical Aspects of Food Policy report of the panel on diet in relation to cardiovascular disease*, Report on Health and Social Subjects, no. 28, HMSO, London, 1984.
[9] H.M.Government *The Health of the Nation: A Strategy for Health in England*, Cm 1986, HMSO, London, 1992.
[10] Rowntree, B.S. *Poverty: A study of town life*, Macmillan & Co, London, 1902.
[11] Rowntree, B.S. *Poverty and Progress: A Second Social Survey of York*, Longmans, Green and Co, London, 1941.
[12] Boyd Orr, J. *Food and the People, Target for Tomorrow no 3, Pilot Press*, London, 1943.
[13] Burnett, J. & Oddy, D.J. (ed) *The Origins and Development of Food Policies in Europe*, University of Leicester Press, London, 1992.
[14] Spring Rice, M. *Working Class Wives*, Virago, London, 1981 (1939).
[15] Graham, H. *Hardship and Health in Women's Lives*, Harvester Wheatsheaf, Hemel Hempstead, 1993.
[16] Conway, J. (ed). *Prescription for Poor Health: the crisis for homeless families*, London Food Commission, Maternity Alliance, SHAC and Shelter, London, 1988.
[17] Nestle, M. (ed). Mediterranean Diets: Science and Policy Implications, *American Journal of Clinical Nutrition*, 61, 6 (S), June.
[18] Blane, D., Brunner, E. & Wilkinson, R. The evolution of public health policy, *Health and Society*, ed Blane, D., Brunner, E. & Wilkinson, R., Routledge, London, 1996.
[19] National Heart Forum. *Report on fruit and vegetables by an expert committee*, National Heart Forum, London, forthcoming.
[20] MAFF *National Food Survey 1992*, HMSO, London, 1993.
[21] WHO *Diet, Nutrition and the Prevention of Chronic Disease*. Technical Series, no 797, World Health Organisation, Geneva, 1990.
[22] Leather, S. Less Money, Less Choice: Poverty and Diet in the UK today. *Your Food, Whose Choice?*, ed National Consumer Council, pp 72-94, HMSO, London, 1992.
[23] National Consumer Council. *Benefits and Diets*, NCC, London, 1995.
[24] Beaumont, J., Lang, T., Leather, S. & Mucklow, C. *Report to Low Income Project Team, Nutrition Taskforce*, Institute of Grocery Distribution, Letchmore Heath, 1995.
[25] Lewis, J. *Food Retailing in London*, London Food Commission and CES Ltd, London, 1985.
[26] Mooney, C. *Cost, Availability, and Choice of Healthy Foods in Some Camden Supermarkets*, Hampstead Health Authority, London, 1987.
[27] Charles, N. & Kerr, M. *Attitudes towards the feeding and nutrition of young children*, Health Education Council, London, 1984.
[28] Brannen, J., Dodd, K., Oakley, A. & Storey, P. *Young people, health and family life*, Open University Press, Buckingham, 1994.

[29] Dowler, E. *Factors affecting nutrient intake and dietary adequacy in lone-parent households*, MAFF, London, 1995.
[30] Caraher, M. , Dixon, P. & Lang, T. *Buying, eating and cooking food: A review of a national data set on food attitudes, skills and behavioural change*, Health Education Authority, London, 1996.
[31] NFA. *Get Cooking*! National Food Alliance, Department of Health and BBC Good Food, London, 1993.
[32] Hunton, B. *A Progress Report on the "Get Cooking" project in Hull, Hull and Holderness Community Health NHS Trust*, Hull, 1996.
[33] Health Promotion Agency for Northern Ireland. *Cook It!: A report on a pilot programme promoting low cost healthy cooking in the community*, Health Promotion Agency for Northern Ireland, Belfast, 1994.
[34] Caraher, M. & Lang, T. *Evaluating Cooking Skills Classes: Towards an Evaluation Strategy for Health Promotion Wales*, Health Promotion Wales, Cardiff, 1995.
[35] Demas, A. *Food Education in the Elementary Classroom as a Means of Gaining acceptance of diverse, low-fat foods in the school lunch program*, PhD thesis, Cornell University, Cornell, 1995.
[36] MORI *Survey for Get Cooking!* National Food Alliance, London, 1993.
[37] DoH *Cooking: Attitudes, Skills and Behaviour*. Department of Health, London, 1995.
[38] Caraher, M., Dixon, P. & Lang, T. *Buying, eating and cooking food: A review of a national data set on food attitudes, skills and behavioural change*, Health Education Authority, London, 1996.
[39] Ritzer, G. *The McDonaldization of Society*, Pine Forge Press, Thousand Oaks CA, 1993
[40] Gabriel, Y. & Lang, T. *The Unmanageable Consumer*, Sage, London, 1995.
[41] Barnet, R. & Cavanagh, J. *Global Dreams*, Simon & Schuster, New York, 1995.
[42] Smith, D. & Nicolson, M. Nutrition, Education and Ignorance and Income: A Twentieth Century Debate, *The Science and Culture of Nutrition: 1840-1940*, ed Kamminga, H. & Cunningham, A., eds, Rodopi, Amsterdam, 1995.
[43] Sheiham, A., Marmot, M., Rawson, D. & Ruck, N. Food values: health and diet, *British Social Attitudes - the 1987 Report*, ed R. Jowell, S. Witherspoon & L. Brook, pp 95-119, Gower Publishing, Aldershot, 1987.
[44] Sheiham, A., Marmot, M., Taylor, B. & Brown, A. Recipes for health and diet. *British Social Attitudes - the 7th Report*, ed R. Jowell, S Witherspoon and L Brook, pp 145-165, Gower Publishing, Aldershot, 1990.

Community meals in the next millennium

J. Kenny, B. Pierson
*Worshipful Company of Cooks Centre for Culinary Research,
Bournemouth University, Poole, Dorset, BH12 5BB, UK*

Abstract

One of the most significant aspects of British demographic trends is the increasing elderly population, that is over 65 years of age. As a result, there is likely to be an increased demand for food services to supply meals to elderly people living in their own homes. Community meals have up until now been the major form of mass catering for the elderly living at home, but as research continues into nutritional requirements of the elderly there is an increasing need to identify and measure the role and importance of food and social contact. Studies conducted on the community meals service in the United Kingdom, United States and Australia have raised various questions about the service. Research being carried out at the Worshipful Company of Cooks Centre for Culinary Research, is using both dietary tables and chemical analysis to assess the nutritional adequacy of meals of those nutrients to which the elderly are most frequently deficient and which are most likely to be destroyed during processing and storage.

Introduction

One of the most significant aspects of British demographic trends is the increasing elderly population, over 65 years of age, HMSO[1]. The elderly now constitute 16% of the population and by the year 2031, the number over 65 is projected to increase to 22%, HMSO[2]. As a result, there is likely to be an increased demand for catering within establishments providing for the elderly, such as residential homes, luncheon clubs and day centres. There may also be an increased requirement for meals to be supplied to private homes for those wishing and able to continue to live outside the system of residential homes.

At present, 96% of those over 65 live in private homes in the community. Most live independently but others may need help with daily tasks such as meals, Dunn[3]. Community meals (previously known as meals-on-wheels) have been the major form of mass catering to the elderly living at home but even this system varies in terms of those eligible to receive meals, delivery, frequency of receipt, source of supply, cost and portion size.

Various studies of the community meals service for the elderly, have provided information on the organisation of the meal service, structure of meals and the acceptability of meals. However, today there are a number of additional factors such as wider food choice, increased disposable income, different perceptions and attitudes, all of which are changing the scope of the meal service to the elderly population.

History

The provision for feeding the elderly at home was initiated by the Women's Royal Voluntary Service (WRVS) in the 1940's. The WRVS provided a post air raid service that supplied hot drinks and snacks to those who had just lost or had their homes damaged during the air raids. The service continued through post war years, helping those most at need due to rationing of food. The WRVS continued its work as small catering units providing meals to the most needy and this service became known as "meals-on-wheels". Today the service is known as, "community meals", and it is the responsibility of Social Service departments to purchase and provide community meals to the elderly. Within which a substructure has developed whereby, Direct Service Organisation (DSO's), private contractors and voluntary organisations bid for contracts to provide prepared meals to the elderly living at home, Kenny, Pierson and Edwards[4].

The criteria for those who receive community meals are:-
"The elderly or handicapped persons living in their own home who cannot provide for themselves a hot main meal daily and cannot be provided with one in any other preferable way", DHSS[5].

There seems to be a lack of clear objectives for the community meals service and calls have been made to define more clearly the objectives of the meal service in relation to the nutritional, social and psychological aspects of the meals, Dunn[3]. This, coupled with the growing research into the nutritional needs of elderly, has highlighted the need for the community meals service to be re-evaluated in order to ensure that it meets an elderly persons actual needs for food. These needs for food have been identified by the late Dr Mogens Jul as inclusive of elements such as nutrition, a tasty meal, social contact, pleasant surroundings, encouragement, a feeling of care and being respected, World Health Organisation[6].

The question today is, how is the meal service, to the elderly person living at home, going to satisfy an elderly persons real needs for food in the next millennium.

Nutrition

Research into nutrition and the elderly is increasing. Advancing age is thought to change requirements for energy, protein and vitamins and minerals due to lean body mass, reduced physical activity and decreased intestinal absorption, Wellman[7]. Until recently, guidelines aimed especially at the elderly have not been available. The Department of Health in 1992 produced a report, "The Nutrition of Elderly People", which made recommendations on the nutrient content of an average day's food for an elderly person aged 75 and over. However, it is acknowledged by the Department of Health that little is known about older people's nutritional requirements and the document makes recommendations for further research into establishing energy, protein, vitamin, and trace element requirements. A need has also been identified, by the Department of Health, for better strategies for measuring nutritional status, developing recognition of clinical features of nutritional deficiencies and gaining a better understanding of nutritional status of older people as a whole.

The first major survey of the nutrition and the elderly in the UK was in Sheffield in 1952, Bransby and Osborne[8]. The King Edward's Hospital Fund Publication of a "Report on an Investigation into the Dietary of Elderly Women Living Alone", was carried out in 1965 and increased awareness in nutrition and the elderly, King Edward's [9]. In 1967, a panel on nutrition of the elderly was set up by the Committee of Medical Aspects (COMA) and reported overt malnutrition rarely occurred, but sub-clinical malnutrition may be a wider incidence, DHSS[10]. In 1972 a large scale government survey of the elderly was set up to understand some of the gaps on nutrition and the elderly. The study looked at the nutritional status of the elderly, and found that diets consumed were much the same as those of the whole population but in smaller quantities, DHSS[10]. This study was followed up five years later and drew similar findings to the first report.

Current research is continuing in this area with such studies as Suter and Russell, "Vitamin requirements of the elderly"[11], Sjogren, Osterberg and Steen, "Intake of energy, nutrients and food items in a ten year cohort comparison and in a six year longitudinal perspective: A population study of 70 and 76 year old Swedish people"[12] and Bunker and Clayton, "Studies in the nutrition of elderly people with particular reference to essential trace elements"[13]. However despite the volumes of research, reports and findings, there is still no clear guidance for the full dietary needs of the elderly.

A detailed survey is presently being carried out by the Ministry of Agriculture Food and Fisheries/Department of Health into the "National Diet and Nutrition survey: People aged 65 years and over", this is due to be published in 1997 and will provide data on people over 65 and their nutritional intake and status.

Nutrition and community meals

The DHSS circular 5/70 states that the community meal service, "inevitably is a second best service, since there is an interval between cooking and service and that it must be clearly understood that community meals can only make a significant contribution to the nutrition of an elderly person if it is supplied for at least five days a week", DHSS[5]. Today nutrition is seen to play a greater role in the health and well being of the entire population including the elderly. Community meals which are delivered to the elderly living in the community, who cannot cook or prepare food for themselves, may make a significant contribution to the nutrition of the elderly person living at home.

It was not until 1985 that the issue of nutritional standards for community meals was first addressed. The London Borough Study Group (LBSG) set a basic standard for community meals that was recognised by caterers and dieticians.
In 1992 the Advisory Body of Social Service Caterers(ABSSC)(who advise caterers) reviewed the recommendations of the LBSG and found a number of anomalies, such as meal weights used were higher than those provided by the producers. As a result the ABSSC set its own guidelines in order to answer the anomalies found in the LBSG guidelines.

The ABSSC guidelines are that each meal should provided:-

600-650 Kilo calories	20g Protein
75-80 Carbohydrates	6g Fibre(Non Starch Polysaccharides)
25-30g Fat	3-4mg Iron
10 mg Vitamin C(after cooking and before holding)	
160mg Calcium	0.3mg Thamin
0.4mg Riboflavin	5mg Niacin
80mg Folate	230mg Retinol

This would mean that the content of the meal would be approximately 13% protein, 37.5% fat and 43.75% carbohydrate.

The recommended meal sizes are:-
- 330-400g - Main Course
- 130-200g - Sweet
- 480-600g = Total

(Minimum total does not reflect on the minimal meal and dessert, because the ABSSC wants the minimum total meal to be more than the sum of the two figures).

It is recommended that these meals should make up a third of the elderly person's recommended daily intake. However the ABSSC are concerned that meals may be thrown away or eaten in stages thus are not providing a third of the daily intake. In addition they are concerned where the other 66.6% of the daily intake is coming from, or if the elderly person is receiving any other food. Consequently they suggest that the service should be looking at providing 100% of the recommended daily intake. Also the issues, such as, lack of iron and folate in meals have been raised, because they are hard to address, due to these nutrients mainly coming from cereal sources that are traditionally eaten at breakfast time, ABSSC[14].

In 1995 The Caroline Walker Trust produced a report "Nutrition guidelines for food prepared for older people in residential and nursing homes and for community meals such as meals on wheels and luncheon clubs", The Caroline Walker Trust[15]. This report highlights the lack of nutrition standards in providing food to the elderly through mass catering. The report has raised the subject of food and nutrition, that is central to the enjoyment of an elderly person's life. The report recommends an increase in energy, folate, vitamin C, calcium and iron. However, this may be difficult as this will require an increase in the size of the meals. The Caroline Walker Trust has moved away from meals providing a third of an elderly person's diet, because not all the recommended intakes represent 33% of daily intake. They have recommended that energy is not less than 40% of estimated average requirement , calcium, iron and folate is not less than 40%. Vitamin C are not less than 50% of reference nutrient intake. The rest of the recommendations are still, not less than 33% of reference nutrient intake, Caroline Walker Trust[15].

Studies have been conducted on community meals in the UK by Stanton[16], Malcolm and Johnson[17], Lawson and Thomsom[18], Davies[19], and Tuner and Glew[20]. Most recent surveys have shown that energy, vitamin C, vitamin D, and calcium are below the recommended levels, Malcolm and Johnson[17] and Davies[19].

Research carried out by Turner and Glew[20], examined the nutritional content of meals from various sources. The protein, energy, calcium and iron contents were calculated from available portion weights and food tables.

Meals varied greatly within regions. Frozen meals may provide greater menu variety, but were lower in weight and hence contained fewer nutrients in total, Turner and Glew[20].

Research carried out by, Lawson and Thomson 1984[18], suggests that ideally community meals must supplement dietary intake of an elderly person and not displace other food sources, Lawson and Thomson[18].

Stanton 1971[16] states that meals vary even from the same region due to portion size, storage, cooking methods and are only of nutritional benefit if delivered five days a week. At weekends meals were provided by relatives and neighbours. Meals should be planned with nutrition in mind, Stanton[16].

Davies 1981[19], analysed the nutrient content of elderly people's weekly diet using computer analysis and carried out chemical analysis of vitamin C and potassium, of delivered meals. Vitamin C was chemically analysed because the food table figures for ascorbic acid often give an inaccurate result due to the methods of production and long periods of storage. Potassium was investigated further due to a link with depression. The vitamin C result through chemical analysis was compared to levels given by food tables. All the mean averages for chemical analysis were substantially below the food table levels. This has identified a difference in chemical analysis and food table results, therefore it would be appropriate to carry out chemical analysis on all the major nutrients required by the elderly. Overall the study concluded that meals were low in vitamin C, protein and did not contain enough energy, Davies [19].

Studies have also been carried out in the United States on their system of community meals called "Home delivered meal programs". Darling 1988[21] suggest that meals seek to improve the food and nutritional intake of the elderly receiving them, but due to variations of sizes of meat, fish, vegetables and desserts and that temperatures were lower than recommended, the food quality was not as good, thus nutrition value was reduced. The meals did however make an important contribution to the diet of the elderly person, Darling[21].

Another American study looked at urban and rural areas and found that geographical location does affect nutrient intake, thus this should be acknowledged in any strategy for community meals, Stevens et al.[22].

Studies in Australia found that community meals did not meet recommendations but with minor changes in menus and appropriate nutrition education could improve food choice and result in increased intake, Bell et al[23]. Also Darnton-Hill[24], found that people receiving community meals were often saving part of the meal and reheating it later. Thus the meals on wheels provided more than a third of the elderly persons daily intake, Darnton-Hill [24].

Some studies Davies 1981 used chemical analysis on the selected nutrients, vitamin C and potassium. However the majority of studies use food tables to establish the nutritional content of the meals provided. There seems to be an excessive reliance on food tables for nutrition analysis. Food tables may be flexible and comprehensive enough to deal with changes that have occurred in food products and production methods, such as cook-freeze, since the food tables were produced. This method also fails to take into account the nutritional status of food between cooking and consumption. These aspects are exaggerated in the case of community meals whereby the majority of meals are prepared using the cook-freeze method, regenerated and then transported, resulting in a time delay, to the consumer.

Conclusion

The study of Davies showed that when meals were analysed using chemical analysis, the levels of vitamin C were substantially lower than results from food tables. If this is the case for vitamin C, then it may also be true for the other nutrients. The move away from prime cooked community meals to frozen meals may also have an effect on the nutritional content of meals at point of consumption. Therefore assumptions are being made that community meals are providing the elderly with some nutritional value. As nutrition plays a major role in the health well being of an elderly person, it is essential to undertake research that can establish the nutritional value, using both dietary tables and chemical analysis, of the meal at point of consumption. Research being carried out at the Worshipful Company of Cooks Centre for Culinary Research, is undertaking analytical procedures to assess the nutritional adequacy of meals including both dietary tables and chemical analysis of those nutrients to which the elderly are most frequently deficient and which are most likely to be destroyed during processing and storage. This work will establish the nutrient value of the community meals and allow strategies to be developed to feed the elderly living at home in the next millennium.

References

1. Committee on Medical Aspects of Food Policy, *The Nutrition of Elderly People*, HMSO, London, 1992.
2. HMSO, *Social Trends 24*, Central Statistical Office, HMSO, London, 1994.
3. Dunn, D, *Food glorious food: A review of meals service for older people*, Centre for Policy on Ageing CPS Reports 11, London, 1987.
4. Kenny, J., Pierson, B., and Edwards, J., Competitive Tendering in Social services: Its implications for the meals on wheels services in the United Kingdom. *Services management New directions and Perspectives*, Cassell, London, 1995.
5. DHSS, *Organisation of meals-on-wheels Circular 5/70*, HMSO, London, 1970.
6. Davies, L, Opportunities for better health in the elderly through mass catering. World Health Organisation, Denmark, 1991.
7. Wellman, N.S., Dietary guidance and nutrient requirements of the elderly, *Primary Care*, 21 March, 1994.
8. Bransby, E.R. and Osbourne, B., *British Journal of Nutrition*, Vol 7, pp160, 1953.
9. King Edward's Hospital fund for London., *Notes on Diets for Old People in Homes and Institutions*, 1965.
10. DHSS., *A nutrition survey of the elderly*, Report on Health and Social Subjects, No 3, London, HMSO, 1972.
11. Suter, P.M. and Russell, R.M., Vitamin requirements of the elderly, *American Journal of Clinical Nutrition*, Vol 45, pp501-512, 1987.
12. Sjogren, A., Osterberg, T., and Steen, B., Intake of energy, nutrients and food items in a ten year cohort comparison and in a six year longitudinal perspective: A population study of 70 and 76 year old Swedish people. *Age and Ageing*, Vol 23(2), pp108-112, 1994.
13. Bunker, V. W. and Clayton, B. E., Studies in the nutrition of elderly people with particular reference to essential trace elements. *Age and ageing*. Vol 18, pp422-429. 1989.
14. Advisory Body for Social Services Catering(ABSSC)., *A recommended standard for community meals*, Advisory Body for Social Services Catering (ABSSC), Bradford, 1990.
15. Caroline Walker Trust., *Eating well for older people*, The Caroline Walker Trust, London, 1995.
16. Stanton, B.R., *Meals for the elderly*. Kings Fund, London, 1971.
17. Malcolm, L and Johnson, M., *Meals services for elderly people: An overview of three studies*, Policy Studies Institute Research Reports, London, 1982.
18. Lawson, F. R and Thomson, J., *The Meals on Wheels services in the UK*, University of Surrey, 1984.
19. Davies, L., *Three score years .... and then. A study of the nutrition and well being of elderly people at home*, Heinemann, London, 1981.

20. Tuner, M and Glew, G. Home delivered meals for the elderly, *Food Technology,* Vol July pp46-50, 1982.
21. Darling, A.S.P., Home delivered meals food quality, nutrient content and characteristics of recipients, *Journal of the American Dietetic Association,* Vol 88(1) pp55-59, 1988.
22. Stevens, D. A, Grivetti, L.E. and McDonald, R.B., Nutrient intake of urban and rural elderly receiving home delivered meals, *Journal of the American Dietetic Association,* Vol June 92(6) pp714-718, 1992.
23. Bell, R. Dunn, P. Whitehead, A. and Xouris, S., The contribution of meals on wheels to the nutrient intake of the elderly, *Australian Journal of nutrition and Dietetics,* Vol 50(2) pp46-50, 1993.
24. Darnton-Hill, I., Psychosocial aspects of nutrition and ageing, *Nutrition reviews,* Vol 50(12) pp476-479, 1992.

A consumer led approach to the development of community meals

N. R. Hemmington
School of Hotel and Catering Management, Oxford Brookes University, Headington, Oxford OX3 OBP, UK

Abstract

Traditional approaches to the provision of community meals have focused on technological aspects of provision, particularly delivery systems. Preoccupation with technological systems has led to an operational focus on meal delivery rather than a more consumer orientated strategic view of the role of community meals as part of an integrated system of community care. This study takes a strategic view of the provision of community meals from a consumer needs perspective and concludes with recommendations for future strategies for the provision of community meals as part of a wider system of community care.

1 Introduction

The community meals service has traditionally focused on the delivery of hot meals through technologically focused delivery systems[1]. Extended delivery times have meant that systems based on insulation have rarely been adequate and heated systems were developed to maintain meal temperatures at an acceptable level. Despite the introduction of heated delivery systems, however, concerns remained about delivery temperatures, food quality, nutrition and hygiene[2].

With these concerns in mind a number of alternative meals on wheels systems have been investigated including systems based on the delivery of partially prepared meals (frozen, pouch sterilised and raw ingredient packs) for finishing and regeneration by the client[3]. Although not suitable for all meals recipients, these systems were found to offer improved quality[4], reduced nutrient loss[5] and improved flexibility. The delivery of chilled meals has also been tried although concerns about the relatively narrow margins for food safety and the reduced flexibility as a result of the shorter storage time have tended to discourage local authorities from this option. A delivery system based on the regeneration of frozen meals during delivery in a specially adapted delivery van has also been investigated[6] but was effectively ruled out when it was found to be around 40% more expensive than traditional delivery systems[7].

Despite these initiatives there have been relatively few advances in meals on wheels systems beyond the use of electric heat store containers and centralised production units[8]. The most innovative development over recent years has been the introduction of freezer meals schemes where frozen meals are delivered to clients who then regenerate them at their own convenience[9]. This system is not suitable for less capable recipients, however, who still have to be provided with meals through a supplementary hot meal delivery system.

All these approaches have focused almost exclusively on technological aspects of the provision of community meals, particularly in terms of delivery systems. Preoccupation with technological systems has tended to lead to an operational focus on meal delivery rather than a more consumer orientated strategic view of the role of community meals as part of an integrated system of community care.

The London Borough of Bromley, conscious of the limitations of traditional community meals systems and aware of increasing demand and continuing financial constraints, commissioned a strategic review of its community meals provision in 1995. The Borough was particularly interested in the development of "innovative, practical and cost effective" strategies which would meet future needs and respond to "service recipients views" whilst "not exceed(ing) the cost of the current service plus 10%"[10].

The study took place in April/May 1995 and was reported in July 1995. Whilst the commissioned study included a detailed critical appraisal of the existing service in Bromley, including costs, this paper focuses on the development of consumer led strategies for the provision of community meals.

Using the London Borough of Bromley as a case study, the objectives of this paper are to; identify consumer attitudes to community meals; identify consumer perceptions of their own needs; identify the dietary implications of consumer views; and, develop innovative, consumer orientated strategies for the future development of community meals.

Whilst the research upon which this paper is based is specific to the London Borough of Bromley, it is anticipated that many of the generic principles, particularly the development of new strategies, will have wider relevance.

2 Methodology

Because of the limited capabilities of meals on wheels clients, their caution over official documents and the need to probe opinions and attitudes beyond any social facade, it was decided to take a qualitative approach to the investigation of clients perceptions and attitudes using in-depth interviews. This approach has the advantage of being flexible to the needs of respondents, is less threatening and stressful, facilitates the identification of important issues as perceived by the client and allows much deeper probing by the interviewer. Although qualitative approaches do not allow the quantification of data it is nevertheless possible to identify those issues that are raised frequently by respondents.

A pilot study was conducted with 8 recipients on 13 and 14 March 1995. These initial interviews were unstructured, free flowing discussions around the subject of eating and drinking, meals and specifically meals on wheels. As a result of the pilot study a number of areas were identified as the basis of discussion for the main survey. These key areas were used primarily to ensure consistency of approach and to provide structure whilst allowing as much personal comment as possible.

A number of additional factual and behavioural questions relating to recipients and their meals were also added at this stage. To ensure consistency between interviews a standard proforma was developed to act as an aide memoire for interviewers. The same interviewing team conducted all interviews.

Multi-layer sampling was used. Three representative areas within the Borough were selected (Bromley, Orpington and Penge), delivery rounds were then selected within these areas and then individual clients were selected from each round. Representative delivery rounds were selected by the voluntary organisations involved (WRVS, Age Concern, Red Cross). The sampling of clients was determined by their willingness to be interviewed and their capability. Some resistance was encountered by those clients who felt the interview may affect their entitlement to the meals service.

The interviewers were introduced to recipients by volunteers when meals were delivered. The background to the project was explained and if recipients were willing to take part appointments were made for either later the same day or for the following day. Interviewers worked in pairs, one conducting the discussion and the other recording comments. Respondents were assured of the confidentiality of the interviews and that their meals provision would not be affected. The interviews took around 30 minutes.

3 Findings

3.1 Previous studies

There have been a number of previous studies[11] investigating the attitudes of clients to meals on wheels. In reviewing these studies a number of significant issues can be identified. The majority of the recipients are either satisfied or very satisfied with the service they receive. This could be the result of fear that criticism could lead to the withdrawal of meals and there is evidence that much lower levels of satisfaction are expressed where less direct questioning techniques have been used[12].

The most frequent sources of complaint are lack of gravy, liver, too much potato, mashed potato, croquette potato, Brussels sprouts, sausages/burgers and a lack of variety. The key determinant on the perceived quality of meals is the quality of the protein element (usually meat or fish). Sausages and beefburgers are usually considered to be inferior substitutes for meat.

The most popular time for delivery is 12.00-12.30 p.m. with most complaints from those who receive their meals before 11.30 a.m. and after 1.30 p.m. Evening meals are not popular presumably because the majority of the elderly view the midday meal as the main meal of the day and perhaps because of fear of opening their front doors on dark evenings.

3.2 The Sample

A total of 30 meals on wheels clients were interviewed. Most of those who volunteered their age were over 80 years of age with several in their 90s. Seventy-six per cent (20) of the sample were women.

The physical capabilities of the sample varied with 40% having severely limited mobility, 39% able to get about their homes and 21% able to go out of their homes independently. The requirements of the interview meant that all those selected were reasonably mentally capable although several were hard of hearing and/or suffering from short-term memory loss.

Most of the sample (60%) had been receiving meals on wheels for less than two years whilst some (7%) had been receiving them for more than six years. Several were vague about the length of time they had been receiving meals, "can't remember, time goes so quickly", two of the sample had no idea how long they had been having meals on wheels, although the majority were able to link their receipt of the service with a stay in hospital or other similar personal landmarks.

Most of the sample (86%) receive meals on wheels five or more days per week. The majority of those who receive less than five meals a week supplement this provision with meals at day centres.

Clients stated that meals are normally delivered between 11.15 a.m. and 1.45 p.m. Sixty percent stated that their meals are normally delivered between 11.15 a.m. and 12.15 p.m. which is significant in light of the fact that the majority were of the opinion that delivery before 12.00 noon is too early. Some clients said that when meals are delivered early it makes the afternoon very long.

Most respondents stated that they eat their meals on wheels immediately on delivery although 24% said they keep it hot for later. Methods of keeping meals hot varied from *"in a low oven"* to *"on top of the gas fire"*.

3.3 Attitudes to the meals on wheels service

Discussion with meals on wheels clients identified a number of key attitudes and perceptions that providers should accommodate in the development of meals provision. In general terms, most clients expressed **satisfaction** with the meals service and there were many positive comments such as, *"They're as good as* *you'd cook yourself"*, *"The meals are generally quite nice"*, *"It's a wonderful organisation"*. There were indications, however, that this was to some extent a result of concern that meals may be withdrawn and these comments appear to be more expressions of gratitude rather than objective critical assessments. This was implied by comments such as, *"The meals aren't brilliant but I put up with them"*, *"Generally the meals are OK, I'm lucky to have them"*, *"I think I'm very lucky to have the service"*, *"I'm thankful for the service"*, and is perhaps best illustrated by the comment, *"I have no complaints, I'm very happy. I wouldn't complain anyway because I have no alternative"*.

The most frequently expressed view was that meals on wheels are good **value for money**, although it was interesting to note that several clients said that they would *"rather pay more for good food"*. This indicates that there is the possibility of offering an enhanced service for those willing to pay for it.

There were many comments about the limited **variety** and the lack of **choice** offered by the meals service. A typical comment was that *"meals are repeated too often, not enough variation"*. An interesting perception was that, *"The meals are the same every day, they're just called different things"* and one client stated that *"I had cottage pie three times out of four last week"* ! Several recipients said they would prefer to be able to choose their meals in advance, *"send a menu round to re-order meals"*, *"I don't like not knowing what I'm going to have. I would like to see a menu to choose my own meals"* and, *"Good idea to send a questionnaire round to ask people exactly what food they want"* were typical comments. Clearly the introduction of choice would also allow for the variety of dietary needs relating to medical conditions (diabetes in particular), religion and beliefs as well as personal preferences.

The **convenience** of meals on wheels is clearly the major benefit for clients. This was expressed by clients in a number of ways from, *"you don't have to do it yourself"*, and, *"it saves all the shopping and washing up"*, to more particular factors such as, *"don't have to worry about the gas stove"*, *"I don't have the energy to cook and clean up"*, and, *"it's convenient to eat from the tray"*.

There were a number of comments about the **quality of meals**. Those commenting on the temperature of meals invariably said that they were delivered hot and there were no complaints of cold meals. There were, however, several comments relating to extended periods of warm-holding, including, *"The meals don't taste fresh, they don't taste like home made"*, *"The meals suffer in that they are frozen and re-heated"*, *"Sometimes they are inedible, they're not cooked properly"*, and ,*"I don't like the smell, it puts me off"*.

The role of the meals within the **diet** was noted by some clients - *"It undoubtedly fills a gap"*, *"most important thing is that you get a main meal*. Views on the size of the portions varied, and although most said things like, *"meals are plenty for me, I don't eat a lot"*, some also said that, *" There's not always enough"*. More specifically some clients stated that they don't have sweets because they are diabetic.

Although not asked specifically, many clients commented on the **social role** of meal delivery. The social contact, however limited, is clearly valued; "*Seeing the meals on wheels volunteers breaks up the day*", "*It's important to see someone every day*", "*They're very nice ladies, they're someone to look forward to*". There were however many comments relating to the fact that delivery staff have very little time to talk; "*don't stay because there's not time for conversation*"; and that they change frequently and clients therefore, "*don't get to know the people who deliver meals*". This has important implications for security as well as for social contact.

3.4 Meal preferences

The interviews identified a number of meals/foods that are liked, including **puddings**, **custard**, **roast dinners**, **chicken casserole** and **steaklet and vegetables**, whilst a number of meal items were disliked for particular reasons. **Vegetables** were most frequently identified because they "*aren't cooked* " surprising in light of the typical warm-holding times ! **Chips** are also disliked because they are "*soggy*", probably a symptom of the regeneration process, and indeed one recipient stated that, "*Even Susie my poodle won't eat the chips*".

Perhaps the most significant finding in terms of meal/food preferences was the wide variety of things that were liked and disliked and that several foods appeared in both categories. This is a clear indication that the tastes of meals on wheels clients are varied and that it is potentially dangerous to generalise about meal preferences. Menus without choice are unlikely, therefore, to adequately satisfy the demands of the present meals on wheels market and choice is clearly becoming a very important issue.

4 Discussion and recommendations

4.1. The implications of choice and diversity

Research has indicated that levels of plate waste are high[13] and it is axiomatic that uneaten food cannot make a contribution to the diet[14]. Thus, despite attempts to provide nutritionally balanced meals it is the tastes and preferences of the client that finally determine the nutrient intake. The meals service must, therefore, focus its efforts on the provision of attractive, appetising meals that are nutritious in order to encourage clients to eat. As the WHO Report[15] states, healthy eating has a "different meaning for the over 75s and its primary aim must be to maintain appetite and keep up enjoyment of food".

The need to encourage people to eat leads to a consideration of the wide range of individual needs relating to cultural, religious, ethnic, therapeutic and ethical factors as well as food preferences. The diversity of these needs makes variety and meal choice fundamental to the provision of a quality meals service. As the Caroline Walker Trust[16] states; "The right to exercise choice -

particularly with regard to their food - is firmly embedded in all United Kingdom codes of practice concerning older people in residential care and is repeated in the principles underlying community care policies... This is particularly important where there may be ethnic, religious and cultural requirements."

The dietary implications of meal choice mean that local authorities have a very indirect control over the individual's diet. It is important therefore that advice is provided to clients, or their helpers, that enables them to make informed choices when selecting meals and helps in their dietary planning. Suggestions for easily prepared, nutritious meals and snacks would be helpful and even the provision of more general lifestyle advice on safety, keeping warm and exercise would be of benefit to older people.

In addition to nutrition, there is no doubt that meals play an important psychological role and can make a significant contribution to the quality of life. As the Advisory Body for Social Services Catering[17] says, "For many of the customers the meal could be the highlight of their day". The social psychological support provided by meals, particularly in terms of security, has been identified by McKenzie[18] who stated that; "Food acts as an aid to security. Thus when everything seems to be going wrong we comfort ourselves that things are 'alright really' by turning to established favourite dishes". This wider psychological role makes the provision of quality meals that reflect a caring society even more critical to the success of the service.

4.2. Strategic aims

It is significant that existing community meals services have tended towards centralisation and have, as mentioned earlier, focused on technological approaches to the problem of meal distribution. Looking at the service as part of a wider system of community care and taking a consumer needs perspective, however, reveals a number of requirements that are unlikely to be met by existing approaches. Specifically, community meals should aim to; be responsive to changes in the volume of demand; be flexible to meet a wide variety of client needs particularly in terms of cultural, ethnic, religious and therapeutic factors as well as preferences and increasingly sophisticated tastes; make an effective contribution to the diets of older people; improve quality, particularly in minimising the deleterious effects of regeneration and distribution on the quality of meals; maximise the social role of community meals; develop robust quality assurance systems which ensure compliance with legislation and enable the demonstration of "due diligence".

4.3. A decentralised local approach

In attempting to move away from a preoccupation with technological approaches towards a more responsive, needs based focus, it is suggested that a local decentralised approach be considered. In this system meal preparation and/or regeneration would take place at the local level, as close as possible to the clients' home, and thus minimise the need for extended delivery.

Two complementary types of service would be offered. For those clients who are capable of regenerating their own meals, a freezer meals scheme would be offered where batches of pre-ordered frozen meals would be delivered to clients on a two weekly basis. Choice would be offered through the provision of four weekly cyclical menus. Research evidence indicates that in general freezer meals schemes are effective in meeting the needs of clients[19].

For those unable to regenerate their own meals, a system based on the regeneration of meals at the local level by meals assistants - volunteer neighbours or friends - would be offered. Frozen meals, sufficient for two weeks, would be delivered to the local meals assistants who would then regenerate the meals at times agreed with their clients.

Choice would be offered as for the freezer meals scheme - 4 week cyclical menus (4 menus), each menu including at least 7 meal choices and each menu lasting a week. In this way clients taking 7 meals per week will be able to select a different meal each day and those with dietary restrictions should have sufficient choice to be able to select meals, they may however have to have the same meal more than once over the week.

Choice forms would be administered by meals assistants who would phone their consolidated order through to the meals office the previous week. Assembled meal packs would then be delivered by freezer vans. Payment for meals would be collected by meals assistants.

Close proximity is the important factor for the recruitment of meals assistants. As a guideline clients should be within a 5 minute walk of the assistants home and each assistant should have no more than 5 clients, preferably 3. It would be necessary to offer meals assistants freezers for meal storage and also to provide hygienic insulated containers for the movement of meals to clients' homes.

All meals assistants would need training particularly in food hygiene. It is recommended that they are trained to the Institute of Environmental Health Officers' Basic Food Hygiene Certificate level or equivalent.

The critical issue for this system is whether meals assistants could be recruited and whether they should be paid or unpaid. With current levels of subsidy there is little scope for any significant payment per meal[20]. A rate of £0.50 per meal for an assistant with the maximum 5 clients having 5 meals per week, would only equate to £50 per month. It is recommended that in the first instance the recruitment of unpaid assistants is tried in order to gauge likely response. It may eventually be necessary, however, to offer a sum in lieu of energy and telephone costs, perhaps £0.25 per meal.

The advantages of this system are as follows;
- regeneration at the local level removes the problems associated with extended warm-holding times, particularly those associated with meal quality, nutrition and hygiene;

- it would enable compliance with the Department of Health[21] guidelines that state that frozen meals should be served within 15 minutes of regeneration;
- regeneration at the local level should lead to a more personal and responsive service in terms of choice and meal times;
- the involvement of local meal assistants should encourage a more social and caring aspect, they should have more time to talk to clients and to monitor their well-being, it will also involve the local community in the care of its older people with other possible social advantages;
- it facilitates the move towards offering variety and meal choice, thereby accommodating cultural, ethnic and religious diversity, therapeutic needs as well as preferences and increasingly sophisticated tastes;
- it allows for increases in demand, particularly at weekends, and moves towards the ideal of a 100% meal policy[22];
- by pushing responsibility down to the local level and empowering both meals assistants and clients, the system moves towards an emphasis on quality assurance rather than quality control.
- evidence from the Bromley study indicates that this system can be delivered within existing net costs plus 10% where meals assistants are paid £0.25 per meal[23].

The potential disadvantages of the system are that;
- the recruitment of meal assistants may be difficult, voluntary organisations are already finding it difficult to recruit volunteers to deliver meals;
- the initial recruitment, training and management of meal assistants will be a major logistical exercise, it is suggested, therefore, that the system be phased in over a period of two years.

4.4. Implementation

As a new, innovative approach to community meals, it was recommended in the Bromley study that the system be fully tested in a pilot study with clients from two delivery rounds. A time scale of 12 months preparation in terms of setting up systems, recruitment and training of meal assistants, 12 month trial and then depending upon the evaluation, 12 months preparation for full implementation was suggested. The Borough received and accepted the report in 1996 and has decided to pilot the system in 1997.

References

1. Lawson, F.R. & Thomson, J., 1981, **The meals-on-wheels service in the United Kingdom**, University of Surrey.
2. Hemmington, N.R., 1987, **Welfare Feeding in the United Kingdom**, University of Surrey.

3. Armstrong, J.F., O'Sullivan, K. & Turner, M., 1980, **The housebound elderly - technical innovations in food service,** Huddersfield Polytechnic.
4. Johnson, M. DiGregorio, S. & Harrison B., 1982, **Ageing, needs and nutrition,** Nuffield Centre for Health Service Studies, University of Leeds, Policy Studies Institute.
5. Glew, G., 1984, Meals-on-Wheels - old and new technology, **Health and Hygiene,** No.5, pp7-11.
6. Armstrong, J.F., 1979, Feeding individual elderly people in their own homes, in Glew, G. **Advances in Catering Technology,** Applied Science Publishers.
7. Lee, K. & Martin, S., 1979, **An economic analysis of the meals on wheels service,** Nuffield Centre for Health Service Studies, University of Leeds.
8. See for example; Advisory Body for Social Services Catering, 1992, **Recommended Standard for Community Meals,** ABSSC.
9. Anon, 1993, Switch to Frozen Urges Social Services Caterer, **Welfare Catering,** April.
10. Hemmington, N.R., Cantwell, J., Wilkes, K. & Burlinson, S., **London Borough of Bromley - Meals Study,** Cheltenham & Gloucester College of Higher Education, July 1995.
11. See: Stanton, B.R.D., 1971, **Meals for the Elderly,** King Edward's Hospital Fund for London: Davies, L., 1981, **Three score years .... and then ?,** Heinemann Medical Books: Ruane, P., Cohen, A. & Heiser, B., 1983, **Survey of Current Meals-on-Wheels Consumers,** Clearing House for Social Sciences Research, No. 9: Tilston, C.H., Gregson, K, Neale, R.J. & Tyne, C., 1992, Meals-on-wheels Service in Leicester: A Marketing Study, **British Food Journal,** 94/2., pp 29-36: Tilston, C.H., Gregson, K, Neale, R.J. & Tyne, C., 1994, The Meals on Wheels Service: A Consumer Survey, **Nutrition and Food Science,** No.2, March/April, pp7-10.
12. Johnson, M. DiGregorio, S. & Harrison B., 1982, op cit.
13. Davies, L., 1981, op cit.
14. The Caroline Walker Trust, 1995, **Eating well for older people,** Report of an Expert Working Group, The Caroline Walker Trust.
15. Davies, L., 1991, Community Meals, **World Health Organisation Report 1991,** WHO.
16. The Caroline Walker Trust, 1995, op cit..
17. Advisory Body for Social Services Catering, 1992, op cit.
18. McKenzie, J., 1979, The Eating Environment, in Glew, G., **Advances in Catering Technology,** Applied Science Publishers.
19. Hemmington, N.R., et al., 1995, op cit.
20. Ibid.
21. Department of Health, 1989, **Guidelines for Cook-Chill & Cook-Freeze,** HMSO.
22. Advisory Body for Social Services Catering, 1992, op cit.
23. Hemmington, N.R., et al., 1995. op cit.

Catering systems as laboratories for studying the consumer

H.L. Meiselman

U.S. Army Natick Research, Development and Engineering Center, Natick, Massachusetts, 01760-5020 USA

Abstract

Research identifying the factors which control food acceptance and food choice has usually been conducted in the laboratory. This has led to difficulty in predicting how people react to real food products in normal eating situations and in predicting product success. Catering environments are ideal places in which to conduct research, as we have demonstrated here at Bournemouth University. In a first study, the effort to obtain food was manipulated in a student refectory by placing the food farther away. With increased effort, students selected less of the test product and substituted other items from the same food class. Second, meal economics was studied in the student refectory, comparing meal pricing and item pricing. Students who chose the complete meal increased their intake of vegetables as compared to non-meal consumers. A third study examined the role of expectations in determining customer satisfaction. The same meal composed of pre-prepared institutional food products was served to students in the refectory and to paying customers in the training restaurant. The paying customers in the up-market training restaurant rated their food better than did the students in the typical student refectory. The customers in the training restaurant expected better food and rated their meal accordingly; the students expected lesser quality food and did the same. It is not what you receive; it is what you **expect** to receive that is important. In a fourth study, we investigated the response to ethnic foods when served amidst different environmental cues. In the training restaurant, we observed a change in food choice patterns when served with an appropriate ethnic theme.

1. Introduction

Understanding food choice and food acceptance has traditionally focused on the properties of the food and on the attributes of individuals. Studying the properties of the food has spawned research into the sensory properties of foods (Piggott[1]; Cardello[2]) and their relationship to food constituents. Studying the attributes of individuals has led to research into the physiology and psychology of the eater, both animal and human (Meiselman and MacFie[3]). Over the past ten years we have begun to balance our interest in food and the individual with a realization that the eating situation or context exerts great control over eating (Rozin and Tuorila[4]; Meiselman[5]). This realization was forced upon us by our attempt to improve military rations to ensure their acceptance in field conditions. (Remember that for the Army the field is actually in a field!) After many attempts to improve the field ration and measure its improved acceptance, we concluded that much of the acceptance of the product depended on the context or environment (Marriott[6]). At the same time, we began to question the traditional way in which research on food choice and food acceptance is conducted. Could laboratory research ever duplicate or simulate the richness of the real world? Could we understand contextual/environmental influences on eating by studying simplified laboratory models? Some argue for realistic settings, while others argue for laboratory control (see *Appetite*, 1992[7]).

In this paper I will present four examples of contextual research carried out in natural eating environments. My goal is not to shift all human research on food choice from the laboratory to the field. Rather, the goal is to increase the percentage of research on food choice which is carried on outside the laboratory, in natural eating environments. These are defined as places in which people normally eat. The catering industry offers a unique potential to contribute to this area of research; in so doing it will support the science of food choice and support its own interests in applied research.

2. Method

2.1. Refectory Studies

The two studies on effort (Meiselman *et al.*[8]) and the study on meal price were carried out in the large student refectory at Bournemouth University. Students were recruited from posters to participate in a study "to rate the canteen" and were paid a modest amount for their participation. The first and second effort studies attracted 43 and 60 participants, respectively. The meal price study included 779 meals. In these three studies each participating student first obtained a meal form upon entering the refectory and then selected a lunch meal and paid for it. After finishing lunch the student completed the questionnaire and returned the form and the entire food tray to our student technician. The technician checked the leftovers against the food ratings to ensure that all foods which had been selected appeared on the rating form. In the first effort study, with chocolates, the leftovers were then weighed in the kitchen, out of sight. In the second effort study, with potato crisps, the leftovers were visually estimated. In the meal price study leftovers were neither weighed nor estimated.

2.1.1. Procedure
The procedures and conditions for the three cafeteria studies were as follows: Each study consisted of a baseline period followed by a treatment condition. The second effort study, on potato crisps, also had a recovery period after treatment. In the first effort study there was daily data collection at lunch for a one-week baseline and a one-week effort condition. During the latter, all chocolate confection was moved from the cash register area to another food line 100 feet away. In order to obtain chocolate confection during the effort condition, students had to go through the food line, pay for each item selected, and then go to the second line, obtain chocolate, and pay a second time. Several changes were made for the second effort study, on potato crisps. Data collection occurred every Tuesday and Thursday only. Baseline lasted for two weeks, effort for three weeks, and recovery for three weeks. Potato crisps were moved more than 100 feet away. During recovery, potato crisps were moved back to their original position on the food line.

2.1.2. Measures
Selection and acceptance were measured in the three cafeteria studies. Acceptance was measured for each food item and for the overall meal; the first study used a 7-point category scale and the second two studies used a 9-point scale. Food selections could be obtained from the food acceptance questionnaire. Individual item weighed food intake was obtained in the chocolate confection effort study, and visual estimation of waste was obtained in the potato crisp effort study, but these data are not presented here.

2.2. Training Restaurant Studies

The effects of ethnic theme and the effects of expectations were studied in the "Grill Room" training restaurant at Bournemouth University. The restaurant is staffed with students and caters to a local clientele of non-university and university patrons. The Grill Room serves casual meals with informal service. Since the students working in the Grill Room occasionally do student projects and ask the customers for feedback, our procedures did not seem very unusual. For the ethnic menu study (Bell *et al.*[9]) subjects consisted of 142 paying patrons of the restaurant. For the expectations study subjects comprised 29 paying patrons of the restaurant and 32 students in the refectory.

2.2.1. Procedure

The procedures and conditions for the ethnic theme study were as follows: For a two-day baseline condition, patrons received a menu with British names and with the usual British decor. For the two-day treatment condition, the menu changed to Italian, and a number of Italian theme items were added to the dining area, including red checked tablecloths, Chianti wine bottles, and Italian travel posters and flags. The food served under the two conditions was the same: only the menu and situational cues changed. As a further test of situational effects, we used four different pasta recipes prepared by chef W. Reeve. He prepared two traditional British pastas and two traditional Italian pastas. The British and Italian pastas were two versions of two dishes. Patrons were asked to rate each food on a 9-point hedonic scale and a 4-point ethnicity scale.

The procedures for the expectations study were as follows: The same meal was offered in both the Grill Room and in the student refectory on different days. The menus consisted of pre-prepared items which are shown in Table 4. In both environments, the meals were served as special set meals. Subjects were asked to rate the foods for both acceptance (9-point hedonic scale) and for appropriateness (7-point scale).

3. Results

3.1. Effort

The two effort studies produced very similar patterns of changes in selection rates. In general, the effort treatment item dropped markedly from baseline level in the effort treatment condition (Table 1). No other food group showed a significant selection rate change with effort. Thus, the effort treatment was specific to the food group manipulated. In addition, selection rates of other foods increased under the effort condition (Table 1). In the chocolate confection study the selection rates for baked desserts, fresh fruit, and packaged accessory foods increased. In the potato crisp study the selection rate for starch foods (potatoes, rice) increased.

Table 1. Significant changes in selection rates for chocolate confection and potato crisps with increased effort.

	Selection Rates		
	Baseline	Effort	Recovery
Study 1			
Chocolate Confection	0.392	0.031	
Desserts, Fruit, Accessory Foods Combined	0.353	0.549	
Study 2			
Potato Crisps	0.718	0.092	0.322
Starch	0.274	0.462	0.398

Selection rates also changed during the recovery period when baseline conditions were reintroduced. Selection rates for potato crisps increased from the effort treatment level but not to the baseline level. Conversely, selection rates for starch decreased from the effort treatment level but not to the baseline level.

Hedonic ratings of the foods were also examined. In the chocolate confection effort study, the hedonic ratings for the foods selected more often were lower than for the chocolate candy. This pattern of lower hedonic ratings for substituted foods did not hold for the potato crisp study.

3.2. Meal Price

The effect of changing from item pricing to a set meal price (for main dish, potato, vegetable) was to increase the selection rate for vegetables (Table 2). Other selection rate changes did not appear to be related to the pricing manipulation.

Table 2. Significant changes in selection rates of items in set meals compared to item-priced meals.

Group:	All Subjects	Set Meals	Item-priced Meals
No. of meals in Week 2	395	69	326
Item	vegetables salads	vegetables salads	salads pizza dessert

When hedonic ratings were examined for the item-priced and meal-priced items, there was no overall effect. Comparing hedonic ratings of students who never selected set meals and ratings of students who selected both set meals and item-priced meals showed no pattern of hedonic changes for overall meal ratings or vegetables. However, when examining just Week 2 it appeared that students selected more preferred items when they were selecting individual items than when they selected the set meal. In other words, hedonic ratings for set meals (overall) and for set meal vegetables were lower than for item-priced meals and item-priced vegetables (Table 3A, 3B).

Table 3A. Hedonic ratings for all set meals and item-priced meals.

		Those Selecting Item-priced and Meals	Those Selecting Only Item-priced Meals
MEAL	Wk 1	7.10 (205)	7.06 (181)
	Wk 2	7.18 (207)	7.06 (186)
VEGETABLE	Wk 1	6.00 (40)	6.94 (16)
	Wk 2	6.21 (80)	7.27 (15)

Table 3B. Hedonic ratings comparing set meals with all other meals in Week 2.

			n
MEAL	Set	6.71	(68)
	Item	7.21	(325)
VEGETABLE	Set	6.15	(69)
	Item	7.00	(26)

3.3. Expectations

Median hedonic ratings and appropriateness ratings from the student refectory and from the training restaurant Grill Room are shown in Table 4. The overall meal rating is higher for the Grill Room (median=8) than for the refectory (median=7), while the appropriateness rating is higher for the refectory (median=7) than for the Grill Room (median=6). Individual food items tended to score 7-8 in the Grill Room and 5-7.5 in the refectory.

Table 4. Ratings of the same pre-prepared foods served in two different settings.

	Student Refectory		Training Restaurant	
	Hedonic	Approp.	Hedonic	Approp.
Beef & green pepper stew	7.0	5.0	7.0	4.5
Beef casserole	7.5	5.5	8.0	6.0
Diced potatoes	6.0	5.0	4.0	3.5
Boiled rice	6.5	6.0	8.0	6.0
Dessert mousse	5.0	5.0	7.0	6.5
Dessert apple	6.5	5.5	7.0	4.5
Overall meal	7.0	5.0	8.0	6.0

Hedonic ratings from 1-9; appropriateness ratings from 1-7.

3.4. Ethnic Theme

Adding an Italian theme to the restaurant produced changes in overall meal ethnicity, item ethnicity, and food selections. The perceived Italian identity of the overall meal increased under the Italian theme condition: more people categorized the meal as Italian (36.7%-69.8%) and fewer categorized it as British, French, or Foreign. Also, perceived Italian identity increased for pasta items.

These changes in ethnic categorization were accompanied by changes in food selections (Table 5). Pasta (both spaghetti and macaroni) was selected more often under the Italian theme, and meat less often. Desserts (both ice cream and zabaglione) were selected more often under the Italian theme.

Table 5. Changes in food selections with an Italian theme as compared to control (no theme).

	% Selection	
	No Theme	Italian Theme
PASTA	16.4	44.0
Spaghetti	10.5	21.3
Macaroni	6.0	22.7
MEAT	77.6	56.0
Chicken	25.4	25.3
Trout	28.4	12.0
Veal	23.9	18.7
None	4.5	4.0
DESSERT		
Fruit	4.5	2.7
Ice Cream	20.0	33.3
Zabaglione	4.9	25.3
None	52.2	37.3

The acceptability for the Spaghetti Beef was lower on Italian theme days; for Spaghetti Bolognese, it was higher.

4. Discussion

These studies have sought to demonstrate that situational variables can have substantial effects in natural eating environments. The nature of the effects and the size of the effects will probably change with different situations, different populations, different foods, and over time. Nevertheless, certain conclusions can be drawn.

Effects of situational variables measured in natural eating situations can be very potent. Laboratory and questionnaire research on food attributes and individual eater attributes often produces subtle or small effects. Situational variables, as reported here, can change response measures by 50% or more.

Measures of food choices in these situational studies produce a clearer picture than do acceptance or hedonic measures. This same conclusion was reached in our earlier studies at Natick on military rations (Meiselman *et al.*[10]).

Finally, these studies demonstrate that situational research on eating in natural environments need not be complicated or difficult. Although some would argue that naturalistic research is limited to observation, one can conduct both manipulative research and nonmanipulative research. The former can intentionally vary some independent variable, such as effort to obtain food. Nonmanipulative research can observe behavior in normal eating environments.

Understanding eating will require greater attention to the eating situation, and some of this research can easily and profitably be conducted in natural eating environments. The caterer has a unique opportunity to contribute to research and to profit from it.

Acknowledgements

The author wishes to acknowledge the active collaboration of the following in the individual studies: R. Bell of Natick RD&E Center; B. Pierson, W. Reeve and J. Edwards of Bournemouth University.

References

1. Piggott, J.R. (ed). *Sensory Analysis of Foods*, Elsevier Applied Science, London and New York, 1988.

2. Cardello, A.V. The role of the human senses in food acceptance, Chapter 1, *Food Choice, Acceptance and Consumption*, eds H.L. Meiselman and H.J.H. MacFie, pp 1-82, Chapman and Hall, London, 1996.

3. Meiselman, H.L. & MacFie, H.J.H. (eds). *Food Choice, Acceptance and Consumption,* Chapman and Hall, London, 1996.

4. Rozin, P. & Tuorila, H. Simultaneous and temporal contextual influences on food acceptance, *Food Quality and Preference,* 1993, **4**, 11-20.

5. Meiselman, H.L. The contextual basis for food acceptance, food choice and food intake: The food, the situation and the individual, Chapter 6, *Food Choice, Acceptance and Consumption,* eds H.L. Meiselman and H.J.H. MacFie, pp 239-263, Chapman and Hall, London, 1996.

6. Marriott, B.M. (ed). *Not Eating Enough, Overcoming Underconsumption of Military Operational Rations,* Committee on Military Nutrition Research, Food and Nutrition Board Institute of Medicine, National Academy Press, Washington, DC, 1995.

7. *Appetite,* **19**(1), 1992, 49-86. Nine discussion papers on methodology and theory in human eating research.

8. Meiselman, H.L., Hedderley, D., Staddon, S.L., Pierson, B.J. & Symonds, C.R. Effect of effort on meal selection and meal acceptability in a student cafeteria, *Appetite,* 1994, **23**, 43-55

9. Bell, R., Meiselman, H.L., Pierson, B.J., & Reeve, W.G. Effects of adding an Italian theme to a restaurant on the perceived ethnicity, acceptability, and selection of foods, *Appetite,* 1994, **22**, 11-24.

10. Meiselman, H.L., Hirsch, E.S. & Popper, R.D. Sensory, hedonic and situational factors in food acceptance and consumption, Chapter 7, *Food Acceptability,* ed D.M.H. Thomson, pp. 77-87, Elsevier Applied Science, London and New York, 1988.

Theories of consumer behaviour - their relation to the patient and the catering service in hospitals in the United Kingdom

J. Ervin, W. Reeve, B. Pierson
The Worshipful Company of Cooks Centre for Culinary Research, School of Service Industries, Bournemouth University, Talbot Campus, Fern Barrow, Poole, Dorset. BH12 5BB

1. Abstract

In the last number of years changes in government policy, market structure and internal competition for funding have ensured that the National Health Service and the units within it, have become increasingly aware of the need to design all services available around the requirements of the consumer, a belief long held in the commercial environment. Patient satisfaction surveys have proliferated, the results of which hold important political, financial and clinical influence. This paper aims to investigate the background to, and methodologies currently being used to measure consumer satisfaction in hospital. It also aims to examine theories of consumer decision making in relation to the hospital patient and the catering service which is available to them.

2. Why is Patient Satisfaction with the Catering Service Important?

High levels of patient satisfaction in hospital are important for a number of reasons. Research carried out in 1976[1] showed that from a clinical perspective, 'satisfaction with care' is a key factor in determining whether a hospital patient will seek medical advice, comply with prescribed treatment, and maintain a continuing professional relationship with their doctor. Food in hospital is part of that total care package.

Satisfaction with food in hospital is important from both a physiological and psychological point of view. Being placed in an alien environment, such as a hospital ward, leads to a feeling of social isolation for the patient. This brings about feelings of bewilderment, which is an 'anxiety provoking and disabling state', (Baron Byrne)[2]. Mealtimes in hospital provide a familiar pattern in an otherwise tedious day and are something which the patient may feel they can relate to, exercise a degree of control over and which may raise patient morale ultimately increasing recovery rate from illness. Scheier et al [3] showed that the more optimistic a patient is the faster his recovery and return to normal and active life. The Kings Fund Report published in 1992[4] discussed the incidence of malnutrition in hospitals and highlighted the importance of the catering service in addressing this issue. Results were seen to be both beneficial to the patient and also the hospital, in terms of financial benefits from increased recovery rates. 97% of patients studied in the Gardner Merchant Food in Healthcare Survey 1995[5] recognised that food and beverages contributed to aiding their recovery. It was shown that when studying those aspects of a meal which are important, sick people, with fragile appetites recognise both the physiological and psychological benefits of food by wanting a tasty, hot meal which is presented attractively.

Levels of patient satisfaction are important from a political point of view. In 1983[6] an NHS Management enquiry stated that management should 'ascertain how well the service is being delivered at a local level by obtaining the experience and perceptions of patients and the community'. This was translated by Health Authorities and the 1989 Griffiths paper 'Working for Patients'[7] as a call for the measurement of patient satisfaction, a view reiterated by the then Health Secretary Virginia Bottomley in 1993[8] when the NHS was urged to give greater weight to patients' perceptions of their care.

In terms of hospital food the commitment of the National Health Service to the market testing of all support services in hospital has ensured that levels of patient satisfaction and the frequency of its measurement have become valuable performance indicators for those in a position to assess the industry. In short, patient satisfaction is perceived as a goal for health care delivery and a measure of its quality, and for the patients sake, services available must be geared towards achieving optimum levels.

3. Current Methodologies for Satisfaction Measurement

Methodologies for the measurement and collection of patient satisfaction data can be divided into two subgroups. These are quantitative and qualitative methods. Both are fundamentally different in their philosophical approach. Quantitative methods approach data collection from a structured pre-defined perspective and methods used are often mechanistic. Questionnaires are an example. Qualitative research methods, however, begin research from a more humanistic standpoint viewing participants as the focal point and not as the recipients of predetermined ideas. Methods used are characterised by small samples and a less structured approach. The research is facilitated rather than controlled by the administrator. There are four main types of qualitative methods. These are In-depth interviews, Group Discussions, Observation Techniques and Projective techniques which explore an individuals unconscious or repressed feelings about a particular issue or product. Both qualitative and quantitative methods of data collection are useful when measuring satisfaction if used correctly. Ideally they should be used to complement each other.

In the 1994 National Audit Office Report[9] it was noted that around 80% of catering operations in hospitals seek feedback from patients on their satisfaction with the catering services provided by asking them to complete a questionnaire. 80-90% of the patients responding to these questionnaires in the hospitals visited, as part of the report, said they were satisfied with the catering services they had received. Of these patients the most dissatisfied group were those aged under 35.

Satisfaction measurement using questionnaires such as these is easy to conduct, and cheap to administer. The proliferation of such surveys in the NHS, since managers were first asked to measure satisfaction, is proof that there is a preference for this type of quantitative research and a relative distrust of qualitative or soft data. To many industry viewers the size of the sample studied is often more important than the material which is elicited from them. The quantitative data which is generated also makes it easier to facilitate recommendations such as those given by the Griffiths report[10] and the National

Audit Office Report[11] which called for the monitoring of quality standards and targets by carrying out regular surveys of patient views.

What do the results generated so far actually tell us? Is the information really of any value? Are the current methodologies, using a questionnaire approach, really the best way to gather information on patient satisfaction, in order to improve the catering service available, or are they more useful in providing proof that satisfaction has been measured across a wide number of patients. High figures, such as those given in the National Audit Office Report[12], are excellent from a public relations point of view, but are they really of meaningful value to the catering manager. The results tell you how many patients are satisfied, but not what the consumer means when they say they are satisfied or how they have come to that conclusion. In practice it has been pointed out, (Wedderburn Tate et al)[13] that while managers may show an interest in seeking feedback from patients there has been no real incentive to build patients views into decision making. Barnes[14] pointed out that the questions which have been asked have, in these surveys, given information which when analysed actually means very little and as a result, past survey's have rarely had any major effect on the care process.

Satisfaction as a concept is defined as a 'post-consumption evaluation that a chosen alternative at least meets or exceeds expectations......that it has done as well as you hoped it would'[15]. Satisfaction is a not a single entity but is a composite of many factors which must all be considered if satisfaction is to be measured realistically. (Scott)[16] pointed out that in order to generate meaningful results from patient satisfaction surveys it is necessary to specify a conceptual base that explicitly recognises the decision making context in which survey results are used. No such conceptual base has yet been designed which applies solely to the catering service available to the patient in hospital. This paper aims to stimulate discussion by examining some factors which show that this exercise is necessary.

4. Theories of Consumer Behaviour and their application to the Patient in Hospital.

Many commercial organisations in the UK are now beginning to understand that it is essential for corporate survival to identify and meet the needs of their consumers. Theories of consumer behaviour have been developed, for application in the commercial environment, which follow the consumers decision making process, identifying those factors which interact and influence how the consumer thinks, evaluates and acts throughout the entire decision process.

The foundations of such theories of consumer behaviour lie in societies where consumers have freedom for individual choice, and to choose is to evaluate benefits perceived from the alternatives available. It is important to realise the limitations on alternatives and the choices available to patients in hospital. Unlike consumers in commercial catering environments, the majority of patients have had little or no choice or control over actually going into hospital - their eating environment, or of the type of treatment they will receive once admitted. They normally do not choose, or have any say in, the bed or ward they will occupy or the company they will have for the majority of their day. Often the only recognisable choices they do have is whether to see visitors or not and alternatives for meals. These two aspects of hospitalisation may increase in importance for a patient because they perceive a degree of control over them. Studies of patients with Heart Disease, Cancer and Aids, found that those who felt they were personally in control of their illness, the care they were receiving and their treatment adjusted much better to their condition than those who did not feel in control. (Taylor et al[17] Thompson et al[18]) The benefits perceived in alternatives available can be viewed as being emotional or hedonic or purely utilitarian. Consumption and preferences are generally a mixture of both. In terms of food the utilitarian aspect is to supply the body with its essential nutrients, but the hedonic benefits are gained in choosing something which we like - having a bar of chocolate instead of a, purely functional, high energy drink. The mix of hedonic and utilitarian benefits attached to consumption does not stay static - what we prefer one mealtime, we may not like the next. In the day to day environment the consumer generally makes a decision that a particular combination of foods, in a particular environment will have the ability to satisfy these preferences. Alternatives can be sought. In hospital this choice is more restricted.

In the past one reason given as to why satisfaction levels with hospital food were so high was that expectations were so low on admission that any improvement resulted in satisfaction.(Feldman)[19] Research into the link between consumer expectations and satisfaction has been underway for the last

30 years. Expectations can be defined as 'beliefs, by patients, about the services they think they are going to receive in hospital.' With reference to surgical or medical procedures, Williams[20] states that 'While there are aspects of service provision for which patients clearly have expectations on which they can base their evaluations (e.g. hotel functions or the amenities of care) a number of situations are apparent where expectations might not exist. This may be the case for a patient who is admitted as an emergency case, and who awakens to find themselves in hospital. They will have had little or no expectations on which to base their reaction to the experience, because they did not necessarily know they were ill, unless their medical diagnosis is a recurring one. A hospital patient who is coming in for a planned operation will usually have communicated with their consultant pre-admission and normally have gone on an 'information search' through conversations with friends, family and acquaintances, all in order to talk about what is going on, to compare perceptions and to decide what to do (Morris et al)[21]. It has been shown that knowing what to expect enhances recovery rates. Preoperative patients prefer roommates who are postoperative for the same complaint (Kulik and Mahler)[22]. Patients whose roommate was post operative walked further each day after surgery, were less unhappy and anxious, needed less pain medication and had a shorter length of stay.

The current evidence for the contribution expectations make to satisfaction is highly questionable and in need of much more theoretical development (Thompson)[23]. A significant amount of work needs to be carried out before expectations can be relied on in surveys as key indicators of patient satisfaction.

There are three different perspectives expounded today as to the way consumers make decisions. These are:

4.1 The Decision Making Perspective
Engel et al [24] have broken down this model of consumer decision making into the following stages:-
Need Recognition, Search for Information, Pre-purchase alternative evaluation, Purchase, Consumption , Post purchase alternative evaluation and Divestment. It is thought that somehow the consumer goes through this process making decisions at each stage which are logical and consistent for them. (Ajzen and Fishbein)[25] state 'Generally speaking.....human beings are usually quite rational and make systematic use of the information available to them. We do not

subscribe to the view that human social behaviour is controlled by unconscious motives or overpowering desires, nor do we believe that it can be characterised as capricious or thoughtless. Rather we argue that people consider the implications of their actions before they decide to engage or not in a given behaviour.' The approach is known as the theory of reasoned action. This model 'fits' some decision making activities, but may not be entirely relevant to the patient in hospital. Wilson-Barnett [26] showed that between a fifth and a quarter of patients in medical wards suffer from 'psychiatric' levels of depression. It is questionable whether these patients will make an entirely rational decision. Mood also has been shown to effect propensity for satisfaction for a hospital patient. Research carried out by Maller et al[27] showed that for patients and staff, the opinion of the meal was significantly correlated with the respondents mood at the time. For decisions in hospital to be entirely rational neither of these factors would have any effect.

4.2 The Experiential Perspective
This perspective recognises that consumers do not always make rational decisions with regard to what they consume. Choices are often made in order to satisfy a whim or an emotional feeling - just because it makes the consumer feel happier or good. To examine consumer behaviour from this perspective would be to examine the feelings, emotions and symbols which accompany consumption. In the commercial environment this perspective can be applied to the leisure industry where the perceived benefits of the products available are in the creation of favourable feelings. This model has some application in the hospital environment as it is apparent, from work such as that carried out by Maller et al above on mood, that the benefits which shape the preferences of hospital patients are not entirely utilitarian, but once again the model is not a true fit.

4.3 The Behavioural Influence Perspective
Behavioural influence occurs when strong environmental influences create circumstances in which the consumer is prepared to choose without developing strong emotional feelings about the product. Choice relies entirely upon extrinsic influences of behaviour by environmental forces. It has been shown that satisfaction with food is related to satisfaction with the entire hospital stay. In

1965 Sheatsley [28] showed that a patient is more likely to be satisfied with hospital food if they are satisfied with the other aspects of their hospitalisation and again in 1990 Deluco and Cremer[29] showed that statements from both patients and family members reflecting the characteristics of food and service quality were significantly related with satisfaction with overall hospital care. Deluco and Cremer also showed how a longer hospitalisation is related to positive perceptions of meal quality. Environmental factors therefore have some relevance in the hospital situation.

Past research has shown that Consumer Behaviour and ultimately satisfaction in hospital is influenced by a number of other external factors and determinants. There has been some discussion as to whether demographic differences of patients actually affects propensity for satisfaction among patients. The figures from the National Audit office Report[30] quoted above clearly show that age of the patient has an effect - patients under 35 are more dissatisfied. (Feldman)[31] suggested that critics of hospital food are likely to be of a high socio-economic status. This was discounted by Sheatsley[32] though who found that patients who criticise hospital food are as likely to be of low socio-economic status as high. Differences in culture have also been shown to have an effect on patient satisfaction with food and other hospital facilities (Madhok et al)[33]. Level of prior experience may affect consumer satisfaction as the patient is more familiar with the hospital experience and the information they have gained previously will have given them a 'cognitive road map' with which to interpret the experience while anticipating fewer frightening surprises. Cohen et al [34] found this was true for kindergarten but Deluco and Cremer[35] found that there was no significant correlation in hospitals.

Conclusion

Based on current literature it would appear that there is no clear cut understanding, at present, of the patient in hospital on which to devise methods of patient satisfaction measurement. The argument for the current use of quantitative surveys is valid, in that they are easy and cheap to implement. The validity of the results currently being generated, predominantly from quantitative research, demands that a comprehensive qualitative study of patients in hospitals is undertaken, which identifies all those factors affecting decision making and consumer behaviour. From this a conceptual base can be designed which will provide the foundation for successful quantitative surveys of the future.

This constitutes part of the ongoing plan of research at the Worshipful Company of Cooks Centre for Culinary Research at Bournemouth University.

---

[1] Larsen D.E. and Rootman I., Physicians' Role Performance and Patient Satisfaction. *Social Science Medicine*. Vol. 10, 1976.

[2] Baron and Byrne. *Social Psychology : Understanding Human Interaction.* Allyn and Bacon. 1994

[3] Scheier et al. Dispositional Optimism and Recovery from Coronary Artery Bypass Surgery : The Beneficial Effects on Physical and Psychological Well-Being. *Journal of Personality and Social Psychology.* 57, 1989.

[4] Kings Fund Centre. *A Positive Approach to Nutrition as Treatment - Report of a working party chaired by Professor J.E. Lennard-Jones on the role of enteral and parenteral feeding in hospital and at home.* January 1992.

[5] Gardner Merchant Research Services. *Food In Healthcare Survey*. 1995.

[6] DHSS. *NHS Management Inquiry.* HMSO, London, 1984.

[7] *Working for Patients*. HMSO, London, 1989.

[8] Bottomley V., *The management of change - achieving quality service.* Speech delivered to the annual conference of the Institute of Health service Management. Birmingham, June 1993.

[9] National Audit Office. *National Health Service: Hospital Catering in England.* London : HMSO. 14th April 1994.

[10] *Working for Patients*. HMSO, London, 1989.

[11] National Audit Office. *National Health Service: Hospital Catering in England.* London : HMSO. 14th April 1994.

[12] National Audit Office. *National Health Service: Hospital Catering in England.* London : HMSO. 14th April 1994.

[13] Wedderburn Tate C. et al. What do patients really think? *Health Service Journal*. 12th January 1995.

[14] Barnes M. Beyond Satisfaction Surveys: Involving People in Research. *Generations Review(Journal of the British Society of Gerontology)* 2(4), 15-17, 1992.

[15] Engel J.F., Blackwell R.D. and Miniard P.W. *Consumer Behavior*. 8th Edition, The Dryden Press 1995.

[16] Scott A., and Smith R.D. Keeping the customer satisfied: Issues in the Interpretation and Use of Patient Satisfaction Surveys. *International Journal for Quality in Health Care*, Vol. 6, No. 4, pp 353-359, 1994.

[17] Taylor et al. Self generated feelings of control and adjustment to physical illness. *Journal of Social Issues*. 47, 91-109, 1991.

[18] Thompson et al Maintaining perceptions of control: Finding perceived control in Low-control Circumstances. *Journal of Personality and Social Psychology* 64, 293-304, 1993.

[19] Feldman J.J. Patients' Opinions of Hospital Food. *Journal of the American Dietetic Association*. Vol. 40, 1961.

[20] Williams B. *Patient Satisfaction: A Valid Concept?* Soc. Sci. Med. Vol. 38, No. 4, 509-516, 1994.

[21] Morris et al. Collective Coping with Stress: Group Reactions to Fear, Anxiety and Ambiguity. *Journal of Personality and Social Psychology*. 33, 674-679, 1976.

[22] Kulik J.A. and Mahler H.I.M. Stress and Affiliation in a Hospital Setting : Preoperative Roommate Preferences. *Personality and Social Psychology Bulletin*, 15, 183-193, 1989.

[23] Thompson A.G.H., and Sunol R. Expectations as determinants of Patient satisfaction: Concepts, Theory and Evidence. *International Journal for Quality in Health Care*. Vol. 7, No. 2. 127-141, 1995.

[24] Engel J.F., Blackwell R.D. and Miniard P.W., *Consumer Behavior*. 8th Edition, The Dryden Press 1995.

[25] Ajzen. I., and Fishbein. M. *Understanding Attitudes and Predicting Social Behaviour*. Prentice Hall. 1980.

[26] Wilson-Barnett. J., and Carrigy A. Factors influencing patients' emotional reactions to hospitalisation. *Journal of Advanced Nursing*. 3, 221-229, 1978.

[27] Maller O., Dubose C.N. and Cardello A.V. Consumer Opinions of Hospital Food and Foodservice. Journal of the American Dietetic Association Vol.76, 1980.

[28] Sheatsley P.B. How total hospital experience shapes patients' opinion of food. *Hospitals* 39, 105, 1965.

[29] Deluco D. and Cremer M. Consumers' Perceptions of Hospital Food and Dietary Services. *Journal of the American Dietetic Association.* Vol. 90, 1990

[30] National Audit Office. *National Health Service: Hospital Catering in England.* London:HMSO. 14th April 1994.

[31] Feldman J.J. Patients' Opinions of Hospital Food. *Journal of the American Dietetic Association.* Vol. 40, 1961.

[32] Sheatsley P.B. How total hospital experience shapes patients' opinion of food. *Hospitals.* 39, 105, 1965.

[33] Madhok R, Bhopal R.S. and Ramaiah R.S.W., Quality of hospital service: A study comparing 'Asian' and 'non-Asian' patients in Middlesborough. *Journal of Public Health Medicine.* Vol. 14, No 3, 271-279, 1992.

[34] Cohen R et al. Easing the Transition to Kindergarten: the Affective and Cognitive Effects of Different Spatial Familiarisation Experiences. *Environment and Behaviour.* 13, 330-345, 1986.

[35] Deluco D. and Cremer M. Consumers' Perceptions of Hospital Food and Dietary Services. *Journal of the American Dietetic Association.* Vol. 90, 1990.

Determinants of repeated lunch choices in a cafeteria situation

L. Lähteenmäki

University of Helsinki, Department of Food Technology, P.O. Box 27, 00014 Helsinki, Finland

Abstract

The aim of the present study was to examine whether the theory of reasoned action can be applied to repeated lunch choices in a cafeteria. The choice intentions rated on the questionnaire and those rated just before lunch were compared. Participants (40 women and 10 men, age 23-52 years) rated the use frequency, liking for, fillingness, internal diversity and use intentions of 53 main lunch dishes before and 31 items after a seven-week choice period. The seven targeted choices included five different main courses prepared from ground meat and served weekly. Prior to lunch, participants rated their intentions to choose the options available that day and recorded their choices after lunch. After the choice period, participants assessed their attitudes and subjective norms towards target foods among other items. All ratings were made on seven-point scales. Hierarchical multiple regression analysis showed that intentions were accurately predicted by attitudes. In addition to attitudes, liking for the dish predicted the intentions rated on the questionnaire but not those rated just before lunch. The data suggest that paper-and-pencil responses obtained in surveys give an optimistic picture of consistency between attitudes, hedonic responses and intentions. To obtain a reliable picture of natural purchase situations several choices are needed, as the model may not be applicable to single-choice occasions.

1 Introduction

In Finland about 30% of the adult population eat their lunches outside the home on workdays, in cafeterias or canteens[3]. These lunches are typically hot meals containing a main dish, starches, fresh salad, bread and a drink. In the university cafeteria where the present study was conducted, the six-week preplanned menus define the main courses, whereas other parts of the meal vary little.

The main course is the crucial factor in acceptance of the entire meal.[2,6,13] The preplanned menus guarantee that the same dishes are not served repetitively, which could decrease acceptance,[9] although Kamen and Peryam[4] found that even a three-day repetitive menu does not decrease acceptance. The interval from the previous serving increases the hedonic responses such that the maximum is attained three months after the last serving.[6]

The theory of reasoned action of Fishbein & Ajzen is the most frequently used attitude model in predicting food choices.[8,10] According to this model, behaviour such as food choices can be predicted by behavioural intentions which result from attitudes and subjective norms towards the behaviour. The hedonic responses have been important in some studies.[11,12] The model has been mostly applied as paper-and-pencil tasks predicting reported behaviour, e.g. reported use frequencies, with attitude and norm ratings of food names.[10] As Meiselman[5] points out, very little research exists on 'real people in real eating situations'. Brinberg and Durand[1] achieved a moderate, but significant correlation ($r = 0.41$) between intentions to eat at a fast-food restaurant within two weeks and the actual behaviour. Correlations between attitudes and single behaviours are weaker than between attitudes and a number of similar behaviours.[14] The choices of several similar lunch items over time should therefore be more closely connected with intentions than single choices, and similarly composite measures of intentions should be better predicted by attitudes and subjective norms.

The aim of the present study was to examine whether the theory of reasoned action, extended with hedonic responses, can explain the repeated choices of similar lunch dishes in a cafeteria environment. The aim was also to compare relationships between choices and intentions measured in a questionnaire or those measured just before lunch. The choices between two alternatives were examined in relation to these intentions.

2 Method

2.1 Participants and target main courses

Participants (40 women and 10 men, age 23-53 years) included university staff and students who ate at least four times during the seven research days at the cafeteria and who were recruited through personal contact in the cafeteria.

The main courses chosen as target foods in the present study included dishes prepared from ground meat and served at least once weekly at the cafeteria (Table 1). During the seven weeks five different dishes were served: meatballs twice, meat patties twice, meat patties combined with red beets or mushrooms, and meat loaf. Although the main courses had five different names they were variations of the same type of dish, and therefore were expected to be perceived similarly.

Table 1 Description of dishes and choices at the cafeteria on each research day

Day	Dish	Target Description	N	Alternative Description	Others N	Total N
1	1	Meat patty with mushroom sauce	19	Fried rainbow trout	20	6	45
2	2	Meat-mushroom patty	17	Turkey casserole	13	16	46
3	3	Meatballs	25	Mushroom pie	8	5	38
4	1	Meat patty with mushroom sauce	15	Baked rainbow trout	18	4	37
5	3	Meatballs	12	Steak	20	5	37
6	4	Meat loaf with olive filling	14	Fried plaice	11	7	42
7	5	Meat-red beet patty	10	Fried rainbow trout	27	2	39
		Σ	112	Σ	127	45	284

2.2. Procedure

The study consisted of three parts: 1) a questionnaire about main dishes in the beginning, 2) lunch choices once weekly for seven consecutive weeks and 3) a questionnaire about attitudes, subjective norms and overall satisfaction of cafeteria services at the eighth week.

In the questionnaire before the lunch choices, respondents rated the use frequency (never - almost daily), liking for (I do not like at all - I like extremely), fillingness (not at all filling - extremely filling), internal diversity (monotonous - diverse) and use intentions (unlikely - likely) of 53 main dishes served for lunch. All ratings were made on seven-point scales which were verbally anchored at both ends. The food names contained ground meat dishes together with other most frequently served lunch items at the cafeteria.

On research days prior to lunch, participants filled in the first page of a two-page form rating their mood, intentions to choose among the alternatives available and liking for them, using the same scales as in the first questionnaire. After lunch, they completed the second page and recorded their actual choices, and pleasantness of separate components of the meal on seven-point scales (unpleasant - pleasant).

After the choice period, participants completed another questionnaire containing items on attitudes, social norms and appropriateness of targeted ground meat dishes. Attitudes were measured by rating the statement 'My attitude towards eating *meatballs (food name)* is negative - positive' on a seven-point scale. The subjective norm was rated by the statement 'In my colleagues' opinion eating *meatballs (food name)* for lunch is negative -

positive'. The importance of the opinion of others was rated on a seven-point 'unlikely - likely' scale as an answer to the question 'How likely are these opinions to have an impact on your choices'? The subjective norm was calculated by multiplying the scores on these two ratings. The same ratings as in the first questionnaire were asked about 31 main dishes which were served as alternatives on research days but not included in the first questionnaire. The questionnaire also contained items on overall satisfaction with the lunch services at the cafeteria.

2.3 Data analysis

The attitude, subjective norm and liking ratings of main dishes were entered into hierarchical multiple regression analysis to predict intentions of single target dishes. The total sum of intentions was predicted by the sum of attitudes, norms and liking ratings. As expected the target dishes were perceived similarly; correlations among dishes varied $r = 0.34 - 0.76$ for attitudes, $r = 0.95 - 0.99$ for subjective norms and $r = 0.17 - 0.77$ for liking ratings. The corresponding Cronbach alphas were 0.84, 0.99 and 0.81 for attitude, norm and liking scores, respectively. Intentions from the questionnaire in the beginning and those rated just before lunch were predicted separately; the differences in these intentions were tested with the paired t-test.

The choices were recoded such that choosing the target food was 1 and any other choices were recoded as 0. Those not eating were treated as missing data for single research days, which implies that the subgroups of participants in the analyses vary among research days. As participants ate four to seven times at the cafeteria on the research days the overall choices were calculated as choices of target foods relative to all choices. The relative choices correlated highly with the absolute number of choices ($r = 0.96$).

The choices between the target food and main alternative were analysed for those who chose either of these dishes. Hierarchical multiple regression analysis was used to test the role of the intentions to choose the target food and intentions to choose the alternative. The differences in the intentions were tested with the paired t-test.

3 Results

3.1 Intentions to choose ground meat dishes

The number of lunches eaten at the cafeteria varied 37 and 46 on the seven research days; about one third of the choices were target dishes (Table 1). The main alternative was slightly more popular than target dish over the seven lunches, but much variation occurred among single days. The ground meat dishes were well liked (mean 4.2 over five dishes), filling (mean 5.4) but not diverse (mean 4.0).

The intentions to choose target dishes were rather high, with the exception of meat-mushroom patties on Day 2 (Figure 1). The means of intentions were similar whether measured in the questionnaire or just before lunch; the only exception was again Day 2 in which intentions just before lunch were more positive than they were in the questionnaire. The intentions asked in the questionnaire and just before lunch correlated with each other moderately ($r = 0.55 - 0.74$), but these two measures were not identical.

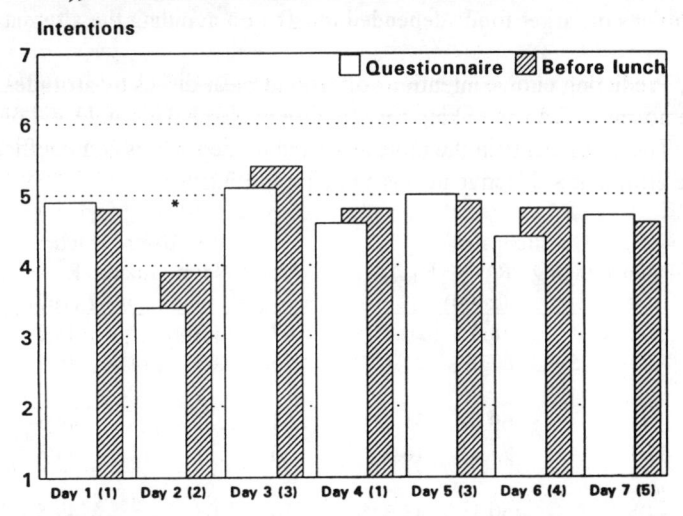

Figure 1. Intentions to choose target dishes on seven research days (numbers in parentheses relate to type of dish). The asterisks (t-test, * $p < 0.05$) show the significant differences between the two types of intention.

3.2 Predicting intentions rated in the questionnaire and just before lunch

In hierarchical multiple regression analysis, choice intentions were predicted by attitudes, whereas subjective norms were not important (Table 2). Liking ratings were good additional predictors of intentions asked in the first questionnaire but not of those asked just before lunch. The overall questionnaire intentions were predicted as well as the best of the single meals, but for intentions rated just before lunch the overall intention was predicted better (80%) than intentions of single meals.

3.3 Intentions and choices between two options

The intentions measured in the questionnaire differed significantly between the target food and the main alternative on most research days (Figure 2a), but some of the differences disappeared from the intentions measured just before lunch (Figure 2b). The choices were more closely connected with the intentions measured just before lunch than with the questionnaire intentions (Table 3). The correlations are not high throughout the seven days and questionnaire intentions could explain at most 31% of the choices and intentions rated just before lunch 44% of the choices. The intention to choose the alternative was almost as sufficient a negative predictor of choices, especially when rated just before lunch. Most choices appeared to depend as much on the intention to choose the alternative as on the intention to choose the target foods, while on Day 5 the choices of target foods depended mostly on avoiding the alternative.

Table 2. Predicting choice intentions of ground meat dishes by attitudes, subjective norms (TRA) and liking for separate research days and total over seven days. The coefficients in the table are standardized regression coefficients for the attitudes, subjective norms and liking ratings.

Day	Att	Norm	Liking	R^2 (x100)	F_{change}	Att	Norm	Liking	R^2 (x100)	F_{change}
			Questionnaire					Before lunch		
1	.69	-.03		48.0	20.8**	.67	.03		45.2	16.9**
	.29	-.05	.59	67.3	26.1**	.56	.03	.17	46.9	1.2
2	.83	.03		69.5	51.4**	.64	.11		44.3	16.7**
	.51	-.01	.44	77.8	16.4**	.56	.11	.11	44.8	0.4
3	.77	-.05		60.1	33.9**	.70	.03		49.5	19.6**
	.53	-.06	.38	68.6	11.8**	.45	.00	.37	57.0	6.8*
4	.69	-.03		48.0	20.8**	.82	.07		69.7	42.5**
	.29	-.05	.59	67.3	26.1**	.76	.08	.08	70.0	0.4
5	.77	-.05		60.1	33.9**	.60	.13		37.1	10.0**
	.53	-.06	.38	68.6	11.8**	.51	.13	.14	38.2	0.6
6	.67	-.10		43.3	15.3**	.85	.07		73.2	54.5**
	.15	-.07	.67	62.5	19.9**	.83	.07	.02	75.2	0.0
7	.70	-.00		48.8	21.5**	.75	-.11		57.6	25.1**
	.24	-.00	.61	64.6	19.5**	.62	-.12	.18	59.1	1.4
Tot	.79	-.04		62.2	32.9**	.90	.01		80.5	33.0**
	.34	-.04	.55	73.1	15.7**	.74	.02	.18	81.4	0.7

* $p < 0.05$; ** $p < 0.01$

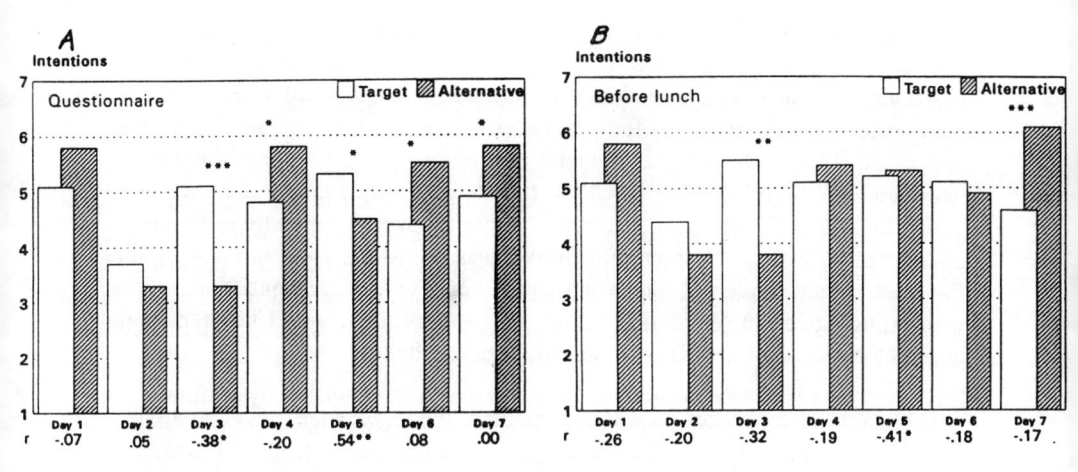

Figure 2. Intentions to choose target dishes and their main alternatives when rated a) in the questionnaire and b) just before lunch on seven research days. The asterisks (t-test, * p < 0.05; *** p < 0.001) show significant differences between the two intentions. The correlations between the intentions are under the columns.

4 Discussion

Studying real eating behaviour in a real eating situation brought many uncontrolled variables into the study. In situations in which individuals experienced a free choice and varying time schedules the number of missing data became relatively large. The results appeared consistent over time, which suggests that the slightly different subpopulations of the 50 participants on each research day did not affect the results systematically. The structure of meals was homogeneous compared with British lunches.[2] Day 5 was, however, inconsistent with the remaining data which suggests that studying an occasional isolated choice is always risky, as several situational factors may influence eating behaviour.[7] Intentions measured in the questionnaire and just before lunch showed similar means and also correlated with each other, suggesting that intentions are relatively stable measures.

As in previous studies attitudes were the most important predictors in the theory of reasoned action, [10,11] but subjective norms showed no influence on intentions. Colleagues at work do not appear to be important influences on choice decisions. Both questionnaire and situational intentions were well predicted by attitudes but the variation among research days was wide, especially when predicting intentions rated just before lunch. The overall composite measure of intentions appeared to improve the prediction of intentions measured just before lunch but not those rated on the questionnaire. The situational factors during research days may explain the greater variation on single days. Since this variation is supposed to be random over time, the composite measure improves the consistency between attitudes and intentions, as was expected.[14] In the questionnaire data all ratings were free of day-to-day variation and therefore the results varied less. Surprisingly, the increased variation in single behaviours resulted in better overall prediction. Attitudes could explain 80% of the overall intention measure, which is exceptionally good for food choice models.[10] Paper-and-pencil tasks give more consistent results than ratings given in real eating situations, implying that in real life behaviour several measures are necessary to improve the reliability.

Table 3 Predicting choices with intentions of the target food and alternative (questionnaire and just before lunch). The coefficients in the table are standardized regression coefficients for the target food and main alternative.

Day	Questionnaire				Before lunch				N
	Target	Altern.	R^2 x100	F_{change}	Target	Altern.	R^2 x100	F_{change}	
1	.39*		15,0	6,5*	.47**		22.3	10.0**	39
	.36*	-.34*	26.2	5.5*	.47*	-.51**	37.7	8.4**	
2	.23		5.2	1.5	.57**		32.7	12.6**	30
	.25	-.34	16.8	3.8	.50**	-.38*	46.5	6.5*	
3	.48**		22.6	9.1**	.64**		41.6	22.1**	33
	.41*	-.18	25.4	1.1	.54**	-.33*	51.5	6.1*	
4	.56**		31.3	14.1**	.67**		44.3	24.6**	33
	.50**	-.32*	41.2	5.1*	.67	-.27	46.7	1.3	
5	-.31		9.9	3.2	.08		0.7	0.2	32
	-.05	-.49	26.6	6.6*	-.12	-.50*	21.7	7.8*	
6	.26		6.8	2.4	.43**		18.5	7.5*	35
	.29	-.37	20.4	5.5*	.36*	-.40*	33.6	7.3*	
7	.47**		22.4	10.1**	.56**		31.4	15.1**	37
	.47**	-.23	27.6	2.5	.50**	-.35*	43.3	6.7*	
Tot	.52**		27.0		.66**		43.6		50

* $p < 0.05$; ** $p < 0.01$

Liking played a different role in intentions measured in the questionnaire and just before lunch. In questionnaire intentions, liking replaced attitudes to some extent as predictors but still added significantly to the entire prediction, whereas liking had no effect on intentions measured just before lunch. Of the questionnaire intentions, 63 - 78% could be explained by attitudes and liking. Liking ratings appear to measure a hedonic component which cannot be assessed by attitude measure. On the other hand, intentions rated just before lunch were more closely connected with attitudes than liking ratings.

Intentions measured just before lunch were on average more closely connected with choices than were those from the questionnaire. This was expected, as the time interval between the intention rating and choice is shorter, and in addition the other alternatives available are known. Intentions, however, could explain only a minor fraction of the choices; those from the questionnaire could explain 5 - 31 %, while those measured just before lunch could explain from 1 to 44%. One of the strong situational factors causing this variability appears to be the other options available.[7] The choice sometimes appears to be driven by avoidance of the other option rather than by a positive decision of choosing the intended option. Although the other option appears to be almost as important in the choice decision the weight of these variables changes from one choice occasion to another. The exceptional results of Day 5 may be explained by the correlations of intentions between the target dish and the alternative. During other days the intentions of the options did not correlate with each other significantly, but on Day 5 they were positively correlated in the questionnaire data. This correlation became negative between intentions measured just before lunch. Menu planning had failed in the sense that the same people liked and disliked both options, which made the choice decision difficult, and in the regression analysis avoiding the alternative became the strong predictor.

Predicting intentions which have been measured as paper-and-pencil entities may result in an overly optimistic view of the relationships between attitudes and intentions, as real life appears to have exceptional days when the choices follow some logic other than the most common. Studying several single-choice occasions therefore gives a more reliable picture of the possible variation over time. In general attitudes are good predictors of intentions, but even in a limited real-choice situation individuals follow their intentions only to a certain extent and at best 30 - 40% of their choices can be explained.

5 References

1. Brinberg, D. & Durand, J. Eating at fast-food restaurants: an analysis using two behavioral intention models, *Journal of Applied Social Psychology*, 1983, 13:459-472.
2. Hedderley, D.I. & Meiselman, H.L. Modelling meal acceptability in a free

choice environment, *Food Quality and Preference*, 1995, 6, 15-26.
3. Helakorpi, S., Berg, M.-A., Uutela, A. & Puska, P. *Health Behaviour among Finnish Adult Population, Spring 1995*. Publications of the National Public Health Institute. B14/ 1995. (In Finnish)
4. Kamen, J.M. & Peryam, D.R. Acceptability of repetitive diets, *Food Technology*, 1961, April, 173-177.
5. Meiselman, H.L. Methodology and theory in human eating research, *Appetite*, 1992, 19, 49-55.
6. Rogozenski, J.G. & Moskowitz, H.R. A system for the preferences evaluation of cyclic menus, *Journal of Food Service Systems*, 1982, 2, 139-161.
7. Rozin, P. & Tuorila, H. Simultaneous and temporal contextual influences on food acceptance, *Food Quality and Preference*, 1993, 4, 11-20.
8. Shepherd, R. & Sparks, P. *Modelling food choice*, Chapter 8, Measuring Food Preferences, eds MacFie HJM & DMH Thomson, Blackie, London, pp.202-226.
9. Siegel, P.S. & Pilgrim, F.J. The effect of monotony on the acceptance of food, *American Journal of Psychology*, 1958, 71, 756-759
10. Stafleu, A., de Graaf, C. & van Staveren W.A. A review of selected studies assessing social-psychological determinants of fat and cholesterol intake, *Food Quality and Preference*, 1991/2, 3, 183-200.
11. Tuorila, H. Selection of milks with varying fat contents and related overall liking, attitudes, norms and intention, *Appetite*, 1987, 8, 1-14.
12. Tuorila-Ollikainen, H., Lähteenmäki, L. & Salovaara, H. Attitudes, norms, intentions and hedonic responses in the selection of low salt bread in a longitudinal choice experiment, *Appetite*, 1986, 7, 127-139.
13. Turner, M. & Collison, R. Consumer acceptance of meals and meal components, *Food Quality and Preference*, 1988, 1, 21-24.
14. Weigel R.H. & Newman L.S. (1976) Increasing attitude-behavior correspondence by broadening the scope of the behavioral measure. *Journal of Personality and Social Psychology*, 33:793-802.

This study was conducted as a part of EU-funded AAIR-project 'The development of models for understanding and predicting consumer food choice'

The effects of exposure to a food odour on food choice, consumption and acceptability.

L. R. Blackwell, B. J. Pierson

Department of Food and Hospitality Management, Bournemouth University, Fern Barrow, Poole, Dorset BH12 5BB, UK

Abstract

Fifty four subjects participated in experiments in which half were exposed to a neutral odour stimuli (the control group) and half were exposed to a bacon odour stimuli (the experimental group). Subjects were then invited to select one of three meats for their lunch and were allowed to consume as much food as they liked. The foods chosen, the amount consumed and the acceptability of the meal for the two groups was analysed and compared. The results indicated that exposure to the odour did not significantly influence food choice, but did significantly effect the acceptability of the meal and the amount of food consumed.

Introduction

Food intake in humans is guided by a variety of factors including physiological, cognitive, social, cultural, economic, religious and environmental influences. The extent to which any single factor or group of factors predominates in this process is presently unknown, but since responsiveness to sensory based food cues represents only one of many mechanisms guiding dietary behaviour, its effects may not be evident in all individuals (Kissileff & Van Itallie[1]). It is, however, possible that the sensory experience associated with a given meal or snack may be a particularly salient cue for influencing intake of that same meal or snack on subsequent days.

Chemosensory experiences associated with food are viewed as important determinants of food choice (Meiselman[2]). Little is known, however, about the mechanisms by which sensory cues influence food selection and ingestion, other than an obvious influence via the innate or acquired hedonic aspects of foods. One mechanism by which sensory cues may exert their influence on food intake is through learned associations. Research suggests that an individual's responses to foods are largely learned not innate, and learned flavour cues have been shown to influence snack food selection in children (Logue & Smith; Hook[3&4]). More recent research has been carried out to investigate the contribution of learned flavour cues to daily patterns of food ingestion and the findings indicated that nutritional meaning may be another factor influencing decisions regarding food selection and portion size (Tepper & Mattes[5]). Booth (1981) noted that once a food's sensory properties acquire nutritional meaning, they might influence future decisions regarding food selection and portion size by providing pre-absorptive information about the probable metabolic implications of ingesting the item (Booth[6]).

It is well known that sight is now the primary human sense and visual cues are, therefore, of primary importance, as the choice or acceptance of a food product may well be initially determined by its visual appearance (Williams, et al.[7]).

This present study investigates both the qualitative and quantitative role of olfactory cues and their effects on food choice and acceptability. It is based on the previous work on hunger perception (Blackwell & Pierson[8]) in which subjects rating their hunger levels after being exposed to a variety of food odours. The results of the hunger perception study indicated that exposure to the odour of cabbage (which had previously been given a low hedonic rating by the subjects) led to a decrease in hunger whilst exposure to the odour of bacon (previously given a high hedonic rating) significantly increased hunger levels (Blackwell & Pierson[8]).

One of the factors which was likely to have effected the results of the hunger perception experiment is expectation of eating. Wooley (1972) found that salivary responses were attenuated when subjects did not expect to eat the food

(Wooley & Wooley[9]). Hence, the hunger rating of the subjects may have been higher if they were expecting to eat the foods they could smell. This present experiment was therefore designed to investigate this potential problem and rather than subjects simply rating their hunger levels on a scale, their hunger was measured in terms of food consumption by allowing subjects to select and eat as much food as they desired.

The experiment had three main aims. The first was to test whether exposure to a food odour influences food choice (a qualitative measure). Secondly, to test whether the exposure leads to an increase in food consumption (a quantitative measure) and finally, to measure whether exposure to the odour prior to consumption effects the acceptability of the meal (a qualitative and quantitative measure).

Experimental Foods

The results of the previous hunger perception experiment showed that the odour of grilled smoked streaky bacon was given a high hedonic rating and significantly increased the hunger levels of all subjects (Blackwell & Pierson[8]) Based on these results the same odour stimuli was selected for this follow-up experiment.

In order to ensure uniformity throughout the test it was necessary that the other foods selected for the experiment were of a similar nature to bacon and therefore sausages and beefburgers were chosen. The three meats have similar nutritional values, are found in similar eating situations, and may be classified under the same 'health' image. Jacket potato, salad, French bread and water were offered for the participants to serve themselves. These accompaniments were chosen as they were claimed to be liked by all of the subjects, were appropriate for the time of year, complemented the meat component and were suitable for 'self-service'.

The meats used were thick link pork sausages with a cooked weight of 20g each, smoked back bacon of which each cooked rasher weighed 5g and beefburgers with a cooked weight of 30g. Smoked streaky back bacon was used to create the odour in the laboratory as the high fat content produced a stronger odour than other bacon cuts.

Subjects

Fifty four members of staff from Bournemouth University aged between 19 and 58 years took part in the experiment. These were thirty four females and twenty males who were selected on the basis that they were non-dieters, non-vegetarians and liked the foods which were involved in the experimentation. An initial pilot questionnaire consisting of 9 different foods (6 of which were to be involved in the experiment) was completed by volunteers to establish their liking for the experimental foods. Subjects who expressed a disliking for any of the foods were not included in the experiment.

Procedure

The experiment took place over a period of six days with nine volunteers taking part on each day (a total of 54 participants). The participants were divided into two quota samples of twenty seven balanced by age and sex. These were labelled as the control group and the experimental group.

Throughout the morning of the tests, subjects completed written tasks in the sensory laboratory. During this time the control group was exposed to a neutral odour stimuli (that is, the odour inherent to the sensory laboratory), whilst the experimental group was exposed to the odour of bacon.

Smoked streaky back bacon was grilled in the sensory laboratory on the experimental test days, and removed just before the experimental group arrived. No visual or auditory food cues were present in the laboratory. Subjects were instructed to complete the written tasks in silence and no reference was made to the odour in the room.

The first written task was completed at 11.30am in the sensory laboratory. This was a questionnaire collecting details about the time the subjects had consumed breakfast that day and what they had eaten. At 12.30pm subjects returned to the sensory laboratory and completed a 'lunch order form' where they were asked to select one of the three meats (bacon, sausage or beefburger) and indicate the amount they would like (ie. number of rashers of bacon, links of sausages, etc.). Subjects were exposed to the odour stimuli for approximately 10 minutes on each of the two occasions.

The participants then went into the restaurant where they were served the meat they had requested and helped themselves to jacket potato, French bread and salad. Water was also provided. Once everyone had finished their meal a second serving of meat was offered. This time subjects were allowed to choose a different meat from their first serving if they so wished. Extra salad, etc. was also available if desired. Plate waste and extra portion sizes were closely monitored.

At the end of the meal subjects were given a 'Meal satisfaction questionnaire' to complete. This asked them to give ratings for the appearance, flavour, smell, texture and overall satisfaction of the meal. Ratings were made on a 100mm line anchored by like extremely and dislike extremely. Finally, they were asked why they had chosen that particular meat.

Results

The results were analysed in order to measure firstly, the effects of odour exposure on food choice, that is, the number of subjects who selected bacon. Secondly to measure the effects of odour exposure on food consumption and finally to analyse the subjects' satisfaction with the meal.

The results were analysed for significance using Analysis of Variance with a significance level of $p<0.05$.

Food Choice

From the control group, 10 subjects chose bacon as their first meat and when offered a second serving, 4 subjects chose bacon. The control group consumed a total of 32 rashers (160g) of bacon. From the experimental group 9 subjects selected bacon as their first meat with 7 subjects choosing it on the second serving. A total of 38 rashers (190g) of bacon was consumed by the experimental group. There was no significant difference between the number of subjects choosing to eat bacon or the amount of bacon consumed between the control group and the experimental group (p>0.05).

Consumption

The amount of meat consumed by the two groups is displayed in tables one and two.

Table 1.   Meat consumption during the first serving

	Control Group	Experimental Group	SE
Bacon	135g	150g	
Sausage	260g	280g	
Beefburger	510g	870g	
Total	905g	1300g	4.67

Table 2.   Meat consumption during the second serving

	Control Group	Experimental Group	SE
Bacon	25g	40g	
Sausage	80g	100g	
Beefburger	120g	120g	
Total	225g	260g	2.64

Although the total weight of meat consumed by the experimental group (1560g) was greater than that of the control group (1130g) it was not significantly greater, F value = 3.67 (p>0.05). The amount of meat consumed by the experimental group on the first serving alone, however, was significantly greater than the amount consumed by the control group on their first serving, F value = 4.91 ($p<0.05$), see table one. This is believed to be a more accurate comparison, as by the time the subjects were offered a second serving of meat, visual, auditory and textural cues had come into force. That is, the subjects had seen the meats and therefore probably had some indication of the crispiness of the bacon, greasiness of the beefburgers, etc. Therefore, a comparison of the first servings only, is more valid as the subjects' anticipated their consumption in the presence of olfactory cues alone.

When analysing the amount of meat consumed on the second serving only (table 2), no significant difference was found between the control group and the experimental group, F value = 0.12 ($p>0.05$). This finding also highlights the effect of the olfactory cues in isolation, as the results indicate that when other cues are present (ie. during the second serving) there is no significant difference in consumption levels.

Acceptability

The mean hedonic ratings given by each group for the appearance, smell, flavour, texture and overall acceptability of the meal are shown in table three.

Table 3.   Mean hedonic ratings given for acceptability

	Control Group	Experimental Group	SE
Appearance	5.87	6.67	
Smell	6.26	6.53	
Flavour	6.66	7.28	
Texture	6.17	6.94	
Overall	6.33	7.00	
Total	31.29	34.42	1.03

These hedonic ratings were compared using analysis of variance and the total ratings given by the experimental group were significantly higher ($p<0.05$) than those of the control group.

A variety of responses were given to the last question on why the particular meat was chosen but only two of the participants made comments relating to the odour. One person commented that she did not choose the bacon because it smelt as though it was smoked and a second participant said she chose the sausages because she could smell them cooking from the sensory room. This comment indicates that those who may have been aware of the odour in the laboratory may not have been able to identify it as bacon, as it would not have been possible to smell the odour of sausages cooking.

Plate waste/extra portions

On the first serving all subjects helped themselves to one jacket potato except for one subject who had two pieces of french bread instead and all subjects had one serving of salad. When offered a second serving of meat, one subject from the experimental group also had extra salad and one subject had an extra jacket potato. From the control group no extra salad or potato were consumed. No meat or salad was left by either group but three members of the control group left half of their jacket potato.

Jacket potato, salad and bread consumption was not measured because these foods were selected using visual cues rather than olfactory cues.

Conclusions

Exposure to the food odour did not significantly influence food choice as there was no significant difference between the number of subjects choosing bacon ($p>0.05$).

Exposure to the food odour was found to significantly influence food consumption as significantly more meat was consumed by those exposed to the odour ($p<0.05$).

Exposure to the food odour significantly influenced the acceptability of the meal as higher hedonic ratings were given for the appearance, smell, flavour, texture and overall satisfaction by the experimental group ($p<0.05$).

When other sensory cues came into force such as the taste and appearance of the food, no significant difference in the consumption levels was found.

Finally it may be concluded that exposure to the odour of bacon prior to eating had both a qualitative and quantitative effect on the meal experience.

Factors to Consider

As discussed in the introduction, nutritional meaning has been found to influence decisions regarding food selection and portion size (Booth[6]). Although non-dieters were specifically selected for this experiment, due to the increased awareness of the implications of a high fat diet it is possible that some of the participants, although maybe not consciously on a low-fat diet, were aware that they should not be eating large amounts of these foods (bacon, sausage and beefburger). Therefore the amount consumed may well have been influenced by the nutritional value of the meats.

Sensory specific satiety is also likely to have had an effect on the amount of meat consumed in the experiment. This can be simply described as a drop in liking for an uneaten food but not for an eaten food (Manthey[10]). Had the participants only been offered a second serving of the *same* meat rather than a choice of the others it is likely the amount consumed during the second serving would have been less, due to sensory specific satiety. In addition to this, when offered a second serving, visual cues had come into force as the participants had seen all of the meats by this stage and hence it is likely that their second choice of meat was largely based on the visual cues.

References

1. Kissileff, H R. & Van Itallie, T B. Physiology of the control of food intake, *Annual Review of Nutrition,* 1982, **2**, 371-418.

2. Meiselman, H L. Determining consumer preference in institutional food service, In *Food Service Systems,* pp127-153, Academic Press, New York, 1979.

3. Logue, A W. & Smith, M E. Predictors of food preferences in adult humans, *Behaviour Research and Therapy,* 1986, **19**, 319.

4. Hook, E G. Dietary cravings and aversions during pregnancy, *American Journal of Clinical Nutrition,* 1978, 1355.

5. Tepper, B J. & Mattes, R D. Learned flavor cues influence food intake in humans, *Journal of Sensory Studies,* 1990, **6**, 2, 89-100.

6. Booth, D A. The physiology of appetite, *British Medical Bulletin.*, 1981, **37**, 135-140.

7. Williams, A A., Langron, S P. & Noble, A C. Influence of appearance on the assessment of aroma in Bordeaux wines by trained assessors, *Journal of the Institute of Brewing,* 1983, **90**, 4, 250-253.

8. Blackwell, L R. & Pierson, B J. The effects of exposure to food odours on hunger perception, *Food Quality and Preference,* 1995, in press.

9. Wooley, S C. & Wooley, O W. Salivation to the sight and thought of food: a new measure of appetite, *Psychosomatic Medicine,* 1972, **35**, 2, 136-142.

10. Manthey, J. Relationship of fibre to sensory specific satiety. *Proceedings of the Second Pangborn Sensory Science Symposium*, University of California, Davis, USA, 1995.

Physiological aspects of beer drinkability

T. Fushiki,[1] H.Kodama,[1] T.Yonezawa,[2] K.Morimoto[2]

[1]*Department of Food Science and Technology, Faculty of Agriculture, Kyoto University, Kyoto, 606-01 Japan*

[2]*Kirin Brewery Company, Ltd. 26-1, Jingumae 6-chome, Shibuya, Tokyo, 150-11, Japan*

Abstract

This research examined beer drinkability from the physiological standpoint. It compared two beers already known to have high and low drinkability. It revealed that the beer with high drinkability has a higher urination rate and is transported faster from the stomach. This result suggests that these factors contribute to better adjusting the blood serum components maintaining dipsetic action. It was also discovered that unpleasant odours and bitterness significantly reduced the urination rate, leading to a loss of drinkability.

1. Introduction

Measures of beer's deliciousness include not only the ability to enjoy the first glass, but also the ability to continue enjoying the beer after consuming several. One way to describe this quality is as beer that the drinker does not tire of. In German, this is called weitertrinken; in English, drinkability. In Europe and other areas where beer is widely drunk, it is a major yardstick for evaluating beer. The division of beers into ones with high and low drinkability takes place not just among brewery technicians, but among beer lovers as well. The constituents that affect a beer's drinkability are, however, virtually unknown.

As beer is consumed, the body gradually becomes saturated with water, a situation that frequently reduces the desire to drink more. A beer that can be drunk without producing this saturation and concomitant reaction against drinking more is obviously high in drinkability. In German beer halls, patrons who drink for hours are a common enough sight. They can do so not simply because they like beer, but also no doubt because their tradition placing high value on beers that can be drunk without losing interest has given birth to beers with high drinkability.

This research, conducted using healthy volunteers, attempted to scientifically evaluate beer drinkability by looking for explanations in such physiological aspects as gastric emptying rate, urination rate, changes in blood mineral composition, and anti-diuretic hormone levels. The beers used in these experiments were bottom-fermented Pilsners from Czechoslovakia (Beer B) and Japan (Beer A).

2. Comparing the compositions of beer and bodily fluids

The massive consumption of beer or other fluid with a composition so radically different from blood has pronounced effects on the control of blood volume and mineral balance. As Table 1 indicates, beer, like the malt from which it is made, is high in potassium and low in sodium. Human blood, however, has these two minerals in the reverse ratio.

Table 1 Mineral balances of human blood serum and beer

Mineral	Human blood serum	Beer A	Beer B
Sodium	3200 mg/L	23	13
Potassium	150 mg/L	286	409
Calcium	100 mg/L	30	18
Osmotic pressure	280 mOsm	75*	172*

*Each value has been subtracted the osmotic pressure originates in alchol.

Cells constantly pump out sodium ions so as to maintain a difference in sodium concentrations inside and outside the cell that is essential to nutrient transport, neuron action, and other processes. Maintaining these relative electrolytic concentrations on both sides of the cell walls is highly important to maintaining life. Since too high a blood potassium level interferes with cell activity, the excess is promptly absorbed by cells or excreted in the urine.

Since beer has these two minerals in the reverse ratio to that found in blood, being able to continue drinking large volumes of it requires prompt mineral metabolism. Animals have highly precise homeostatic systems for maintaining such mineral balances, so the minerals in the beer consumed are swiftly processed. The beer's composition can therefore greatly affect whether people can drink great quantities of it.

3. Urination rate and drinkability

Buday and Denis [1] have reported different urination rates for beer and water. We therefore conducted experiments to explore whether there was any difference in urination rate between beers of high and low drinkability.

Figure 1 compares urination rate results obtained with a beer, B, considered to have high drinkability by a preliminary sensory test with brewery technicians, and one, A, with low drinkability. In these experiments, 15 healthy human volunteers drank 200 ml of beer every 15 minutes for two hours. Urine output

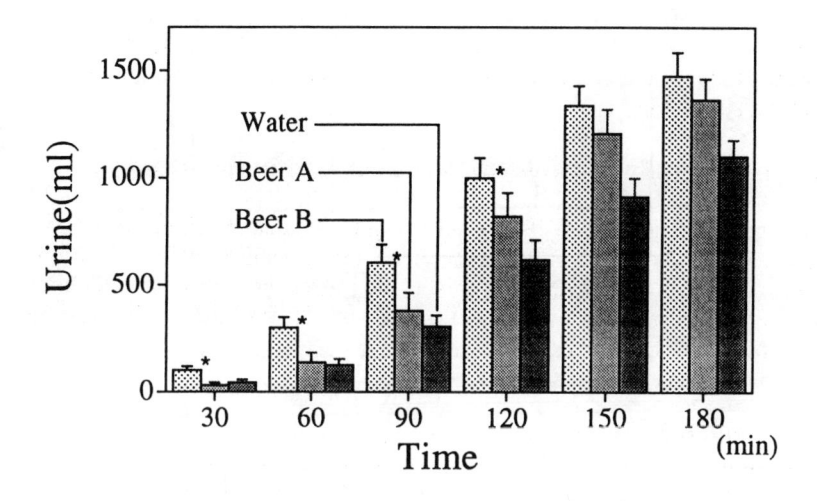

Figure 1: Urination rates of water, beer A, and beer B

was measured every 30 minutes for three hours. That these volunteers all produced the proper aldehyde dehydrogenase isozymes was previously checked by analyzing their white blood cell DNA. This Figure shows how the beer with high drinkability produced higher urination rates than water and the other beer over the first 90 minutes. These results are consistent with the findings by Buday and Denis [1] that beer produces higher urination rates than water. This experimental result indicates that it is possible to investigate the factors influencing beer drinkability using urination rate as an indicator.

4. Gastric emptying rate and gastric transfer speed

Oral questioning of the volunteers revealed that drinking the beer with low urination rate and low drinkability produced a feeling of stomach fullness. This result suggested that the gastric transfer speed for this beer was lower. Analysis of scans obtained with ultrasonic diagnostic equipment (SSA-250A, Toshiba, Tokyo, Japan) of the volunteers' stomachs revealed, for the beer with the high drinkability, a trend toward faster transport out of the stomach through accelerated gastric emptying rates (Get), Gasitic motility (MI), and Gastric frequency (Figure 2). Although the statistical sample is too small to draw significant conclusions, this result suggests a causal relationship between drinkability and gastric activity.

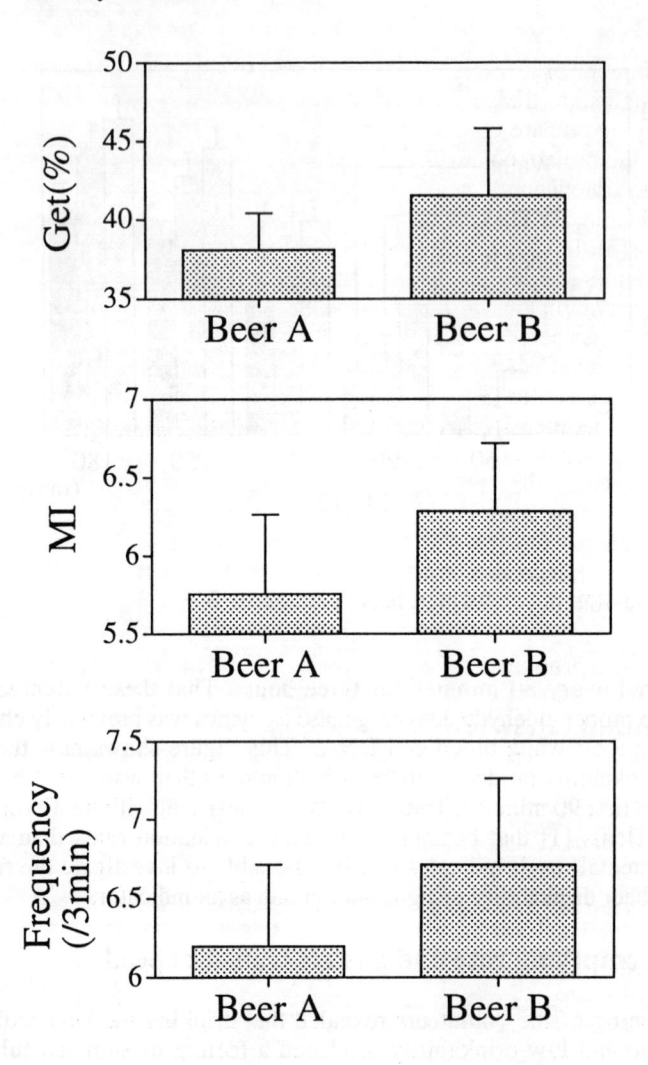

Figure 2: Gastric emtying rate (Get), motility index (MI) and frequency of gastric contraction (Frequency) whith beer intake.

Beer that leaves the stomach faster is naturally excreted faster. The fact that the beer with the higher drinkability also had higher urination rates would seem to be the result of its leaving the stomach faster. Similar results were obtained in experiments with laboratory rats force-fed beers of different drinkability through a gastric tube. Weighing the stomach contents five minutes later revealed that the beer with the higher drinkability was transported out of the stomach faster. (Data not shown.)

A large number of researchers have investigated the mechanisms by which stomach contents are transported out. Among the factors known to affect the gastric emptying rate are the food/drink volumes, the osmotic pressure, the fat content, and the protein content [2]. The common explanation is that chemical receptors in the duodenum control the secretion of such digestive tract hormones as cholecystokinin, PYY, and secretin and thus regulate the opening and closing of the pylorus[3]. In addition to these humoral factors, there are also known to be control factors exerted by the nervous system. Reports note, for example, that vagus denervation lowers stomach activity and that applying a mild electrical current to the vagus nerves increases stomach activity [4, 5]. The beer ingredient that produces these effects on gastric emptying rate is, however, as yet unknown and remains a subject for further study.

In human experiments, Mori [6] concludes that beer's diuretic effect is due to alcohol. It should be noted, however, that the diuretic effect investigated by Mori is somewhat different from the urination rates measured in these experiments. Mori measured the total urine output for five hours after drinking beer and the excess moisture excreted once urination had leveled off. What was measured is thus totally different from the speed of urination measured in these experiments.

5. Effect of unpleasant taste on beer drinkability

5.1 Unpleasant odors reduce drinkability

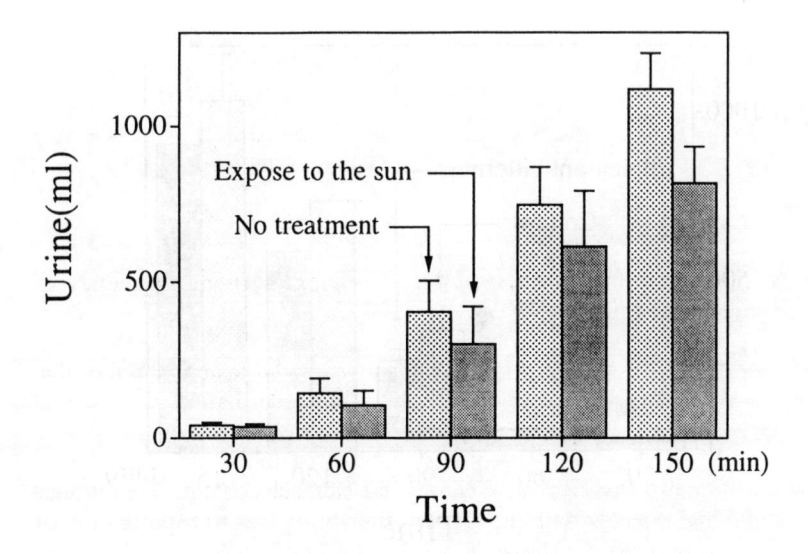

Figure 3: Urination rates of the beer with the unpleasant odor

One observation made during the course of these beer drinking experiments was that the volunteers soon tired of and thus could not drink as much of beers smelling of oxidized hops as a result of poor storage. We therefore attempted to confirm this observation in our experiments measuring urination rates. Beer stored in a clear bottle left in the sun for two hours acquires an unpleasant odour due to hop oxidation. This phenomenon has long been known and is due to the breakdown, under the influence of light, of isohumulone, the bitter ingredients of hops, into 3-methyl-2-butenethiol. This compound has a sulphury odor reminiscent of skunk and is numbered among the contributors to lowered beer quality. The sulphury beer made this way was equivalent to beer from a store with improper storage conditions.

This beer was given to 15 volunteers at the rate of 200 ml every 15 minutes with urine output then measured at 30-minute intervals. A control group was given the same brand of beer, but properly stored. The beer with the unpleasant odor produced markedly lower urination rates than those from the control group (Figure 3). Many volunteers reported that the beer with the unpleasant odor gave them a sensation of a full stomach. There was a tendency for inebriation to start later yet proceed faster. The overwhelming opinion was that the odor gradually became more noticeable, hindering the volunteers' ability to drink vast quantities.

5.2 Effect of unpleasant bitterness on urination rate

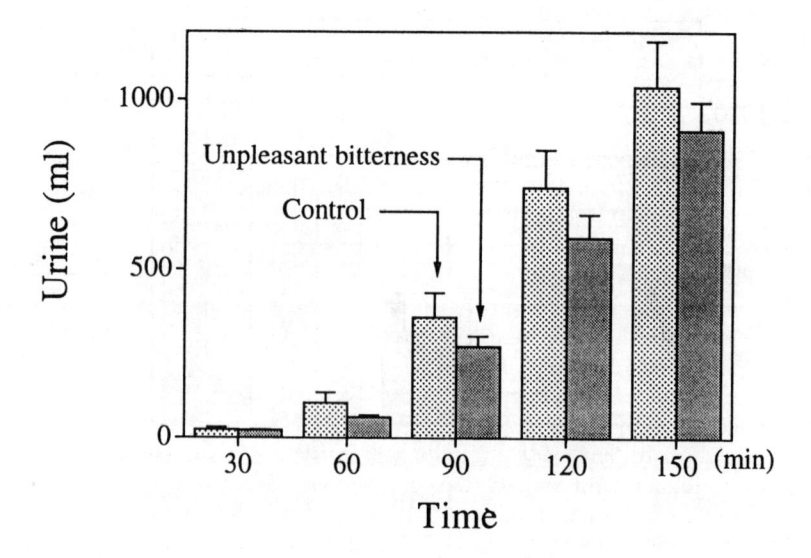

Figure4: Urination rates of the beer with the unpleasant bitterness

Our experiments revealed that unpleasant bitterness as well as unpleasant odours affected the gastric emptying rate and the urination rate. Extensive boiling of hops releases a red pigment and increases the unpleasant bitterness component. Although the task of identifying the specific component is complicated by the fact that the bitter components of hops are said to number over 900 [7], adding this extract produced markedly lower urination rates than those from the control group drinking unadulterated beer (Figure 4). The majority of the subjects reported that the bitterness became all the more noticeable as they drank, thus lessening their desire to drink more.

One possible explanation for these results is that the unpleasant hop odor or bitterness triggered the human alarm system for warning against suspect foods and thus caused the nervous system to close the pylorus, lowering the gastric emptying rate and the urination rate. The human body leaves potentially hazardous substances in the stomach instead of immediately passing them on to the intestines. The result is that an unpleasant taste markedly lowers drinkability.

6. Changes in fluid composition due to beer ingestion and their effects on drinkability

To shed further light on the close relationships between drinkability and such physiological aspects as urination, we investigated the differences that ingesting drinkable and less drinkable beers produced on the adjustment of liquid volume and mineral composition. In this experiment, ten volunteers drank 20 ml of beer for each kilogram of body weight at a fixed rate over two hours. Blood samples were taken at the beginning of the drinking session and then at the 1- and 2-hour marks. Although beer B, the one with high drinkability, was extremely high in potassium, there was little change in the potassium and sodium content of the blood. This result indicates that the homeostatic system for maintaining such mineral balances takes priority and that consuming large quantities of beer does not affect this aspect of blood composition.

There was, however, a significant increase in serum protein content with beer B (Figure 5). Blood and other extracellular fluids rapidly move into the cell so as to maintain mineral balances within highly specific tolerances. The serum protein content was higher apparently because, in order to promptly remove the excessive potassium originally present in beer B, the water moved into the cells or was removed in the form of urine. In other words, drinking beer B seemed to reduce the water content of the blood, producing a thirsty condition that delayed saturation of the body with water. This effect no doubt plays a role in accelerating both gastric emptying and urination rates.

The concentration of arginine vasopressin (ADH), a hormone involved in mineral metabolism that plays a crucial role in thirst and saturation [8], also tended to be higher with beer B (data not shown). Since the release of this hormone signals thirst and produces a desire to drink, its presence indicates that the body is not saturated with liquid, and the subject is therefore able to drink more beer.

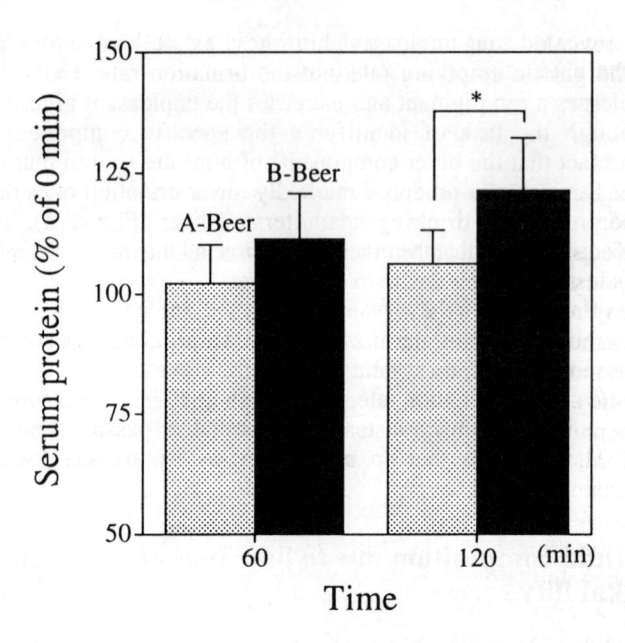

Figure 5: Change of serum protein content with beer intake.

7. Conclusions

The chemical composition of beer is totally different from that of blood, so drinking large quantities therefore represents major stress for the homeostatic system for maintaining blood mineral balance. These experiments reconfirmed that there are differences in drinkability between brands and concluded that it is possible to at least partially explain this fact with reference to physiological factors that is, that the body uses gastric emptying and urination to counteract this stress. The faster these two rates, the greater the drinkability of the beer. These experiments also discovered that an unpleasant odor or bitterness delayed passage of the beer through the pylorus, reducing the drinkability of the beer. Analysis of the changes in blood composition following the consumption of large amounts of beer revealed that differences in drinkability are closely linked to the metabolism of minerals and water.

Special thanks are extended to Dr. Haruma, Hirosima University school of Medicine, for measuring human gastric emptying.

References
1. Buday, A.Z. & Denis, G. The diuretic effect of beer, *The Brewers Digest*, 1974, **49**, 56-58.
2. Kelly, K.A. Motility of the stomach and gastroduodenal junction, Chapter 12, *Physiology of the gastrointestinal tract*, ed L.Johnson, vol.1, pp 393-410, Raven Press, New York, 1981
3. Fisher, R.S., Lipshutz, W. & Cohen, S., The hormonal regulation of pyloric sphincter function, *J. Clin. Invest.*, 1973, **52**, 1289-1296.
4. Anuras, S., Cooke, A.R. and Christensen J., An inhibitory innervation of the gastroduodenal junction. *J. Clin. Invest..*, 1973, **54**, 529-535.
5. Mir, S.S., Telford, G.L., Mason, G.R. & Ormsbee, H.S. III, Noncholinergic nonadrenergic inhibitory innervation of the canine pyrorus, *Gastroenterology*, 1979, **76**, 1443-1448.
6. Mori, R. Uber die diurestische Wirkung des Bieres, Archiv fur Hygiene Bd. VII, Munchen und Leipzig, 1887, **7**, S354
7. Meilgaard, M.C. *Beer flavor*, Doctoral dissertation, The royal Technical Univ. of Denmark, 1981
8. Robertson, G.L. & Berl T. Water Metabolism. *in The kidney*, ed B.M.Brenner and F.C. Rector, Jr. , vol.1, pp. 385-432, W.B.Saunders, Philadelphia,1986.

Section 3.2: Marketing and Media

Service and guilt - a Norwegian view on service
S. Larsen,[1] I.S. Folgerø[2]
[1]*Finnmark College, Dept. of Tourism N-9500 Alta, Norway*
[2]*Stavanger University, Centre, N-4004 Stavanger, Norway*

Abstract

The paper examines general aspects of the Hotel and Restaurant scene. Aspects of outstanding individual service ARE presented, and the concept of guilt is introduced as a psychologically relevant element in managing clients' experience of excellent service.

The paper concludes that the guilt factor should be managed thorough increasing personnel awareness of variations in client expectations. Further, it is argued, personnel training should focus on the individual employees' tendencies to project his or her feelings onto others.

1   General background

In spite of - or perhaps because of - its relative ambiguity, the service concept is much debated. People working within the hotel- and restaurant sector, as well as the general public, seem to hold strong views about what constitutes "good" and "bad" service. One often gets the impression that the script for designing excellent service is easily learned, however difficult it may seem to pursue this script. Norwegian celebrities, for example, particularly after a jaunt abroad, seem to know exactly what is the problem with Norwegian service quality. After a trip to southern Europe, or even to the USA, they excitedly report what excellent service experiences they have had in other countries, where hotel and restaurant staff behave correctly, are service minded and friendly - as opposed to what these celebrities claim to have experienced at home. It has frequently been suggested that it is, and always will be, difficult to produce outstanding service in Norway due to the "fact" that Norwegian hospitality workers do not know, and may never learn (!), how to integrate their own freedom ideals with customers' needs and demands. The implications of such a view - that people in other countries should be less freedom oriented and have a greater potential for learning - seems to slip comfortably past the minds of these critics.

A closer examination of the statements of these strong opinion holders, leaves the impression that their service concept is fairly undefined. Needless to say, many of these opinions are based on vague feelings that abroad one is taken better care of for a lesser amount of money, than one is at home. In the present paper we describe the basic contents of some aspects of human interaction related to the service concept. In addition the paper presents an analysis of opinions currently expressed in the Norwegian debate about service and service quality.

2   General Aspects of the Hotel and Restaurant scene

The production of service is fundamentally different from other forms of production. However true it may be that all enterprises contain some degree of service, the service employee is in a unique position; having to interact with clients more often, more directly and more intensively than employees in other industries. In addition, there is an almost inverse relationship between the level of academic training and the level of direct contact with clients in the hospitality industries. An unskilled production employee, working on an assembly line or otherwise involved in industrial production, very rarely has to face the customer; while the (comparatively as unskilled) hospitality worker constantly has to face his clients, his guests. Furthermore, each individual hospitality employee seems to be perceived by the clients as a representative for the company. One may very well say that the hospitality worker continuously plays a game - acts a part - in his or her interaction with the clients[1]. But the play is different from an ordinary stage performance in a very important way: The script is not given, each and every client represents an individual challenge to the hospitality worker. Every client brings his or her own expectations, cognitions, and personality into the play.

The implication of these aspects are of course plentiful. The most central and obvious consequence, however, is that the interaction with clients places a strong pressure on the individual service employee. Consequently, it is quite paradoxical that the lower the employee's theoretical background is, the closer s/he has to relate to clients in the daily work. This is a major challenge to every employee in the hospitality industrey; but even more, it underlines the heavy responsibility of personnel departments in recruiting and developing their staff. Waiters, front desk staff and housekeepers must be given the opportunity to learn about human behaviour, particularly to develop their skills in diagnosing clients demands and expectations, and in responding accordingly and adequately to these.

It is very likely that the employee in the hospitality industry needs an off-stage area where s/he can be relatively free from critique and conflicts. Most service workers will at times be in conflict with customers. These conflicts may not be openly acted out, which again may lead to a build-up of frustrations and aggressions in the employee. If the employee is to be able to play his or her role adequately, handle conflicts professionally, and deal with the unavoidable feelings of shortcoming and - sometimes - defeat, s/he needs to be taken care of off stage by colleagues and supervisors. A "safe" area, free from critique and conflict, is necessary if the employees are to regain their strength and perform at their best. This represents another major challenge for supervisors and personnel managers in the hospitality industries.

3 Aspects of outstanding Service

So what is "good service", and how can we achieve it ? The Danish philosopher Søren Kierkegaard wrote (in the 1840's !) that excellent service is characterised by the fact that the client gets his or her way 1. It is not the enjoyment itself, Kierkegaard underlined, but the experience of having had ones wish come true, that makes one experience one is being "taken care of". We accede this point, and consequently agree with other writers who have underlined that the main attribute of excellent service is that the client's experience exceeds his/her expectations. Let's examine a few conditions that may increase the likelihood of this happening.

Conjecturing from theories of achievement motivation, our opinion is that the *first* condition is that the service employee experiences absence of, or a low degree of anxiety for clients, colleagues and supervisors. An anxious employee cannot perform to his/her best standards. A psychosocial setting that does not provide opportunities for the service employee to be him/herself, inevitably leads to burn-out, frustration and anger. From a customer point of view, it is like going to the opera; the quality of the musical experience improves if one does not have to worry about the sopranos ability to perform the difficult C'' in the final act. From a managerial point of view, a major goal must be to create an atmosphere of "absence of fear" in the organization.

Our *second* conjecture draws on theories of social perception. Our opinion is that the attitudes of the individual service employee are of importance. Does s/he permit the clients to be individual ? Does s/he accept the diversity of human nature, taste and preference, or does s/he feel that some clients are "good" and some "bad", based on stereotypes and prejudices? Is the hospitality worker prepared to do something extra for the client, is s/he helpful and friendly ? In-house professional curricula should aim at teaching employees basic psychological processes that condition personal behaviours, feelings and cognitions. A reliable measure for individual service attitudes, as well as programmes aimed at developing more positive attitudes among staff, is often called for. This represents another major challenge for hospitality managers.

The *third* condition to be met in order to improve service quality is related to the question of guilt. In Western societies guilt, in its many shapes and disguises (consciousness, superego, morale), plays a major role in determining the emotional structure of customers and employees alike. Even so, the guilt aspect is overlooked in current literature; also among the many who enjoy criticising the hospitality industries. Many clients have experienced that the hotel or restaurant they visited did not respond adequately to their demands and expectations. The wine was too warm or too cold, the hotel room was not available upon arrival, or the food was badly prepared or presented. The worst possible response by the organization/ employee is to add insult to injury; to imply that the *client* is to blame for the (perception of) poor quality of the product ("Do *you* know anything about wine ?", "Why did *you* arrive so early ?"). Logically, nothing could decrease the likelihood of resale more than the conveying of such views to the client. Again, the management has a job to do.

4 The guilt question analyzed

Yet: Is this the full story, or are there complicating factors even in this obvious observation? It seems that people in general enjoy a feeling of moderate guilt. It may be an important aspect of our (Norwegian protestant) culture that when we enjoy, we feel obliged to experience a suitable, corresponding guilt (Confer Folgerø and Larsen[2], Larsen[3]). Several authors have emphasised that protestant ethics tend to combine restrictive attitudes towards the use of adiafora[4,5] with an economic ethical codex highlighting diligence, industriousness and hard work[6]. Let's, for example, imagine a married Norwegian couple, who have - for once - arranged for a baby-sitter, so that they may go out for an intimate, romantic dinner. They may feel guilty for "abandoning" their children (who knows what the baby-sitter is really like ?); they may also feel guilty about spending some £50-100 on an evening out, money that could and "should" have been used more sensibly according to their Protestant ethics. This may be lurking in the back of their minds as they approach the restaurant. Then, at least in some settings, they experience a boost of guilt as they enter the dining room, where the staff seems far too occupied to handle more clients than they already have. It is most likely that our couple feel that their presence is seen as an annoying intrusion into the private lives of the restaurant staff. This may partly explain why at least Norwegians, when they wine and dine, tend to leave a rather handsome tip regardless of the service quality experienced. A Danish chef decuisine, who worked in our country for a number of years, said that Norwegian guests were the easiest of clients: Uncomplaining, they accept poor service and poor food, and leave a nice tip regardless of how they are treated. This heedless tipping may be an atonement, an effort to achieve peace of mind even though one has indulged - behaved sinfully - by spending so much money in the first place. Some clients seem to enjoy being punished by the staff, since they then experience that their inner feelings correspond to their outer reality. At least subconsciously, they may experience that some of the punishment that they rightfully deserve, has thus been served. Logically, even if seemingly contradictory, this may increase their experience of good service.

These observations point towards a better understanding of why some people feel that the service quality is much higher in foreign countries. There is probably a linear relation between price and guilt, as there very often is between price and quality. The higher the price, the more intense the feeling of guilt. In our country the price level seems higher than in countries where the Norwegians like to spend their holidays (adjusted for income levels, the prices are fairly similar; but for the Norwegian wallet, they seem a lot better). The shame and guilt related to spending money on food and wine will then be less intense, as a function of the price/guilt relation. All other factors held constant (type of restaurant; objective quality of wine; etc.), the customer's enjoyment (and thus the perceived quality) of food, wine and service would seem poorer in Norway, because the higher price of the meal would imply more guilt. Our simple hypothesis is that service quality is closely related to the staff's ability to handle this guilt. In short, if hotel and restaurant employees manage to alleviate their customers' feelings of guilt, the experienced service quality will increase.

Results of a recent study of service attitudes[7] indicated that employees in the public sector (registered nurses) were more likely to blame clients for poor service quality, than were employees in hotels and restaurants. Nurses were more likely to agree with statements like "If we have much to do, I always tell clients that we can not provide a full service", "Clients have to understand that we can not give full service when we are busy", and "When we have a lot to do, I often feel that clients understand that we are unable to provide full service". Clearly, hotel and restaurant staff are conscious that the customers pay their salaries; also, they are well aware that their own part in the customer/employee interaction is vital for the prospect of future business opportunities. In Norwegian hospitals the clients lack this power; they have no choice but to come back should they need more of the service, no matter how they have been treated. It is, however, promising that hotel and restaurant staff are able to modify their attitudes (and, presumably, their behaviour) accordingly, and quell the possible urge to blame customers for poor service in stressful situations.

5   May the Guilt problem be dealt with efficiently ?

Initially we indicated that the service concept is a complex one. The present discussion has not offered any definition of the concept, but some important aspects have been underlined. Let's see what conclusions may be drawn from our analysis in terms of practical employee/customer interactions.

First of all service employees need to be aware that sometimes enjoyment and pleasure arre in conflict with commonly held ethics. S/he should understand that some customers feel shame for spending their money on something as unproductive and selfish as their own pleasure. The front-line employee should aim to fool - in the very best meaning - these customers, to make them believe that they really deserve what they pay for. The customers should be encouraged to enjoy the moment, and his/her real, guilty feelings should be covered. Sometimes it may be necessary to actively join the resistance; at other times a more superficial treatment will be enough. The goal is simple: Help the customer enjoy him/herself and for a short time feel free from ethical and other demands placed on him/her by post-industrial society. S/he will soon enough discover the fraud (if not before, when the credit card bill arrives), but not before s/he has left the establishment; and so the restaurant seems a pure, blessed haven. A revisit will be guaranteed by the subconscious need we all have once in a while to feel free from our collective, and at the same time highly private guilt.

Secondly the hospitality employee needs to develop his/her abilities to carefully distinguish different clients. S/he should be aware that every market is comprised of different segments, people with different needs. Some clients are perfectly able to enjoy, their ethics permit it. Others may not. A careful training in communication skills will help the service employee to diagnose customers from different segments accordingly, and to treat each group adequately in the practical everyday setting.

The *third* step represents a major personal challenge to all employees in service organisations. The staff must be aware of their own feelings of guilt and their tendencies to project this feeling onto others. A careful and honest analysis of one's own tendencies to condemn clients for "interrupting" one's work, for "misbehaving" and for ordering "wrongly", will lead to an increased service awareness and hence an increased service quality. We should (or at least try to) understand the dynamics of our own feelings in relation to our clients. For managers in hospitality organizations it is of paramount importance to practice the same unprejudiced attitude towards staff members. Managers should actively avoid using employees guilt as a management "tool". Such an effort would lead to generally less condemning attitudes towards clients and staff, a more friendly atmosphere, a fairer treatment of clients by the staff, and fairer treatment of the staff by the management. In these authors' view this is one of the central issues in the struggle to improve service quality in hospitality organizations. Our bottom line is: Know yourself, and do something about it !

3   References

1. Larsen, S. & Aske, L. On Stage in the Service Theatre. *International Journal of Contemporary Hospitality Management*, 1992, **4**(4), pp 12-15.

2. Folgerø, I.S. & Larsen, S. Take the Plunge - get the breakfast right. Paper presented at the *ICCAS*, Bornemouth, June 25-28, 1996.

3. Larsen, S. Norsk jul - dårlig mat og bitter drikke (Norwegian Christmas - bad food and bitter drinks), *The Norseman*, 1992, **32**(6), pp. 26-29.

4. Larsen, S. The origin of alcohol related social norms in the Saami minority. *Addiction*, 1993, **88**, pp. 501-508.

5. Weber, M. *Den protestantiske etikk og kapitalismens ånd* (Protestant Ethics and the spirit of capitalism), Gyldendal, Oslo, 1971.

6. Fivelsdal, E. Om Max Webers sosiologi (On Max Weber's sociology). Introduction to M.Weber's *Makt og Byråkrati* (Power and Bureaucracy), Gyldendal, Oslo, 1971.

7. Larsen, S. & Bastiansen, T. Service attitudes in Hotel and Restaurant Staff and Nurses. *International Journal of Contemporary Hospitality Management*, 1992, **4**(2), pp 27-31.

The marketing of the wine experience: an innovative approach

J. Fattorini

The Scottish Hotel School, University of Strathclyde, 94 Cathedral Street, Glasgow, G4 0LG, Scotland

Abstract

This paper proposes that restaurants fail to make the most of their wine list as a marketing tool. Hidebound by notions of the 'correct' order of service of wines, most restaurant wine lists are unimaginative and ill suited to the desires of the modern dining public. Moreover they fail to give the customer anything more than the bottle on the table; they fail to give a 'wine experience'. This paper points out the benefits that many restaurants already find from selling wines from less well known regions of the world. More fundamentally though, the paper highlights the benefits that restaurants might find in following the example of the licenced retail trade. The harsh commercial environment in which off-licences and wine merchants operate has led them to look for ever more innovative ways of marketing their product. This paper discusses how many of these ideas; celebrity endorsements, product sampling, product information and specialist staff training, might be applied to a restaurant setting. The paper proposes a concept of a 'wine experience' that restaurateurs might employ, giving 'added value' to the customer's wine purchases. The paper shows that by using the concept of a 'wine experience' restaurant wine lists cease to be the 'cash cows' that restaurateurs have so long depended on for gross profit, and customers have so long resented, but become an innovative and flexible marketing tool integral to the success of the restaurant.

1. Introduction

Customers are too often given a poor deal when they buy wine from restaurants. They face high prices for everyday wines, costs which are sometimes out of all proportion to the costs involved for the restaurateur. Wine lists are dull and ill thought out, merely presenting a ledger of available products. The staff, even if they are keen to give advice, are hampered by a lack of relevant training and wine knowledge.

Yet this apathy by many restaurateurs towards the wine in their restaurants is hardly mirrored by a lack of interest towards wine from customers. They want to read about it, talk about it, take courses in its appreciation and eventually become 'experts' in its consumption. All restaurateurs need to do is harness this burgeoning interest and convert it into sales.

Many may well say that this is a case of easier said than done. Yet many restaurateurs have done little other than to latch onto high profile examples of consumer wine interest and sell their wine, excessively marked-up, as before. An example of this has been the vast increase in the number of so called 'New World' wines on many restaurant wine lists. Much consumer interest in wine can be attributed to easy drinking styles from countries who have only recently come into the fold of 'quality producers' Unhampered by laws or tradition that demand wine is made in a set-piece style, producers in Australia, the USA, South America, New Zealand and elsewhere, produce wine designed to satisfy consumer demand. Not surprisingly consumers 'untrained' in traditional wine styles have found these fruity, 'up-front' wines very appealing. More so certainly than the acidic, bitter and tannic flavours that are frequently the 'hallmark' of the poor and moderate quality wines from traditional Western European producer countries that used to appear on restaurant wine lists.

Changing the wines listed in restaurants to styles consumers find more appealing is encouraging. However customers still feel aggrieved if they purchase a bottle that they could buy for half the price at home, merely for the pleasure of someone else opening it; no matter how much they like the wine inside. The very wide availability of many of the most popular wines on the high street is a constant problem for restaurateurs.

The problem many restaurateurs seem to have is not so much choosing good wines but convincing the wine interested customer that they have bought more than a bottle. Furthermore much of the groundwork in converting latent wine interest into wine sales has already been done. The licensed retail trade has spent recent years contending with an increasingly hostile and competitive environment. Those retailers who have dealt with the harsh trading conditions most successfully were not the ones with the broadest range or necessarily most competitively priced wines. The ultimate victors have been those who draw customers into a much broader 'wine experience'. Their customers do not just buy wine, but buy into a less tangible yet more powerful set of values.

In the licensed retail example these values can be divided into two broad categories. On the one hand the convenient and unstuffy nature that characterises supermarket wine buying. On the other, the total immersion into a Disneyesque world of wine that is found in many High Street wine chains and independent wine merchants. Both these retailing styles have their parallels when constructing a 'wine experience' in a restaurant, although the second is perhaps the more obvious and widely applicable.

2. The Marketing of the Meal Experience

The title of this paper is unashamedly adapted from Campbell-Smith's (4) influential 1967 text 'The Marketing of the Meal Experience: A Fundamental Approach'. As identified by Wood (10), 'From being an explicit concept, the "meal experience" has now become a largely implicit one in catering and hospitality education'. Although it appeared radical at the time, Campbell-Smith pointed out that when eating out people were not simply concerned with the food and drink they consumed but with the 'total environmental experience of dining-out' (10). Campbell-Smith identifies the three components of this total environmental experience as food and drink, service and atmosphere.

The implications of Campbell-Smith's text have been far reaching. The discipline of hospitality marketing and positions created within hospitality firms and consultancies to accommodate these specialists, revolve around creating 'meal experiences'. Specialists then spend their time developing restaurant 'concepts'. Built around a particular theme, the restaurant will use the three elements of food and drink, service and atmosphere, to present this 'meal experience' to the customer.

3. Consumer Dissatisfaction With The 'Wine Experience'

The genius of Campbell-Smith's 'meal experience' was that it merely made explicit something that was already true. As Barr (1) points out 'People go out for the occasion more than the food - this was even more true in the 1950's, when restaurant food was far worse than it is today'. But as Barr continues 'Part of that occasion is the drinking of wine, even if this is not something people normally do'. Invariably restaurateurs and customers concentrate on the food rather than the wine when planning or evaluating a restaurant. Wine is an afterthought and only ever plays a supporting role.

In the years after the Second World War to the start of the modern wine 'boom' in the 1970's, wine's role as both a rarity and support to food was a benefit to restaurateurs and a bind to customers. Customers felt obliged to buy wine in restaurants because that was part of the experience. Yet as few ever drank it at home they had nothing to compare it to, either in terms of price or quality. Restaurateurs knew that customers would choose them on the basis of their menu, never their wine list. By underpricing the menu customers would be drawn in to a restaurant, where they would feel obliged to buy overpriced wine that paid for the overheads.

Not surprisingly, for as long as dining-out and drinking wine retained their rarity value few complained, for nobody knew any better. The problem for restaurateurs came when customers did start drinking wine at home and eating out more frequently. Naturally, if someone enjoys an Australian Cabernet/Shiraz at home that they bought in a supermarket for £4.50, they will feel cheated at having to pay £12.00 to £15.00 for a similar wine when eating out; or at least they will feel cheated if the restaurateur has not perceptibly 'added value' to the wine in some way.

At this point many restaurateurs would be keen to point out that they have to make money somehow. If they reduce the mark-ups on wine, then the price of food must rise. As customers still choose restaurants on the basis of their menus as opposed to their wine lists then this would be foolish. Yet creating a wine experience is not about selling wine cheaper, it is to do with making the high price for wine appear better value, or in other words augmenting the product. From the point of view of the customer, no matter how well chosen a wine list is, no matter how knowledgeable the boss, if the member of staff taking the order has no wine knowledge or selling skills, then the customer will be left feeling sold short. Equally, if the list shows no imagination with anodyne descriptions that could apply to fizzy pop as much as they apply to a particular wine, again the 'experience' is incomplete and poor.

3.1 The Consumer as Expert

The lack-lustre nature of the wine experience in restaurants is for many consumers exacerbated as their own knowledge and familiarity with wine develops. The familiarity with wine that many modern customers enjoy is not just bought about by drinking it, but by reading and talking about it as well. There is now a small industry turning out books and periodicals on wine. Asa Briggs (3) notes the increasing number of books written on the subject. He refers to a article written by Auberon Waugh, who in 1984 felt that 'the production of books about wine had outstripped even the growth in wine consumption'.

Several other writers have noted the development of new types of wine consumers who particularly seek out the wine experience either in restaurants or wine shops. Often these are the same people who invest large amounts of time in reading about wine. Robert Joseph (8) described this customer as 'The Pagoda Man', using a title first coined by Steve Daniel, Head Buyer at Oddbins wine merchants. Daniel says how 'The Pagoda Man' 'likes to pretend that he knows all about wine'. Fattorini (6) alternatively uses the term 'Professional Consumers'. Usually middle class, they are wine consumers who 'wish to be seen to have affiliations with a business they perceive as romantic and glamorous whilst retaining the security of their other life [in better paid jobs than the wine trade]'. Fattorini identifies how many of these 'Professional Consumers' are very keen to take trade examinations. This would appear to serve a double purpose of both engendering respect amongst their peers, and giving them a sense of knowledgeable authority when they enter a wine shop.

4. The Retailing Experience

Critically, both these articles identify the great commercial potential in this variously termed consumer type. Joseph for instance suggests that 'Pagoda man is one of the wine trade's 'most tempting prey'. They also identify the strategy that the licensed retail trade has developed in order to target them. Joseph (7) simply identifies this strategy as creating shops that provide a '"comfortable environment" liberally baited with lime-washed wood and canvas'. Fattorini is more specific, commenting 'False, curved wood ceilings give a more enclosed, cellar-like feel, whilst pictures of vineyards and wineries constantly seek to remind the customer where the wine came from. The shop thus becomes a wine-lover's theme park'. Not surprisingly different retailers have very different interpretations on this theme, although bottle green colour schemes, bare wood and half barrels and claret cases used in the displays are recurrent items.

It is not only the decor in modern wine shops that contributes to the 'wine experience' that customers enjoy there. The 1996 Which? Wine Guide (5) says of Oddbins (Best High Street Chain Award) 'staff who not only are interested in wine but, despite working more hours than a junior doctor for rather less money, also want to sell it to you' (Eyres, 5). In fact the quality of the staff is a frequent theme in the Guide. Within the multi-brand Thresher empire the top-end Bottoms Up stores benefit from 'the wines on the shelves and the usually keen bright staff [which] combine to make the stores more appealing', although the middle market Wine Shop brand suffers from staff that 'do not seem as enthusiastic as those in Bottoms Up'. At the fast growing Wine Cellar group 'The staff at head office may be bright, but those in the shops still have a long way to go to catch up with their counterparts at Oddbins, Wine Rack and Bottoms Up'. However Victoria Wine is damned with faint praise for its 'rigorous instruction programme' for counter staff, as 'the standard of the shop staff, even in the Victoria Wine Cellars [their prestige brand], still needs to rise to reach the Bottoms Up and Oddbins level'.

The 'Oenological Disneyland' (Fattorini, 6) is not the only 'wine experience' that wine retailers invite consumers to buy into. Although frequently regarded as a purely functional experience (and an inverted snobbery frequently causes customers to express it as such) supermarket wine buying is a 'wine experience' every bit as complete as the theme park experience of the high street retailer. The mystique that surrounds wine and wine knowledge is extremely powerful. For many the only way to buy wine is as another grocery and away from the prying eyes of sales assistants and wine buffs who might mock your choice. The importance of the 'service' element in supermarket wine buying is that there is none, thus removing the problem of embarrassment through ignorance. The 'atmosphere' in a supermarket is the same whether you buy wine or something as anodyne as potatoes. But for many that is a vital part of the experience. In fact one could argue that this 'non-wine experience' is in fact the powerhouse behind the modern wine boom. As Barr (2) points out, supermarkets 'have removed any stigma that might be attached to wine-buying by treating it as just another grocery product, and have therefore attracted respectable women who would no more have thought of entering an off-licence than they would have thought of entering a bookmaker's or a pornographic bookshop'. Supermarkets have not so much expanded the existing market or developed their market share, as introduced a whole new market sector to wine buying. Furthermore by virtue of their 'wine experience' supermarkets have ensured they maintain a 100% market share of this important if previously untapped sector

Thus, within all arms of the of the retail wine trade one can see how a model for the 'wine experience' is formed. In the years immediately after the last War wine retailers were in much the same situation as restaurateurs. Only a very small proportion of the population used them frequently, whilst the majority of consumers had little knowledge of what to expect when they occasionally bought from them. Now though, customers eat out and buy wine far more frequently. Even if they are not 'Pagoda Man or a 'Professional Consumer' they can reasonably estimate what a bottle of reasonable quality wine should cost them. This has led to a far harsher retailing environment with many more shops chasing the (admittedly far larger) wine buying market. In this situation competing firms can gain competitive advantage either by selling wine cheap, or selling it differently. Many have chosen the latter, augmenting their product (wine) by inviting customers to buy into an experience. The shop looks like a wine shop because it is filled with icons of wine production and rustic agriculture; it has the right atmosphere. The staff seem to know what they are talking about and they are polite, they provide a good service. Then there is the wine itself, the product, arranged in a ordered way on the shelves (although slight air of clutter is fashionable in many stores).

Alternatively retailers, notably supermarkets, create a 'non-experience'. Here the three elements of Campbell-Smith's experience, atmosphere, service and product are moulded into an experience that is as close as possible to ordinary grocery retailing. The key to consumer satisfaction when selling wine in a restaurant would appear to be to adapt one of these proven models to the restaurant environment.

5. The Meal Experience vs The Wine Experience

The root problem with restaurant wine sales is that they are only ever seen as a subset of restaurant meal sales. The first rule to successful wine selling is to see food sales and wine sales as totally separate entities (Jordan, 7). There is far too great a temptation to use one to subsidise the other. Restaurateur Chris Hartley in interview with Brian Jordan points out, 'kitchen inefficiencies are often masked by wine profits, especially as it's usually easier to increase wine mark-ups than to locate a kitchen problem... forcing a kitchen to act as an independent profit centre (preferably with profitability bonuses) must result in better food margins'.

By reducing the pressure on wine to act as a cash generator the restaurateur has two possibilities, one is to reduce prices on the range, the other is to maintain prices but introduce better quality wine. The choice ultimately decides what sort of 'wine experience' the restaurant is going to offer. Unfortunately many restaurateurs currently choose a middle-way. Moderately high margins on mid-market, medium quality wines satisfies many, but excite none.

The first strategy leads to a 'wine as ordinary commodity' approach as found in supermarkets. This mode of wine selling is epitomised by very informal French-style brasseries. Only the most basic choice is offered: red, white and possibly rosé. The wine is served in unlabelled bottles which are refilled from a much larger container. This allows the restaurateur to buy in bulk and thus reduce costs. The bottle is simply put on the table and is not discussed; it is just a drink. There is no need for a wine list, staff simply need to ask if the customer would like wine and whether they would like red or white.

This is the most extreme form of this model. But still in the same vein is a wine list that is light-hearted, but informative enough to minimise interaction between staff and customer. To do this it needs to be very short and only contain wines that customers are likely to know. Customers are not going to be desperately adventurous, but then they don't want to be. The wines on the list should have names that are easy to pronounce. Nothing makes an uneasy wine consumer less happy than tripping over complicated names that they feel they ought to know. Finally, the wines should be easy to drink. Very powerful or pronounced flavours are more likely to clash with menu items. By choosing a menu and wine list where most foods go with most wines, customers are not going to have the meal of a lifetime but equally they are not likely to be embarrassed at choosing food and wine that clashes.

Thus if this style of restaurant wine selling is placed in the context of Campbell-Smith's three elements of a 'meal experience' the wine or product is kept as simple as possible. The list should rarely number more than two dozen wines, and preferably be half that. The wine is chosen entirely with what the customer knows they like in mind. The service element is minimised. Staff should be able to describe in the most simple terms where the wine comes from and what it tastes like, but they should be discouraged from doing so unless asked. Simplicity is also the guide when actually serving wine. The sort of young wines this list would contain are rarely faulty, and so tasting, bottle presentation and specialist glasses for particular wines are all unnecessary. A 'no quibble' returns policy deals with any customer complaints and gives customers a sense of security when ordering. As for atmosphere, any sense that wine is treated specially or with great respect should be avoided. This means not having specialist wine staff, decanters or very obvious wine racks. If there are pictures of vines or wine production they should simply be part of the general rustic milieu, and not a special feature. The same goes for barrels, claret cases and any other wine memorabilia. The atmosphere must afford wine no special status, just like the supermarket.

The second strategy tackles the three elements very differently. Here the idea is to use the technique of many high street merchants and flatter the customer's possibly quite limited wine knowledge and immerse them into a sea of vinous images. Large numbers of wine drinkers find popular images of the wine trade endlessly fascinating. It is still perceived as 'trade' that is pursued by 'gentlemen' (notwithstanding the large number of women employed in the wine trade). It has a long history in the UK. Encyclopaedic wine knowledge is still a hallmark of the 'connoisseur', a sign of sophistication. However more people aspire to such encyclopaedic knowledge than will ever attain it, so there is money for restaurateurs (and wine merchants) who can give customers the impression that they are wine buffs without the effort of hard study.

It is this sense of flattering the customer that distinguishes this form of wine selling, from the imposing wine lists and superior sommeliers that have for so long characterised up-market wine sales in restaurants. Ultimately the customer (and their guests) must feel as though they chose the wine, and they chose it because they knew what they were talking about.

Once again therefore, if one considers the three elements of Campbell-Smith's meal experience. The product, or the wine list, should be manageable. That is, it should be as short as possible. Nico Ladenis (9) writes in 'My Gastronomy', 'I have never been able to understand why people go into rapture and ecstasy over the wine list of a restaurant which is pages long... Why not a short, well-thought-out wine list?'. Ladenis's intention is that it is better if the restaurateur can concentrate on managing a concise and considered wine list than losing track of wines on a much longer one. In the context of this paper though, it is much easier for the customer to make up their mind if their choice is limited to fewer wines.

5.1 The Wine Product

The winelist should be easy to read and understandable. Information put on the wine list should be useful to the 'ordinary customer'. Information about producers and production methods, so highly prized by wine experts, is no use when choosing wine in restaurants and for most it is not that interesting. The best technique rather, is to tell customers what the wine goes with and then tell them a story, give the wine a personality. For example, take Côte Rotie, a full-bodied red wine from the northern Rhone Valley in Southern France. A reasonable example might be listed at about £30.00 or so, a good one though could cost very much more. For as long as it is simply wine in a bottle with a name they do not know it has little appeal to most customers. But first tell them what it goes with, it is robust and powerful and is superb with game, wild duck and venison. Then give it a personality. The term 'Côte Rotie' means 'roasted slope'. It is still home to many small scale producers who depend for a living on fruit harvesting as well as wine. Tell them the story of its two distinct vineyard areas, the Côte Blonde and the Côte Brune that were bequeathed to two sisters by their father and that the slopes embodied the characteristics of the girls. The Côte Blonde with its lighter soil produced lighter, alluring wines to be consumed early, the more robust, austere wines of the Côte Brune needed longer to come into their best though. So by building up this image, a mental picture, the restaurateur is adding value; turning an ordinary bottle of wine into an icon of rural, rustic (mythical) Frenchness.

5.2 Wine Service

The service offered by the restaurant adds to the images given in the list and built up around the product. Staff should feel free to discuss wines, and should be capable of doing so. An advantage of the short wine list is that it is easier for staff to be able to talk knowledgeably about all the wines on it, being able to sell every wine with a story. In training though the emphasis is not on developing wine knowledge but on developing selling skills. A waiter/ess who comes out with a great deal of technical information can make the customer feel belittled. If they are asked about wines they should keep the stories short and introduce them with words like 'apparently...' or 'I believe that...' so that their knowledge is no greater than the customer's, just different. When actually serving wine, let customers play the role of expert. Let them taste it, provide special glasses if the wine is over a certain value, decant old bottles, in other words, give them value for money by putting on a show. The only thing to avoid is any feeling that they are anything less than an expert. Rather than the archetypal and insincere 'And a very good choice if I may say so Sir/Madam', staff should mention what a good match it would be with one of the main courses. It is a far more meaningful comment and sounds far more sincere, even if the customer chose it because the waiter/ess told them it would go well with their main course.

5.2.1 The Fear of Getting it Wrong

For many consumers one of the biggest fears and worries when buying wine is that they will not like it. This fear is exacerbated when the wine is new or expensive. They are aware that they cannot simply reject it on the basis that either they (or their companions) do not like it. Overcoming this fear is useful for the restaurateur because they can steer customers towards less well known wines. Less competitive market sectors not only offer the consumer better value but also offer the opportunity for increased flexibility in marking-up.

The two main strategies are to either let customers try wine before purchase or show them that someone else liked it. Product sampling has become much easier recently with the advent of a number of preservation systems for the commercial wine vendor. By having a range of profitable wines available by the glass, staff can offer customers the chance to try these wines before they buy a whole bottle. These are also useful for wines rarely bought by the bottle such as sweet pudding wines.

Endorsements tell the customer that another person, particularly 'one who should know' enjoyed a particular wine. In its most basic form staff should feel free to recommend wines that they particularly enjoyed. This involves the restaurateur giving staff these wines to taste, something that is still not universally done. Alternatively wines can be recommended by somebody 'famous'. This could go from the chef of the restaurant to a television celebrity. Some winelists now contain quotations from wine writers about particular wines or even generalist comments about the region. The most important part is that if the consumer recognises the name of the endorser it makes them more confident about ordering.

5.3 Wine Atmosphere

This is where restaurateurs can use their imaginations to the fullest. Being known for a good wine list can only ever add to a restaurant's reputation and given the public's fascination with wine it is worth creating a 'winey' atmosphere. There are a great many examples of ways that different restaurateurs have done this and travelling to new restaurants will always generate new ones. A reputation for knowledgeable staff is never a bad thing and framing certificates of exams passed and putting them up on the walls always looks good. Equally, membership of wine guilds and societies can be displayed. Earlier in the context of wine retail, icons of wine production were mentioned. Building up a collection of these often requires a friendly wine merchant but the effect can be superb. The key here is names. The ends of claret cases where the names and pictures of the chateau have been branded look particularly good. Over time whole walls can be covered in them and varnished to very good effect. Sometimes people worry that they have merchant's writing on them and old cellar stickers. In fact the shabbier the better as it provides a link with the romanticism of the wine trade. Even old bottles (although the collection should be limited to particularly good, rare or famous wines) can be used, particularly high up out of reach. Other little touches depend on what is available. A small, lined, half barrel with iced water makes a very good cooler for house white wine. The range of possibilities is vast.

6. Concluding Remarks

The concept of the 'meal experience' has revolutionised dining-out in the UK by selling customers more than just a meal. By interlinking service and atmosphere to the meal, the restaurateur 'adds value' to the food and wine that is their core product. Unfortunately, whilst the 'meal experience' is still valid, owing to changes in the way the public perceives wine, it can no longer be sold simply as a subset of the 'meal' but must be treated as a separate entity. Equally, greater knowledge in the public about the value of wine means that it can no longer be used as a cash cow to make up for underpriced or inefficiently produced food.

In order to make the most out of wine restaurateurs must provide a 'wine experience' quite distinct from the 'meal experience'. This gives the consumer obvious added value for a product they are aware is heavily marked-up. The inspiration for different types of 'wine experience' is taken from the retail wine trade, both in its high street and supermarket forms. Here strategies to deal with increased competition and squeezed margins have developed in recent years.

The supermarket and high street wine chain have inspired the two main forms of the 'wine experience'. The supermarket style, treating wine as an ordinary commodity satisfies those customers who want to avoid any association with the snobbery/mystique of wine that is often found in the UK. The high street style of wine theme park is suitable for those customers who value associations with wine and the wine trade but lack the encyclopaedic knowledge required to pass as a connoisseur.

References

1. Barr, A. (1988), Wine Snobbery, London: Faber and Faber.

2. Barr, A. (1995), Drink, London: Bantam Press.

3. Briggs, A. (1985), Wine for Sale, Chicago: University of Chicago Press.

4. Campbell-Smith, G. (1967), The Marketing of the Meal Experience, Guildford: Surrey University Press.

5. Eyres, H. (ed.) (1996), The Which? Wine Guide 1996, London: Which? Books.

6. Fattorini, J (1994), 'Professional Consumers: Themes in High Street Wine Marketing', International Journal of Wine Marketing, 6, 2, pp. 5-13.

7. Jordan, B. (1996), ''Me and My Cellar: Chris Hartley', Hotel and Restaurant, December/January, pp. 35-37.

8. Joseph, R. (1994), 'The Pagoda Man cometh', The Sunday Times Review, 6 November, p.17.

9. Ladenis, N. (1987), My Gastronomy, London: Headline.

10. Wood, R.C. (1995), The Sociology of the Meal, Edinburgh: Edinburgh University Press.

Poster

Destination North Cape: are German tour bus operators' needs and expectations understood by the North Norwegian hotel business?

T. Gustavsen, L. Lervik and S. Larsen.
Department of Tourism and Hotel Administration, Finnmark College, N-9500, Norway.

The hotel industry in the very north of Norway has a significant variation in occupancy throughout the year. The winter season is extremely slow, while the summer season is equally as busy. The most sustainable segment in the summertime is the German bus tour operator. This poster addresses the issue of how well hotel operators in western Finnmark, the northernmost county in Norway, understand the expectations and demands of the German bus tour operators. The results show a gap between what the hoteliers think about their own product, compared to the bus tour operators' rating. Five of the seven hotels had a significant gap on the whole hotel product, one only on part of the product, while one hotel had a good understanding of how the German bus tour operators evaluated the hotel. There was also an incoherence between the hoteliers and the bus operators concerning how the hotels were ranked. The hotel that ranked itself highest, came out with the lowest grade from the bus operators, whereas the hotel that ranked itself lowest, came out with the best grade from the operators.

INDEX OF AUTHORS

Alexieva, I. 266,267,315
Almeida, M.D.V. 387
Aujezdská, A. 307
Beer, S.C. 351
Békássy-Molnár, E. 241
Biller, E. 269
Blackwell, L.R. 479
Borowski, J.G. 397
Caraher, M. 415
Carr-Hill, R. 415
Colquhoun, A. 173,251
Corney, M.J. 59
Creed, P. 221
Cui, W. 263
Danowska, M. 262
Dixon, P. 415
Dodson, H.I. 199
Eastham, J. 49
Edmondson, A.S. 199
Edwards, J.S.A. 3,117,335
Ervin, J. 457
Eskin, N.A.M. 263,264,397
Eves, A. 59
Fattorini, J. 509
Ferrone, L.R. 186
Folgero, I.S. 81,501
Fushiki, T. 489
Gofman, A. 101
Goodinge, A.W 3
Grysman, C. 397
Gustavsen T. 91,522
Hann, P. 173
Hemmington, N.R. 437
Herbage, P.F. 3
Jones, P. 129,,161,186
Karg, G. 285
Karpinska, M. 262
Kállay, M. 241
Keeling, H. 211

Kenny, J. 427
Kipps, M. 59
Kivela, J. 25,184,185
Kleynhaus, I.C. 270
Kodama, H. 489
Kreutzmeier, S. 285
Lang, T. 415
Larsen, S. 81,91,501,522
Lawson, J. 325
Lähteenmäki, L. 469
Leighton, C. 405
Lervik, L. 1,522
Liu, H. 263,264
Lyon, P. 251
Manninger, K. 241
McGlade, M. 405
Meiselman, H.L. 447
Mitchell, J. 375
Morimoto, K. 489
Moskowitz, H. 101
Müller, L. 307
Müllerová, D. 307
Neryng, A. 268,269
Noble, C. 59
Peacock, M. 275
Pereira, R. 263
Pierson, B. 427,457,479
Póltorak, A. 268
Przybylski, R. 264
Redman, M.H. 51
Reeve, W.G. 117,457
Reid, A. 251
Ridley, S. 129
Rodrigues, S.S.P. 387
Schafheitle, J.M. 161
Seaman, C.E.A. 294,405
Sharples, A.E. 69
Sheard, M.A. 199,211,231
Sutton, J. 363

Tungaturthy, P. 101
Ulbricht, G. 297
Van Rensburg, D.M.J. 271
van Westering, J.M. 15
Vatai, Gy. 241
Wade, J.A. 151
Walker, A. 325
West, A. 139,325
Wierzbicka, A. 269
Worsfold, D. 191
Xie, G. 211,231
Yonezawa, T. 489
Zalewski, S. 265